CANCER ETIOLOGY, DIAGNOSIS AND TREATMENTS

RESEARCH ON NEW GENERATION TUMOR MARKERS

CANCER ETIOLOGY, DIAGNOSIS AND TREATMENTS

Additional books and e-books in this series can be found on Nova's website under the Series tab.

CANCER ETIOLOGY, DIAGNOSIS AND TREATMENTS

RESEARCH ON NEW GENERATION TUMOR MARKERS

YILDIZ DINCER
EDITOR

nova
Medicine & Health
New York

Library of Congress Cataloging-in-Publication Data

ISBN: 978-1-53614-367-6

Published by Nova Science Publishers, Inc. † New York

CONTENTS

PREFACE

Tumor markers providing information about biological characteristics of the tumor have been used in the clinical setting for many years. As described in a Egyptian papyrus, the first known attempt to find markers for malignancy (breast cancer) was performed 2000 years ago. The first tumor marker was incidentally determined in 1846 by Bence-Jones. In modern medicine, tumor markers were discovered after the demonstration of tumor-specific transplantation antigens in experimentally-induced tumors in syngenic rodents. The history of currently used tumor markers began with discovery of alpha-fetoprotein in 1956, followed by identification of carcinoembryonic antigen in 1965. Since then the number of tumor markers has increased continuously. The potential uses of tumor markers are cancer screening in general population, differential diagnosis of symptomatic patients, clinical staging of cancer, estimating tumor volume, monitoring response to therapy and detecting the recurrence of cancer. Unfortunately, there is no ideal tumor marker meeting all the criteria. Currently used tumor markers are far from to be used in screening and diagnosis due to the low sensitivity and specificity of the tests, thus new tumor markers which have high sensitivity and specificity are continuously investigated. Detection of cancer stem cells, circulating tumor cells and cell-free tumor DNA in the bloodstream (the liquid biopsy) launched a new era in this field. In the past decades developments in molecular techniques have provided an opportunity to detect genetic and epigenetic alterations in circulating tumor DNA. The number of circulating tumor cells; mutations and epigenetic alterations in circulating cell-free tumor DNA; profile of circulating microRNAs and tRNAs; the signature of long non-coding RNAs expression in tumors have been introduced as new generation molecular tumor markers and are currently under investigation. In this book, starting from a definition and limitations of commonly used tumor markers, recent investigations on cancer stem cells, FXYD proteins, circulating tumor DNA, long non-coding RNAs, microRNAs and tRNAs were introduced; liquid biopsy examinations in the hepatobiliary and gastro intestinal cancers and molecular markers in thyroid cancers were discussed. I hope this book will be helpful for students and clinicians and provides a baseline for future investigations.

April 2018,
Istanbul

In: Research on New Generation …
Editor: Yildiz Dincer
ISBN: 978-1-53614-367-6

Chapter 1

Commonly Used Tumor Markers and Their Limitations

Ali A. Tuz[1], MD Student, SelinYuksel[2,3], MSc and Yildiz Dincer[2,*], PhD

[1]Istanbul University Cerrahpasa Medical Faculty, Istanbul, Turkey
[2]Istanbul University- Cerrahpasa, Cerrahpasa Medical Faculty, Department of Medical Biochemistry, Istanbul, Turkey
[3]Biruni University, Medical Faculty Department of Medical Biochemistry, Istanbul, Turkey

Abstract

Early detection is a crucial factor for successful therapy and increased survival rate in cancer patients. Many cancers are associated with the abnormal production of some tumor markers which can be detected in higher than normal levels in the blood, urine, or tissues. Despite some limitations many tumor markers have been used in prediction of stage/tumor grade, monitoring the therapy, and in the detection of recurrences in cancer patients. An ideal tumor marker should be released from only tumor cells and tissue-specific. Unfortunately, there is no ideal tumor marker. All tumor markers are present in normal, benign and cancer tissues hence not spesific enough to be used for screening or early diagnosis of cancer. Even levels of some tumor markers may not rise in every cancer patient. In this chapter, limitations of the commonly used tumor markers will be discussed.

Keywords: Conventional tumor markers, limitations of tumor markers, ALK, BRAF V600 mutation, KRAS mutation, BRCA1 and BRCA2, C-kit/CD117

---

[*] Corresponding Author Email: yldz.dincer@gmail.com.

INTRODUCTION

A tumor marker is any substance that increases in blood, body fluids or tissues of patients with certain types of cancer. In practise, the tumor markers can be used in diagnosis, prognosis, detection of recurrences but they are mostly beneficial to monitor treatment and follow-up. The tumor marker may be an enzyme, a hormone, an intracellular protein, a cell surface antigen or a receptor. Regardless of its structure, there is great need for sensitive tumor markers for screening, risk identification, differential diagnosis of symptomatic patients and early detection. For this purpose, an ideal tumor marker should meet some criteria. The tumor marker must be sensitive enough for screening and early diagnosis; it must be synthesized only by the tumor, must not be found in healthy tissues and must be tissue-specific. If it is synthesized in other tissues, it must exhibit different immunological properties. Its serum level must be proportional to the size and activity of the tumor, at a detectable level in serum even if the tumor is still very small; and its serum level must be very low in benign tumors.

Unfortunately, many known tumor markers are not specific to a single type of cancer, and has an insufficient sensitivity for cancer screening. Certain tumor markers are also produced in healthy tissues in small amounts. Although high levels of the tumor markers are mostly seen in cancer, they are also elevated in some benign diseases, in many hormonal and metabolic changes. In addition, many cancers do not have blood tumor markers that are useful for guiding care. For this reason, their use in diagnosis and screening is extremely limited.

Conventional tumor markers are frequently useful in follow-up of previously diagnosed cancer patients. The plasma level of the tumor marker at the time of diagnosis is associated with the prognosis and survival in cancer patients.

Although a temporary increase in the level of the tumor marker may be seen during the therapy (due to cell lysis), the plasma level decreases after a successful treatment. The rate of the decrease in the plasma level of the tumor marker reflects the success of the treatment. A consistent high level of the tumor marker level indicates poor response to the therapy. An increase in the diminished tumor marker level after a period of successful treatment is usually associated with recurrence.

Testing for Tumor Markers

Since the tumor marker tests are used to confirm or refute the presence of cancer, they should have the lowest error rates and the highest accuracy. An ideal test must correctly differentiate malignant cases from physiological or benign disease states.

Accuracy

The accuracy of a test is its ability to differentiate the patients and healthy cases correctly, and is mathematically stated as:

$$\text{Accuracy} = \frac{\text{True positives} + \text{True negatives}}{\text{True positives} + \text{True negatives} + \text{False positives} + \text{False negatives}}$$

The 100% accuracy does not happen in reality since the accuracy of a test varies in different diseases and in different situations.

Sensitivity

The sensitivity of a test is its ability to determine the patient cases correctly. To estimate it, the proportion of the true positives within patient cases should be calculated. It is mathematically stated as:

$$\text{Sensitivity} = \frac{\text{True positives}}{\text{True positives} + \text{False negatives}}$$

If the tumor marker test is not sensitive enough, the results may suggest a false negative. This means that persons should not have a tumor regarding the tumor marker levels but they actually do, or tumor marker levels suggest cancer treatment is working although it is not [1].

Specificity

The specificity of a test is its ability to determine the healthy cases correctly. To estimate it, the proportion of the true negatives within the healthy cases should be calculated. It is mathematically stated as:

$$\text{Specificity} = \frac{\text{True negatives}}{\text{True negatives} + \text{False positives}}$$

If the tumor marker test is not specific enough, the results could suggest a tumor is present or growing despite treatment but they actually do not [1].

No tumor marker is 100% sensitive and 100% specific, but optimal clinical use can be ensured by applying appropriate cutoff points. Different tumor markers may be used for different cancers, how and when they are used varies quite a lot [2].

Table 1. Commonly used tumor markers and clinical situations where they have been found to be useful

Tumor Marker	Cancer Types	Tissue Analyzed	Diagnosis	Staging	Response to Treatment	Prognosis	Determine Targeted Therapy	Assess Recurrence
ALK gene	Non-small cell lung cancer, Anaplastic large cell lymphoma	Tumor			x	x		
Alpha-fetoprotein (AFP)	Liver cancer	Blood	x		x			
Alpha-fetoprotein (AFP)	Germ cell tumor	Blood		x	x	x		
BCR-ABL fusion gene (Philadelphia chromosome)	Chronic myeloid leukemia, Acute lymphoblastic leukemia, Acute myelogenous leukemia	Blood and/or bone marrow	x	x	x			
Beta-2-microglobulin (B2M)	Multiple myeloma, Chronic lymphocytic leukemia	Blood, urine, cerebrospinal fluid			x	x		
Beta-human chorionic gonadotropin (beta-hCG)	Ch choriocarcinoma, Germ cell tumor	Blood, urine		x	x	x		
BRAF V600 mutations	Cutaneous melanoma, Colorectal cancer	Tumor					x	
BRCA1, BRCA2 gene mutations	Ovarian cancer	Blood					x	
CA 125	Ovarian cancer	Blood	x		x			x
CA 15-3/ CA27.29	Breast cancer	Blood			x			
CA 19-9	Pancreatic cancer, Gallbladder cancer, Bile duct cancer,Gastric cancer	Blood			x			
Calcitonin	Medullary thyroid cancer	Blood	x		x			x
Carcinoembryonic antigen (CEA)	Colorectal cancer	Blood			x			x
CD20	Non-Hodgkin lymphoma	Blood					x	
Chromogranin A (CgA)	Neuroendocrine tumors	Blood	x		x			x
Chromosomes 3, 7, 17, 9p21	Bladder cancer	Urine						x

Tumor Marker	Cancer Types	Tissue Analyzed	Diagnosis	Staging	Response to Treatment	Prognosis	Determine Targeted Therapy	Assess Recurrence
Circulating tumor cells of epithelial origin	Metastatic breast, prostate, colorectal cancers	Blood				x		
C-kit/CD117	Gastrointestinal stromal tumor, Mucosal melanoma	Tumor	x					
Cytokeratin fragment 21-1	Lung cancer	Blood						x
EGFR gene mutation	Non-small cell lung cancer	Tumor			x			
Estrogen receptor (ER) / Progesteron receptor (PR)	Breast cancer	Tumor					x	
Fibrin/fibrinogen	Bladder cancer	Urine						
HE4	Ovarian cancer	Blood		x				x
HER2/neu gene amplification or protein overexpression	Breast cancer, Gastric cancer, Gastroesophageal junction adenocarcinoma	Tumor					x	
Immunoglobulins	Multiple Myeloma, Waldenstrom macroglobulinemia	Blood and urine	x		x			x
KRAS gene mutation analysis	Colorectal cancer, Non-small cell lung cancer	Tumor					x	
Lactate dehydrogenase	Germ cell tumor, lyphome, leukemia, melanoma, neuroblastoma	Blood		x	x	x		
Neruon-specific enolase (NSE)	Small cell lung cancer, Neuroblastoma	Blood	x		x			

Table 1 (Continued)

Tumor Marker	Cancer Types	Tissue Analyzed	Diagnosis	Staging	Response to Treatment	Prognosis	Determine Targeted Therapy	Assess Recurrence
Nuclear matrix protein 22	Bladder cancer	Urine			x			
Programmed death ligand 1 (PD-L1)	Non-small cell lung cancer	Tumor					x	
Prostate-specific antigen (PSA)	Prostate cancer	Blood	x		x			x
Thyroglobulin	Thyroid cancer	Blood	x		x			x
Urokinase plaminogen activator (uPA)	Breast cancer	Tumor		x				

Anaplastic Lymphoma Kinase Gene (ALK) Fusions

ALK fusions have been identified and used to detect the prognosis, response to treatment in non-small cell lung cancer (NSCLC) and anaplastic large cell lymphoma. Stage IV patients can be detected with higher sensitivity compared to stage I patients. However, the ALK fusions in NSCLC are seen in a small subpopulation of patients, approximately 5%, who are responsive to the ALK kinase inhibitors, and ALK fusions in NSCLC are expressed at low levels and tumor heterogeneity may result in variability in replicate samples [3]. In addition, different techniques are used to detect the ALK rearrangements, fluorescent in situ hybridisation (FISH) and immunohistochemistry (IHC) are the most commonly used techniques, and their results are inconsistent. Frequency of the ALK rearrangements based on FISH is lower than the previous reports [4] and IHC sensitivity is low [4]. In order to obtain more accurate results, different new methods are being tested, such as using antibody clone D5F3 with a sensitivity ranging between 90-100% for IHC [5]. That result may replace ALK-FISH, as stated before the most common technique used. Furthermore, platelet RNA is tested as a source to detect the ALK rearrangement with a sensitivity of 65% and specifity of 100%. This method also identified the ALK rearrangements 2 months before the radiologic detection of tumor progression [6].

Alpha-Fetoprotein (AFP)

AFP levels are used for the diagnosis of liver cancer and monitoring response to treatment. However, elevated AFP levels may be a result of pregnancy, chronic hepatitis and cirrhosis, in addition to the liver cancer. AFP has a sensitivity of 39% to 65%, a specificity of 65% to 94%, and approximately one-third of the early stage liver cancer patients with small tumors (>3cm) have normal AFP levels [7]. Moreover, different biomarkers such as Des-Gamma-Carboxy Prothrombin (DCP) can be used to evaluate small sized liver cancer, as the levels of DCP in malignancy is significantly higher than in healthy adults and patients with nonmalignant hepatopathy, chronic hepatitis and cirrhosis. Thus, using only the AFP level is not effective at differentiating healthy subjects from the liver cancer patients, and clinicians are dissatisfied with the results. Using DCP and AFP in combination can be a good predictor in both small and large size liver tumors, as DCP is better than AFP in detecting the small size liver tumors [8].

AFP is also used in germ cell tumor staging, prognosis and monitoring the treatment. It may be used with β-hCG, as at least one of them is elevated in 85% of nonseminomatous germ cell tumors and in which only 20% is of early stage. However, similar to liver cancer patients, elevated AFP results may be confused with alcohol abuse, hepatitis, cirrhosis, biliary tract obstruction and Fanconi anemia [9, 10]. Some patients may have hereditary, mildly elevated serum AFP levels which remain same during the course of chemotherapy and follow -up. This condition causes confusion in progressive cancer patients who are expected to have increasing values.

Beta-2 Microglobulin

Beta-2 Microglobulin is a marker for multiple myeloma and chronic lymphocytic leukemia. However, it is a cell surface protein found in many cells and its serum level also increases in conditions that increase cell production and/or destruction, or that activate the immune system and in renal failure. Thus, the beta-2 Microglobulin testing can not be used for the diagnosis; instead, it provides additional information about the prognosis. The beta-2 microglobulin testing is typically conducted in conjunction with a urine test to help determine the prognosis of multiple myeloma [11].

Furthermore, the combination of beta-2-microglobulin with serum albumin concentrations is used to determine the prognosis of multiple myeloma patients [12]. However, these parameters may be misleading for tumor load in patients with substantial myeloma infiltration in the bone marrow. Most patients in this situation have normal β_2-microglobulin levels with stage I disease. Patients with stage IV disease show increased β_2-microglobulin levels but this may be late for the treatment. Moreover, even though there is a reference material for the β_2-microglobulin measurements, it is not universally utilized for the detection kits by the manufacturers. Different techniques and reference ranges are currently in use. This may result in difference in the β_2-microglobulin assay performance [13].

Beta-Human Chorionic Gonadotropin (β-hCG)

β-hCG is tested for pregnancy under normal conditions. It also elevates in choriocarcinoma and germ cell tumors. Reference ranges of β-hCG in these pathological conditions are suggested to be updated in recent researches. β-hCG can be elevated in benign conditions such as primary hypogonadism and giant cell tumors of bone [14]. Thus, it is important to eliminate these conditions in order to avoid misdiagnosis.

BRCA1 and BRCA2 Gene Mutations

BRCA1 and BRCA2 are human *genes* that encode DNA repair proteins. BRCA1 and BRCA2 mutations commonly increase the risk of female breast and ovarian cancers. Women who have inherited mutations in *BRCA1* and *BRCA2* tend to develop breast and ovarian cancers at younger ages than people who do not have these mutations. Women diagnosed with breast and ovarian cancers are considered for genetic testing and if positive, for targeted therapy. These tests are common in Europe and in the USA, but their usage is very limited in the Eastern countries. Moreover, regarding the overall survival, there is no significant difference between the patients with mutated BRCA and wild-type BRCA [15]. Although serious adverse events are reported for the higher grades, in general, results are inconsistent for the *BRCA1 and BRCA2* targeted therapy. For example, 18% patients who received treatment and 9% cases who received placebo have been reported to exhibit serious side effects [15]. More investigation is required to evaluate the effects of the BRCA targeted therapies.

BCR-ABL Fusion Gene

BCR-ABL fusion gene is used in diagnosis, staging and assessment of response to treatment for leukemia types. Patients with chronic myeloid leukemia have the BCR-ABL fusion in their genome. However, there are some patients who do not have any sign of cancer although they have the BCR-ABL fusion [16], that situation leads to some incorrect assumptions in clinics. It is not clear whether patients with the BCR-ABL fusion gene will develop leukemia in the future or they have some kind of resistance to oncogenesis. Further animal and clinical data are needed to answer these questions.

BRAF V600 Mutation

BRAF V600 mutation testing and targeted therapy is indicated for cutaneous melanoma and colorectal cancer. The sensitivity of BRAF V600 mutation detection is different in several papers, changing from Chinese, European and USA patients, which may be related its clinicopathological characteristics. The sensitivity is between 38-79% in melanoma patients. Moreover, biopsy is made from tumor tissues and tumor heterogeneity is one of the problems in analysis. Circulating tumor cells (CTC) may be used instead of the tumor biopsy, but success rate of polymerase chain reactions (PCR) in CTCs to detect the BRAF mutations is relatively low. Higher concentrations and newer PCR techniques may be used to have better results, but in clinics these techniques are not easily affordable [17].

C-Kit/CD117

C-kit/CD117 test is used to diagnose gastrointestinal stromal tumor and mucosal melanoma. This test has relatively high false-negative results [18]. In order to eliminate these results, different antibodies such as DOG1 are being investigated. DOG1 can be used for suspicious or negative stainings of CD117.

CA 15-3 and CA 27.29

CA 15-3 blood test is used for breast cancer. However, early stage breast cancers and some tumors do not produce CA 15-3, thus this test is not useful for diagnosis. About 30% of breast cancers cause an increase in the CA 15-3 level, however, increased CA 15-3 levels may also be related to other cancers (uterine, lung, pancreas), cirrhosis of liver, tuberculosis, benign breast conditions, pregnancy, breastfeeding, endometriosis, sarcoidosis and pelvic inflammatory disease [19].

CA 27.29 is also used for breast cancer. As CA 15-3, elevated levels of the CA 27.29 may be a result of ovarian cysts, liver disease and other cancers such as lung, liver, colon, pancreas, ovary and prostate cancer. It is elevated in about 33% of early stage breast cancers and 67% of late stage breast cancers. Furthermore, CA 27.29 is more sensitive but less specific than CA 15-3, but in clinics they are used equivalent in most of the cases.

CA19-9

CA 19-9 is elevated in different cancer types. It is elevated in 80-90% of pancreatic and 60-70% of biliary tract cancers. It is also elevated by gallstones, blockage of bile duct, pancreatitis, cystic fibrosis and liver disease [20]. In advanced cancers, the CA 19-9 level should be combined with imaging techniques to understand the localization of the cancer.

CA-125

Although CA-125 is a marker for ovarian and endometrial carcinomas, it is not specific. The National Cancer Institute-sponsored Prostate, Lung, Colorectal and Ovarian Cancer Screening Trial (PLCO) conducted research on CA-125 which is a tumor marker used to diagnose ovarian cancer. It is elevated in about 85% of ovarian cancer but only in 50% of early stage ovarian cancers. PLCO data showed that, it is not sufficiently specific or sensitive to be used with transvaginal ultrasound to screen for ovarian cancer in average-risk women. It also elevates in benign conditions such as menstruation, pregnancy, fibroids, ovarian cysts, pelvic inflammation, cirrhosis, ascites, pleural and pericardial effusions and endometriosis. These benign conditions are relatively common, so it is important to eliminate other possibilities before diagnosing patients with ovarian cancer [21].

Calcitonin

High levels of Calcitonin in blood may indicate medullary thyroid cancer. Different methods are used in different clinics, which leads to a difference in thresholds and stimulation values. Since it is difficult to evaluate the results, different laboratories should have their own basal and stimulation test values [22]. Hypercalcemia, hypergastrinemia, neuroendocrine tumors, renal insufficiency, goiter and papillary and follicular thyroid carcinomas may lead to high levels of calcitonin in blood. Some drugs, such as omeprazole, beta-blockers and glucocorticoids also increase calcitonin. After a threshold, calcitonin level may have a risk for medullary thyroid carcinoma [23].

Carcinoembryonic Antigen (CEA)

CEA in blood is used to assess the recurrence and response to treatment in colorectal cancer (CRC). It may also be elevated in benign conditions such as a peptic ulcer, ulcerative colitis, rectal polyps, emphysema, breast disease, inflammatory situations and smoking. Interlaboratory differences are possible, so different tests in the same laboratory are recommended in order to assess the recurrence [2]. Serial CEA measurements can detect the recurrence in colorectal cancer with a sensitivity of 80% and specificity of 70% [24]. In addition, the glycosylation patterns on CEA differ in different stages of CRC. Tumor adjacent normal tissues can be detected regarding to the glycosylation patterns of CEA, usage of these patterns may increase the usefulness of CEA in the future for detection of recurrence and assessment of response to treatment in CRC [25]. Some tumors don't produce CEA, thus the CEA testing cannot be used for the cancer screening.

B-Lymphocyte Antigen CD20

CD20 targeted therapy is one of the most important drugs that are preferred in non-Hodgkin lymphoma. However, even though the lymphoma cells express CD20, some patients are resistant to targeted antibody therapy. Some patients also show initial response to therapy followed with a resistance in short duration. Heterogeneity of the cells may also cause false test results [26].

Chromogranin A (CgA)

CgA is elevated in neuroendocrine tumors (NETs) and used in the diagnosis, recurrence assessment and treatment response evaluation. Some other non-neoplastic diseases, such as inflammatory bowel disease, renal failure, heart failure, rheumatoid arthritis; other cancer types such as breast, digestive gland, female genital tract, head and neck, hematological and usage of proton pump inhibitors may also lead to the increased CgA levels, therefore a cutoff range is necessary to diagnose the NETs. The cutoff range for CgA can be different in different laboratories. Several techniques are used with different antibodies and epitopes which in turn leads to significant variability in sensitivity and specificity from 60% to 93%. 86% of the patients with recurrence showed the elevated CgA levels, and all patients who had liver metastasis have high CgA levels. However, the CgA concentration may be normal in the mild proliferative NETs, as most of the appendix and insulinoma. Rapidly proliferating and poorly differentiated NETs have smaller vesicles, resulting in false-negative test results [27].

Chromosomes 3, 7, 17, 9p21

Targeting the centromeres of some chromosomes may be used in bladder cancer. Researches show promising results on diagnosis and recurrence assessment. Centromers of the chromosomes 3, 7, 17 and the locus of 9p21 may have such a potential. However, these tests are time consuming, cost intensive and not common. They require manual investigation by trained investigators and large scale epidemiologic studies are needed. Computer assisted FISH spot analysis is also introduced in some papers, but this technique is also relatively new [28].

Circulating Tumor Cells of Epithelial Origin (CTCs)

CTCs are used for prognosis of the metastatic breast, prostate and colorectal cancer. Different techniques are used and 75% of the prostate and 80% of the breast cancer CTCs were detected in some of them, but sample numbers were not satisfying in these experiments [29]. Most widely used techniques are immunology-based ones, with nucleic acid-based alternatives. The nucleic acid-based techniques may detect cell free DNA in the blood from apoptotic cells as well as CTCs. However, method standardisation is required [30].

Cytokeratin Fragment 21-1 (CYFRA 21-1)

Blood level of the CYFRA 21-1 can be used to detect the recurrence in lung cancer. However, combination of different markers including this fragment has better sensitivity and specifity, 64%, 90% respectively [31]. Thus, the CYFRA 21-1 testing is not enough itself to assess recurrence or prognosis. Some papers have shown its prognostic value after surgery, but these research results were not satisfying because of their limited number of cases [32, 33].

EGFR Mutation

The EGFR gene mutation is seen in about 40% of the NSCLC patients. Different exons may be mutated in different patients. Detecting these different exons may not be cost-effective in all countries. Tyrosine kinase inhibitors (TKIs) are used in these patients with good outcomes. As these mutation is not seen all the NSCLC patients, detecting the right gene and targeting that gene is important in early therapy [34].

Estrogen Receptor (ER)/Progesterone Receptor (PR)

By binding to its specific receptors, ER influences tumor growth. The PR testing is a useful adjunt to the ER testing, since the PR synthesis appears to be dependent on ER action. Breast cancer tumors may express ER and PR. 2 of the 3 breast cancer patients are ER/PRreceptor-positive, so the presence of these receptors is used to determine the targeted therapy. Approximately two thirds of patients with the estrogen receptor (+) tumors respond to the hormonal therapy. However, in a small percentage of the ER-positive patients (2.1% to 5% in different papers), the tumor lacked expression of the estrogen-responsive genes [35]. Immunohistochemistry for the estrogen-responsive gene products is used in most of the laboratories, leading to false diagnosis in these patients. Different gene expression-based techniques may be used to clear these issues. Furthermore, side effects such as fatigue, low blood cell count, nausea, vomiting, diarrhea are relatively common in patients receiving the hormone receptor-targeted therapy.

Fibrin/Fibrinogen Degradation Products

Fibrin/fibrinogen degradation products in urine are used in urinary bladder cancer. They may be seen at low levels in urine of healthy people. Tumor grade and stage is directly proportional with the fibrin/fibrinogendegradation products level in urine, which improves sensitivity in detecting aggressive tumors. However, the accuracy of test is around 75%, which is not enough to be used on its own. Specificity of the test is low, they also increase in non-specific inflammatory conditions, such as urinary tract infections [36]. Small tumors may not be detected with this test. Combined tumor markers are needed to detect the bladder tumors.

HE4 Protein

The HE4 protein is frequently overexpressed in ovarian cancers [37]. Sensitivity of the HE4 protein testing is higher than those of other single markers used for both the early and late stage ovarian cancers, 62.4-82.7%, 74.6-92.5% respectively. If HE4 and CA-125 is used in combination, the rates are better [38]. Even though these results seem promising, in order to have a positive predictive value of 10% for the detection of all the ovarian cancer stage, sensitivity of 75% and specificity of 99.7% is needed in the USA, which is significantly higher than the current data of the HE4 protein testing. Late stage cancers may be detected easier with the current test, however, the combination of transvaginal sonography, even though it is expensive, is required to screen and detect the early stage ovarian cancers [39].

HER2/Neu Gene Amplification

The HER2/neu is a gene that may be amplified in breast and gastrointestinal cancers. HER2/neu-positive patients are prescribed with special gene therapies. It is important to detect the HER2-positivity, since the drugs used as anti-HER2/neu are highly toxic [40]. Cell- surface proteins related to the overexpression of HER2/neu gene are routinely tested in clinics for this indication. However, immunohistochemistry scoring for the breast and gastric cancers are different, basolateral and lateral staining alone in gastric cancer is positive and treated. Tumor heterogeneity is crucial in this aspect, because some patients may exhibit the HER2/neu positivity in one of the sites even though most of the tumor is HER2-negative [41]. Different potential biomarkers such as HER1, HER3 which are also important for the pathways of angiogenesis and may have clinical benefits are currently under investigation [42].

Immunoglobulins

Bence-Jones proteins are abnormal proteins produced by plasma cells. They are free monoclonal immunoglobulin light chains found in the urine. No Bence-Jones proteins are detected in the urine of healthy individuals but 75% of myeloma patients have the Bence-Jones proteins in their urine samples [43]. The Bence-Jones proteins are considered as a marker for multiple myeloma. In addition, Waldenstorm-macroglobulinemia shows similar characteristics with multiple myeloma. Differential diagnosis is hard for these two diseases, because biopsy results and clinical features are similar in some cases. Organomegaly, macroglobulinemia are the main similar features and new markers are needed to differentiate these situations [44].

KRAS Mutations

The KRAS mutations are detected in almost 25% of lung adenocarcinoma patients in the USA. Drugs targeting the KRAS mutation sites are in use in the clinics, but mutations that cause the resistance to the Ras-binding drugs decrease the effectiveness of the therapy. The extent of the mutations is unknown, as new mutations occur in different patients. Besides, the effect of the KRAS mutations on survival of the patients with early-stage NSCLC is unknown, even though some older data showed worse prognosis in these patients [45-47].

Lactate Dehydrogenase (LDH)

The LDH activities are elevated in germ cell tumors, lymphoma, leukemia, melanoma and neuroblastoma. Patients with four times higher LDH activity are expected to have shorter life after diagnosis. High LDH activities are also related with poor therapy response, but in some studies, it is reported that some patients (1/3 of the patients receiving chemotherapy) had decreased LDH activities after the full dose chemotherapy [48]. It is important to have a large scale research about these patients to have good responses to therapy.

Neuron-Specific Enolase (NSE)

In neuroblastoma patients, NSE is used to diagnose and evaluate the response to treatment. It has a moderate sensitivity and specificity to diagnose pediatric neuroblastoma patients, but it has good values as a marker in prognosis and follow-up after therapy. LDH and NSE are used in some patients and these values indicate a better outcome when used together [49]. NSE is also used in small cell lung cancer (SCLC) patients. However, in prospective studies, since survival of the patients in the SCLC is relatively short, it is hard to indicate an exact survival rate regarding the NSE results.

Nuclear Matrix Protein 22 (NMP22)

NMP22 is used as a marker in bladder cancer patients. It has a sensitivity of 56% and specificity of 88%. It is a protein which can also elevate in inflammation and in some other benign conditions. It has better potential for high-grade and high-stage tumors. Different markers and algorithms are under investigation to be combined with NMP22 in order to have better outcomes, because NMP22 itself is not sufficient for screening [50].

Programmed Death Ligand 1 (PD-L1)

The PD-L1targeted therapy is used in NSCLC. The PD-L1 expression within the tumor parts may be different, even metastatic sites can differ over time. The PD-L1 expression change can occur spontaneously and also induced by cancer therapies [51]. Multiple and repeated biopsies are needed to have a valuable result. Furthermore, immunochemistry testing process differs in different laboratories. Regrading methodology and tests, different cutoff values are needed for different laboratories.

Thyroglobulin (TG)

TG is elevated in thyroid pathologies. However, this elevations does not differ according to the type of the pathology, it shows the mass of the pathology within thyroid gland [52]. The serum TG level is elevated in approximately 2/3 of differentiated thyroid cancers (sensitivity of 70% and specificity of 80%), nearly 50% of the poorly differentiated and anaplastic carcinomas of thyroid. TG is also elevated in benign conditions, such as Graves' disease. In contrast, approximately 12% of thyroid cancer patients has low or normal TG levels. Mutations in the TG gene may cause no-expression of the TG gene even though the TSH levels are high [53].

Urokinase Plasminogen Activator (uPA)

uPA is considered as a tumor marker in breast cancer staging. However, in order to detect this protein within tumor tissue, large amounts are needed. Thus, this marker cannot be used in patients with small tumors. Moreover, this test requires fresh or freshly frozen tumor tissue, which is not applicable in some clinics [54].

Prostate-Specific Antigen (PSA)

Blood PSA testing is a common screening method for prostate cancer and it is elevated in more than 75% of organ confined prostate cancers. However, most men with an elevated PSA level don't have prostate cancer, as elevated PSA may also be a result of benign prostate conditions such as prostatitis, benign prostatic hypertrophy, prostatic trauma and ejaculation.

The NCI-sponsored Prostate, Lung, Colorectal, and Ovarian Cancer Screening Trial (PLCO), and the European Randomized Study of Screening for Prostate Cancershowed that the PSA testing at best leads to a small number of decrease in deaths caused by prostate cancer. Furthermore, it is not clear whether the benefits of the PSA screening outweigh the harms of following diagnostic tests and treatments for cancers that in many cases would never have threatened a man's life [55, 56].

Conclusion

Despite the improvements in cancer treatment during the past decade, cancer is still the most frightening disease of this century. The lack of sensitivity for early stages of cancer and specificity for malignancy limit the use of current serum/urine/tissue-based tumor markers in screening and diagnosis.

Detection of cancer stem cells and cell-free tumor DNA in the bloodstream launched a new era in the tumor marker researches. In the past decades, developments in molecular techniques led to detection of genetic and epigenetic alterations in the circulating tumor DNA that is a new concept termed as liquid biopsy. The number of circulating tumor cells; mutations, epigenetic alterations in the circulating cell-free tumor DNA; levels of circulating microRNAs and tRNAs; the signature of long non-coding RNAs expression in tumors have been introduced as new generation molecular tumor markers and are currently under investigation.

References

[1] https://www.cancer.net/navigating-cancer-care/diagnosing-cancer/tests-and-procedures/tumor-marker-tests.

[2] https://www.cancer.gov/about-cancer/diagnosis-staging/diagnosis/%20tumor-markers-fact-sheet#q5

[3] Lira, M. E., Kim, T. M., Huang, D., Deng, S., Koh, Y., Jang, B., et al. (2013). Multiplexed gene expression and fusion transcript analysis to detect ALK fusions in lung cancer. *J Mol Diagn;* 15(1):51-61.

[4] Mattsson, J. S. M., Brunnström, H., Jabs, V., Edlund, K., Jirström, K., Mindus, S., et al.(2016). Inconsistent results in the analysis of ALK rearrangements in non-small cell lung cancer. *BMC Cancer.* https://bmccancer.biomedcentral.com/articles/10.1186/s12885-016-2646-x .

[5] Savic, S., Diebold, J., Zimmermann, A. K., Jochum, W., Baschiera, B., Grieshaber, S., et al. (2015). Screening for ALK in non-small cell lung carcinomas: 5A4 and D5F3 antibodies perform equally well, but combined use with FISH is recommended. *Lung Cancer;* 89(2):104-9.

[6] Nilsson, R. J., Karachaliou, N., Berenguer, J., Gimenez-Capitan, A., Schellen, P., Teixido, C., et al. (2016). Rearranged EML4-ALK fusion transcripts sequester in circulating blood platelets and enable blood-based crizotinib response monitoring in non-small-cell lung cancer. *Oncotarget;* 7(1):1066-75.

[7] Daniele, B., Bencivenga, A., Megna, A. S., Tinessa, V. (2004). Alpha-fetoprotein and ultrasonography screening for hepatocellular carcinoma. *Gastroenterology;* 127(5):108-12.

[8] Liu, C., Xiao, G. Q., Yan, L. N., Li, B., Jiang, L., Wen, T. F., et al. (2013). Value of α-fetoprotein in association with clinicopathological features of hepatocellular carcinoma. *World J Gastroenterol;* 19(11):1811–1819.

[9] Germà, J. R., Llanos, M., Tabernero, J. M., Mora, J. (1993). False elevations of alpha-fetoprotein associated with liver dysfunction in germ cell tumors. *Cancer;* 72(8):2491-4.

[10] Cassinat, B., Guardiola, P., Chevret, S., Schlageter, M. H., Toubert, M. E., Rain, J. D., et al. (2000). Constitutive elevation of serum alpha-fetoprotein in Fanconi anemia. *Blood;* 96(3):859-63.

[11] https://themmrf.org/multiple-myeloma/multiple-myeloma-tests/blood-tests/beta-2-microglobulin/.

[12] Bataille, R., Annweiler, C., Beauchet, O. (2013). Multiple myeloma international staging system: "staging" or simply "aging" system?. *Clin Lymphoma Myeloma Leuk;* 13(6):635-7.

[13] Fedele, P. L., Choy, K., Doery, J. C. G., Grigoriadis, G., Shortt, J., Lu, Z. X. (2014). Inter-laboratory discordance of beta-2 microglobulin results: impact on the

validity of the international staging system for multiple myeloma. *Br. J. Haematol;* 166(6):951-3.

[14] Lawless, M. E., Jour, G., Hoch, B. L., Rendi, M. H. (2014). Beta-human chorionic gonadotropin expression in recurrent and metastatic giant cell tumors of bone: a potential mimicker of germ cell tumor. *Int J Surg Pathol;* 22(7):617-22.

[15] Ledermann, J., Harter, P., Gourley, C., Friedlander, M., Vergote, I., Rustin, G., et al. (2014). Olaparib maintenance therapy in patients with platinum-sensitive relapsed serous ovarian cancer: a preplanned retrospective analysis of outcomes by BRCA status in a randomised phase 2 trial. *Lancet Oncol;* 15(8):852-61.

[16] Ross, T. S., Mgbemena, V. E. (2014). Re-evaluating the role of BCR/ABL in chronic myelogenous leukemia. *Mol Cell Oncol*; 1(3):450-8.

[17] Tang, H., Kong, Y., Si, L., Cui, C., Sheng, X., Chi, Z., et al. (2018). Clinical significance of BRAFV600E mutation in circulating tumor DNA in Chinese patients with melanoma. *Oncol Lett;* 15(2):1839–1844.

[18] Güler, B., Özyılmaz, F., Tokuç, B., Can, N., Taştekin, E. (2015). Histopathological Features of Gastrointestinal Stromal Tumors and the Contribution of DOG1 Expression to the Diagnosis. *Balkan Med J*; 32(4):388–396.

[19] Colomer, R., Ruibal, A., Genollà, J., Salvador, L. (1989). Circulating CA 15-3 antigen levels in non-mammary malignancies. *Br J Cancer*; 59(2):283-6.

[20] Bedi, M. M., Gandhi, M. D., Jacob, G., Lekha, V., Venugopal, A., Ramesh, H. (2009). CA 19-9 to differentiate benign and malignant masses in chronic pancreatitis: is there any benefit?. *Indian J Gastroenterol;* 28(1):24-7.

[21] Perkins, G. L., Slater, E. D., Sanders, G. K., Prichard, J. G. (2003). Serum tumor markers. *Am Fam Physician*; 68(6):1075-82.

[22] Daumerie, C., Maiter, D., Gruson, D. (2013). Serum calcitonin estimation in medullary thyroid cancer: basal or stimulated levels? *Thyroid Res;* 6(1):S4.

[23] Toledo, S. P., Lourenço, D. M., Santos, M. A., Tavares, M. R., Toledo, R. A., de Menezes Correia-Deur, J. E. (2009). Hypercalcitoninemia is not Pathognomonic of Medullary Thyroid Carcinoma. *Clinics* (Sao Paulo); 64(7):699–706.

[24] Duffy, M. J. (2001). Carcinoembryonic antigen as a marker for colorectal cancer: is it clinically useful?. *Clin Chem;* 47(4):624-30.

[25] Zhao, Q., Zhan, T., Deng, Z., Li, Q., Liu, Y., Yang, S., et al. (2018). Glycan analysis of colorectal cancer samples reveals stage-dependent changes in CEA glycosylation patterns. *Clin Proteomics*; 15, 9.

[26] Motta, G., Cea, M., Moran, E., Carbone, F., Augusti, V., Patrone, F. (2010). Monoclonal Antibodies for Non-Hodgkin's Lymphoma: State of the Art and Perspectives. *Clin Dev Immunol;* 2010:14.

[27] Gut, P., Czarnywojtek, A., Fischbach, J., Bączyk, M., Ziemnicka, K., Wrotkowska, E. (2016). Chromogranin A – unspecific neuroendocrine marker. Clinical utility and potential diagnostic pitfalls. *Arch Med Sci;* 12(1):1–9.

[28] Köhler, C. U., Martin, L., Bonberg, N., Behrens, T., Deix, T., Braun, K., et al. (2014). Automated quantification of FISH signals in urinary cells enables the assessment of chromosomal aberration patterns characteristic for bladder cancer. *Biochem Bioph Res Co*; 448(4):467-72.

[29] Autebert, J., Coudert, B., Champ, J., Saias, L., Guneri, E. T., Lebofsky, R., et al. (2015). High purity microfluidic sorting and analysis of circulating tumor cells: towards routine mutation detection. *Lab Chip;* 15(9):2090-101.

[30] Gerges, N., Rak, J., Jabado, N. (2010). New technologies for the detection of circulating tumor cells. *Br Med Bull;* 94:49-64.

[31] Mazzone, P. J., Wang, X. F., Han, X., Choi, H., Seeley, M., Scherer, R., et al. (2018). Evaluation of a Serum Lung Cancer Biomarker Panel. *Biomark Insights;* 13:1-5.

[32] Xu, Y., Xu, L., Qiu, M., Wang, J., Zhou, Q., Xu, L., et al. (2015). Prognostic value of serum cytokeratin 19 fragments (Cyfra 21-1) in patients with non-small cell lung cancer. *Sci Rep;* 5:9444. doi: 10.1038/srep09444.

[33] Pujol, J. L., Molinier, O., Ebert, W., Daurès, J. P., Barlesi, F., Buccheri, G., et al. (2004). CYFRA 21-1 is a prognostic determinant in non-small-cell lung cancer: results of a meta-analysis in 2063 patients. *Br J Cancer;* 90(11):2097-105.

[34] Zhao, D., Chen, X., Qin, N., Su, D., Zhou, L., Zhang, Q., et al. (2017). The prognostic role of EGFR-TKIs for patients with advanced non-small cell lung cancer. *Sci Rep;* 7:40374. doi: 10.1038/srep40374.

[35] Groenendijk, F. H., Zwart, W., Floore, A., Akbari, S., Bernards, R. (2013). Estrogen receptor splice variants as a potential source of false-positive estrogen receptor status in breast cancer diagnostics. *Breast Cancer Res Treat;* 140(3):475–484.

[36] Tsihlias, J., Grossman, H. B. (2000). The utility of fibrin/fibrinogen degradation products in superficial bladder cancer. *Urol Clin North Am*; 27(1):39-46.

[37] Molina, R., Escudero, J. M., Augé, J. M., Filella, X., Foj, L., Torné, A., et al. (2011). HE4 a novel tumour marker for ovarian cancer: comparison with CA 125 and ROMA algorithm in patients with gynaecological diseases. *Tumor Biol;* 32(6):1087–1095.

[38] Li, J., Dowdy, S., Tipton, T., Podratz, K., Lu, W. G., Xie, X., et al. (2009). HE4 as a biomarker for ovarian and endometrial cancer management. *Expert Rev Mol Diagn;* 9(6):555–566.

[39] Gadducci, A., Cosio, S., Carpi, A., Nicolini, A., Genazzani, A. R. (2004). Serum tumor markers in the management of ovarian, endometrial and cervical cancer. *Biomed Pharmacother;* 58:24–38.

[40] Onitilo, A. A., Engel, J. M., Stankowski, R. V. (2014). Cardiovascular toxicity associated with adjuvant trastuzumab therapy: prevalence, patient characteristics, and risk factors. *Ther Adv Drug Saf;* 5:154–66.

[41] Hofmann, M., Stoss, O., Shi, D., Büttner, R., van de Vijver, M., Kim, W., et al. (2008). Assessment of a HER2 scoring system for gastric cancer: results from a validation study. *Histopatholog;* 52:797–805.

[42] Incorvati, J. A., Shah, S., Mu, Y., Lu, J. (2013). Targeted therapy for HER2 positive breast cancer. *J Hematol Oncol;* 6:38. doi: 10.1186/1756-8722-6-38.

[43] Caldini, A., Nozzoli, C., Terreni, A., Staderini, M., Berardi, M., Biagioli, T., et al. (2016). New patterns of relapse in multiple myeloma: a case of "light chain escape" in which FLC predicted relapse earlier than urine and serum immunofixation. *Clin Chem Lab Med;* 54(6):991-5.

[44] Schuster, S. R., Rajkumar, S. V., Dispenzieri, A., Morice, W., Aspitia, A. M., Ansell, S., et al. (2010). IgM multiple myeloma: Disease definition, prognosis, and differentiation from Waldenstrom's macroglobulinemia. *Am J Hematol;* 85(11):853–855.

[45] McCormick, F. (2015). KRAS as a Therapeutic Target. *Clin Cancer Res;* 21(8):1797–1801.

[46] O'Byrne, K. J., Gatzemeier, U., Bondarenko, I., Barrios, C., Eschbach, C., Martens, U. M., et al. (2011). Molecular biomarkers in non-small-cell lung cancer: a retrospective analysis of data from the phase 3 FLEX study. *Lancet Oncol;* 12(8):795-805.

[47] Shepherd, F. A., Domerg, C., Hainaut, P., Jänne, P. A., Pignon, J. P., Graziano, S., et al. (2013). Pooled analysis of the prognostic and predictive effects of KRAS mutation status and KRAS mutation subtype in early-stage resected non-small-cell lung cancer in four trials of adjuvant chemotherapy. *J Clin Oncol;* 31(17):2173-81.

[48] Liu, R., Cao, J., Gao, X., Zhang, J., Wang, L., Wang, B., et al. (2016). Overall survival of cancer patients with serum lactate dehydrogenase greater than 1000 IU/L. *Tumour Biol;* 37(10):14083–14088.

[49] Liu, X., Zhang, W., Yin, W., Xiao, Y., Zhou, C., Hu, Y., et al. (2017). The prognostic value of the serum neuron specific enolase and lactate dehydrogenase in small cell lung cancer patients receiving first-line platinum-based chemotherapy. *Medicine.* https://www.ncbi.nlm.nih.gov/pubmed/29145241 .

[50] Wang, Z., Que, H., Suo, C., Han, Z., Tao, J., Huang, Z., et al. (2017). Evaluation of the NMP22 BladderChek test for detecting bladder cancer: a systematic review and meta-analysis. *Oncotarget;* 8(59):648-56.

[51] Seetharamu, N., Preeshagul, I. R., Sullivan, K. M. (2017). New PD-L1 inhibitors in non-small cell lung cancer – impact of atezolizumab. *Lung Cancer;* 8:67–78.

[52] Knudsen, N., Bülow, I., Jørgensen, T., Perrild, H., Ovesen, L., Laurberg, P. (2001). Serum Tg--a sensitive marker of thyroid abnormalities and iodine deficiency in epidemiological studies. *J Clin Endocrinol Metab;* 86:3599–3603.

[53] Indrasena, B. S. H. (2017). Use of thyroglobulin as a tumour marker. *World J Biol Chem;* 8(1):81–85.

[54] Duffy, M. J., McGowan, P. M., Harbeck, N., Thomssen, C., Schmitt, M. (2014). uPA and PAI-1 as biomarkers in breast cancer: validated for clinical use in level-of-evidence-1 studies. *Breast Cancer Res;* 16(4):428.

[55] Andriole, G. L., Crawford, E. D., Grubb, R. L. 3rd, Buys, S. S., Chia, D., Church, T. R., et al. (2009). Mortality results from a randomized prostate-cancer screening trial. *N Engl J Med;* 360(13):1310-9.

[56] Schröder, F. H., Hugosson, J., Roobol, M. J., Tammela, T. L., Ciatto, S., Nelen, V., et al. (2009). Screening and prostate-cancer mortality in a randomized European study. *N Engl J Med;* 360(13):1320-8.

In: Research on New Generation …
Editor: Yildiz Dincer
ISBN: 978-1-53614-367-6

Chapter 2

FXYD3: A PROMISING BIOMARKER FOR CANCER TREATMENT

Chia-chi Liu**, *PhD
Sydney Medical School,
University of Sydney, NSW, Australia

ABSTRACT

Easy access to the Na^+-K^+ pump in the cell surface membrane and the critical dependence of cell survival on the pump has made it is an attractive therapeutic target in cancer. Use of cardiac glycosides has been explored but has been found to have limited effectiveness due to cardiac toxicity of the drugs. As an alternative we have examined whether FXYD proteins that associate closely with the Na^+-K^+ pump molecular complex might be a useful target. FXYD3 is of particular interest because it is markedly over-expressed in several common cancers and we have shown that several FXYD proteins, including FXYD3, are critical for reversal of glutathionylation, an oxidative modification that inhibits activity of the β1 Na^+-K^+ pump subunit.

We hypothesized that FXYD3 protein overexpression protects pump function against inhibition by the high levels of oxidative stress cancer cells typically display and that a reduction in FXYD3 expression levels would sensitize cells to chemotherapy and radiotherapy that largely induce cell death by increasing oxidative stress.

In light of the reported treatment resistance of overexpressing FXYD3 cancers, results suggest silencing wild type proteins may greatly strengthen the efficacy of treatments that increase oxidative stress within tumors. Such increases are commonly seen with radiotherapy and chemotherapeutic agents. This ongoing study endeavors to develop amalgamated novel treatments for cancer patients while alleviating side-effects associated with traditional therapy; advance diagnosis and improve overall patient well-being.

* Corresponding Author Email: chiachi.liu@sydney.edu.au.

Keywords: Na^+-K^+ pump, FXYD3 protein, oxidative stress, cancer treatment, biomarker

INTRODUCTION

Cancer is an extremely challenging disease due to its high rates of resistance to traditional chemotherapy and radiotherapy, highlighting the crucial need for novel anticancer drugs or more effective therapeutic approaches [1]. The Na^+-K^+ pump mediates counter transport of intracellular Na^+ in exchange for extracellular K^+. Since the Na^+-K^+ pump is easily accessible in the cell surface membrane and critical for cell survival, it is an attractive therapeutic target in cancer. Pump-inhibiting cardiac glycosides are cytotoxic for cancer cells *in vitro* but their toxicity for normal cells limits use in doses high enough for *in vivo* clinical use against cancers.

FXYD domain-containing ion transport regulator 3 (FXYD3), also known as mammary tumour 8, belongs to the FXYD protein family and is localized in the cell membrane and cytoplasm. The proteins are known for their role in regulation of the Na^+-K^+ pump and are expressed in a tissue-dependent manner [2]. Studies demonstrated that this integral cell membrane protein was differentially expressed between tumour cells and normal cells, regulated cell proliferation, cell apoptosis and tumour metastasis, influenced the cell cycle, and participated in tumour angiogenesis and progression [3, 4]. However, FXYD3 expression is particularly high in various types of cancer, including breast cancer [5], prostate cancer [4] and pancreatic cancer [5]. The predominant expression of FXYD3 in cancer has led to its use as a biomarker of malignancy. Recent developments in our understanding of the role FXYD proteins have in cellular function have suggested that FXYD3 up-regulation could also have a functional role that might be targeted therapeutically [6].

NA^+-K^+ PUMP

The Na^+-K^+ pump is a ubiquitous enzyme active in the membrane of eukaryotic cells. It is responsible for creation and maintenance of the Na^+ and K^+ gradients across the cell. This membrane glycoprotein complex utilizes the energy released by hydrolysis of adenosine-5'-triphosphate (ATP) to transport 3 Na^+ ions out and 2 K^+ ions into the cell against the electrochemical gradient of each ion. [7, 8]. The Na^+-K^+ pump is primarily a sodium pump as the K^+ can be substituted with other monovalent metal and inorganic cations [9]. This function is crucial for cell survival and body homeostasis because the Na^+ gradient is used as an energy source to transport ions or solutes [10].

STRUCTURE OF THE NA^{+}-K^{+} PUMP AND FXYD PROTEIN

The Na^{+}-K^{+} pump is a heterodimer comprised of a large α subunit and a smaller β subunit. The α subunit has ten trans-membrane helices [11] and includes the cytoplasmic catalytic domain (ATP-binding and phosphorylation sites) and the extracellular ouabain binding site [12]. The major isoform, α1, is expressed in nearly every tissue. In contrast, the expression of α2, α3 and α4 isoforms is restricted to specific tissues. The β subunit is a single membrane-crossing protein with its amino terminus in the cytoplasm. It is required for folding, membrane insertion and membrane delivery of the catalytically active α subunits and has highly conserved glycosylation sites and disulfide bridge-forming cysteine residues. In addition, the β subunit is essential for enzyme activity and influences the transport properties of mature sodium pumps. Three different Na^{+}-K^{+} ATPase β isoforms have been identified, which are distributed in a tissue-dependent manner. Like the α1 subunit, the β1 subunit is expressed in nearly every tissue [13].

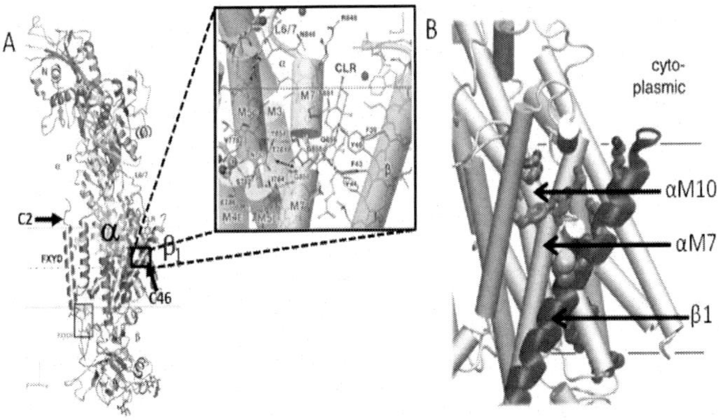

Figure 1. The structure of the Na^{+}-K^{+} pump. 3D structure of α/β/FXYD complex. Transmembrane domains are between the 2 green lines. The β subunit and FXYD proteins are single-transmembrane spanning while 10 helices of the α subunit, numbered 1-10, span the membrane. Cysteine residues in the β1 subunit (C46) and FXYD proteins (C2) are indicated. The inset shows hydrogen bonds from tyrosine residues near C46 in the β subunit to αM713. B. Location of C46 in the β1 subunit is shown with its Sulphur atom in yellow. Polar residues lining the interface between α and β subunits are indicated (oxygen in red, nitrogen in blue). Adapted from Liu et al. [29].

A third subunit is part of the FXYD family of small, single transmembrane proteins, named after their signature extracellular FXYD sequence [14]. They are expressed in a tissue-specific manner [15] and associate with the Na^{+}-K^{+} pump in all tissues. Despite a close association, FXYD proteins are not an integral part of the catalytic and ion transporting function of the Na^{+}-K^{+} pump. However, it is firmly established that the FXYD proteins modulates Na^{+}-K^{+} pump function [16]. The three-dimensional structure of the α/β/FXYD complex is shown in Figure 1A [17]. The crystal structure indicates multiple hydrogen bonds to α- and β subunits in the extracellular domain [18] but the FXYD/α association in the transmembrane domain can only be attributed to weak van der

Walls interactions and possibly to a single hydrogen bond, which has not been firmly identified. However, the basic amino acids lysine and arginine of FXYD proteins and of the α subunit cluster near the cytosol/membrane interface (Figure 1A). They are positively charged at a physiological pH and the electrostatic attraction they share to negative charges at the inner membrane leaflet [19] may stabilize FXYD/α interaction. Alternatively, the cytoplasmic tail of FXYD proteins might interact with the cytoplasmic domain of the subunit but this cannot be evaluated from the crystal structure since the cytoplasmic tail of FXYD proteins has not been resolved.

PARADIGMS FOR REGULATION OF THE NA^+-K^+ PUMP

Protein Phosphorylation Signalling Pathways

Effects of the adrenergic receptor-coupled activation of protein kinases A and C (PKA and PKC) on pump function have been extensively studied, but remarkably, the functional consequences of such signaling on the pump have remained a matter of controversy [10]. The most recent studies have attributed the stimulation to phosphorylation of Ser68 in the cytoplasmic terminal of FXYD1, which relieves an inhibitory effect of FXYD1 on the pump [10]. In striking contrast to this consensus view, the work from our Laboratory has shown that PKA activation with exposure of isolated cardiac myocytes to the adenylyl cyclase activator Forskolin *in vitro* causes Na^+- K^+ pump inhibition. This has been found to be mediated by the activation of PKC and NADPH oxidase in a downstream pathway that is shared with Ang II [20, 21].

Moreover, reconciling the effects of protein kinase-dependent signaling on the pump that are attributed to phosphorylation with the structure of the pump and the expression pattern of its subunits in different tissues poses a challenge. A regulatory role for phosphorylation of FXYD is difficult to reconcile with the fact that only FXYD1, expressed in the heart, has functional phosphorylation sites [16] which cannot account for the regulation of the pump by this mechanism in other tissues. An ideal consensus motif for PKA phosphorylation on the α1 pump subunit exists, and despite an expected poor access of PKA to the site [22], it can undergo phosphorylation.

However, functional effects and molecular mechanisms have been difficult to establish. An amino acid mutation designed to mimic phosphorylation of the PKA motif seemed to have no functional effect in *Xenopus oocytes*, a result that was consistent with molecular dynamics simulation. In contrast, simulated phosphorylation of the wild-type amino acid suggested that phosphorylation decreases Na^+ binding affinity and hence should inhibit the pump [23]. PKC can phosphorylate subunits despite the absence of a consensus motif for this family of kinases, but the functional significance is uncertain [23].

Protein Kinase-Dependent Redox Regulation of the Na^+-K^+ Pump

Since phosphorylation of the Na^+-K^+ pump molecular complex cannot readily account for effects of the protein kinase-dependent regulation of the Na^+-K^+ pump function, the work in our Laboratory has explored the role of oxidative posttranslational modifications in kinase-dependent pump regulation. Oxidative modifications can affect the structure and function of proteins in a manner analogous to phosphorylation [24]. Of these, the formation of a disulfide bond between a protein cysteine and the cytosolic tri-peptide glutathione (Protein S-glutathionylation, P-GSS) is of particular interest for signaling because glutathionylation is a stable, yet reversible modification, with effects on protein structure and function reminiscent to that of phosphorylation [25]. Redox regulation seemed a plausible alternative because chemical oxidants can inhibit Na^+-K^+ ATPase as pump activity [26].

Because of the classic role of Ang II in activation of NADPH oxidase, which induced reactive oxygen species (ROS) generation [27], we examined the effects of Ang II on pump function and a role for protein-GSS [28]. The Ang II-induced activation of oxidative signaling was associated with glutathionylation of the β1 subunit of the Na^+-K^+ pump (β1-GSS) [26]. Mutational studies of Na^+-K^+ pumps expressed in *Xenopus oocytes* identified Cys46 as the reactive residue in the β1 subunit [26] and, consistent with the NADPH oxidase-dependence of Ang II-induced inhibition of pump [20], there was a causal relationship between $ONOO^-$-induced β1 subunit glutathionylation and pump inhibition [26]. Access of the cytosolic GSH to cysteine 46 was counterintuitive in view of the location of the cysteine in the lipid phase of the transmembrane segment (Figure 1B). However, we subsequently showed that structural changes of the pump molecule during its catalytic cycle accounts for access. Polar residues line the interface between the β subunit and transmembrane segments 7 (αM7) and 10 of the α subunit from the cytoplasm to C46 (Figure 1B) and molecular dynamic simulations suggest that that minor structural changes in the pump complex cause a membrane deformation that yields a hydrophilic environment for C46 [29].

Na^+-K^+ Pump as a Target for Cancer

Skou first described Na/K-ATPase and its primary function in 1957. Years later this pump has been shown to play a role in cell signaling and has been considered a target for cancer treatment [30]. A growing number of discoveries of Na^+/K^+-ATPase inhibitors (e.g., cardiotonic steroids, CTS) can be used to effectively treat certain cancers including breast cancer, leukemia, prostate cancer and tumors of the kidney/urinary tract [31].

However, the cytotoxic effects of CTS significantly differ depending on how it is applied. It seems that a significant problem that must be resolved is achieving therapeutic levels of the CTS in the target tissues of the tumour [31]. Therefore, more evidence to support the clinical implications of the use of CTS in patients with cancer still needs to be addressed.

FXYD3 Proteins and Reversal of B1 Pump Subunit Glutathionylation and Pump Inhibition

A reactive cysteine in the C2 position of FXYD3 proteins was critical for the reversal of glutathionylation of Cys46 of the β1 subunit and Na^+-K^+ pump inhibition induced by chemical oxidants [32]. We mutated all 4 cysteines in FXYD3 to serine, thus synthesizing a "Cysless" FXYD3. Mutation of the FXYD3 cysteines in the membrane domain does not affect association with the Na^+-K^+ pump α subunit [33]. WT and Cysless FXYD3 can also cause competitive displacement of the native FXYD protein (Figure 2A). We found that β1 subunit glutathionylation and pump inhibition could be reversed by exposing myocytes to exogenous wild-type FXYD3. A cysteine-free FXYD3 derivative had no effect (Figure 2B) [32, 34]. This suggested that FXYD3 protein protects Na^+-K^+ pump against oxidative damage hence repairs pump function.

Role of FXYD Proteins Involved in the Direction of Na^+/K^+ Pump in Cancer

FXYD proteins are a 7-member mammalian family of single transmembrane-spanning proteins named after an invariant FXYD sequence in their extracellular domain (Figure 3) [2]. They are mostly known for associating with the membrane Na^+-K^+ pump and their ability to modulate its characteristics [35]. Only FXYD1 has phosphorylation sites that might allow it to mediate pump regulation through protein kinases. FXYD2-7 were thought to modify pump function according to their presence or absence and have no role in regulation that is dependent on cell signalling [35]. This left regulation poorly unexplained in the many tissues/cells that predominantly express FXYD2-7.

Two cysteine residues in the cytoplasmic terminal, named C1 and C2 in Figure 2, are conserved in the 7-member mammalian FXYD family (Figure 3). While most cysteine residues in proteins are not susceptible to glutathionylation ("reactive"), reactivity is increased by neighbouring basic amino acids [25]. Figure 1B shows that C1 and C2 are mostly flanked by the basic lysine (K) and arginine (R), and studies on cardiac myocytes indicated glutathionylation of FXYD1 [32].

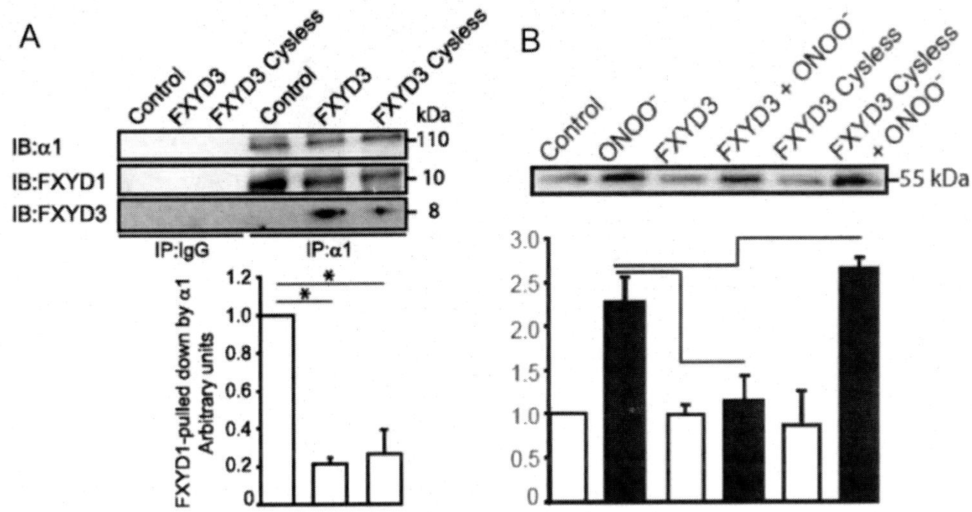

Figure 2. A. Displacement of FXYD1 from α1 subunit by incubation of myocytes with 500 nM WT- or Cysless FXYD3 for 15 min. FXYD3 was used to allow distinction from FXYD1 with antibodies. B. Pre-incubation of myocytes with WT FXYD3 eliminated the increase in glutathionylation of the β_1 subunit detected in myocytes exposed to $ONOO^-$, while Cysless FXYD3 had no effect. From Figure 3A & 3C in [31].

Mutagenesis identified C2 but not C1 as reactive, with the reactivity of C2 dependent on flanking basic amino acids. C2 is flanked by two basic amino acids in FXYD3 but only one in FXYD2 (Figure 3) and mutation of the nonbasic amino acid to a basic one made FXYD2 susceptible to glutathionylation, similar to other members of the FXYD family [32]. In addition to the importance of surrounding basic amino acids in the 2D protein structure, surrounding residues in the 3D structure also play a critical role. Access to C2 in FXYD proteins is more straightforward in view of its location in the cytosol near the inner membrane leaflet (Figure 4). Indeed, the 3D structure of the Na^+-K^+ pump suggests that proximity to basic amino acids in the α subunit might account for the differences in reactivity between C1 and C2 [32]. This reactivity was critical for receptor and protein kinase-dependent Na^+-K^+ pump regulation, a role that must be seen in the context of the new paradigm for pump regulation we have described.

The paradigm for Na^+-K^+ pump regulation we have described preserves established roles of protein kinases in receptor-coupled Na^+-K^+ pump control but it invokes kinase dependent regulation of redox signaling and oxidative modification [20, 36] rather than phosphorylation of the pump molecular complex invoked previously. It also greatly expands the known role of FXYD proteins. Widespread interest in it is indicated by recent reviews across a broad range of disciplines [24, 37-40], and it is well integrated with pathophysiology and pharmacotherapy [20, 36].

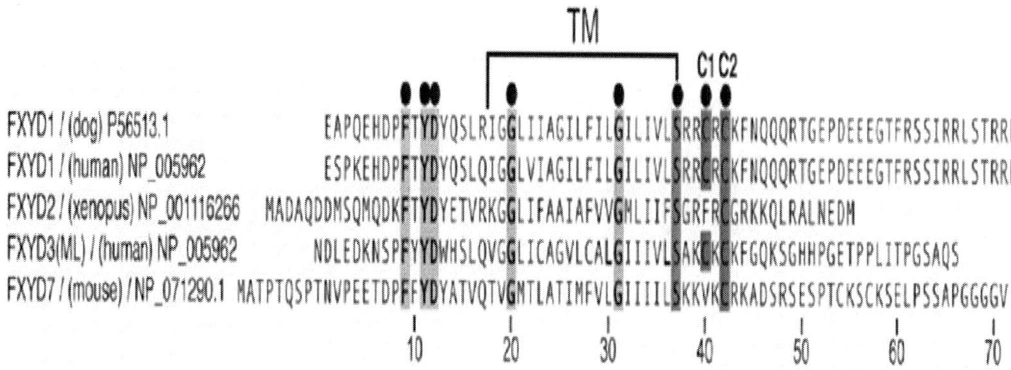

Figure 3. Amino acid sequence alignment in mammalian FXYD protein family. Numbering corresponds to FXYD's sequence and begins at 1 after the signal peptide (not shown). Conserved residues are marked with filled circles. Transmembrane domain is indicated by TM. Susceptibility to glutathionylation of FXYD1, 3 and 7 depends on two basic amino acid residues that flank C2. FXYD2 has one immediately adjacent basic amino acid only and is not susceptible to glutathionylation [31].

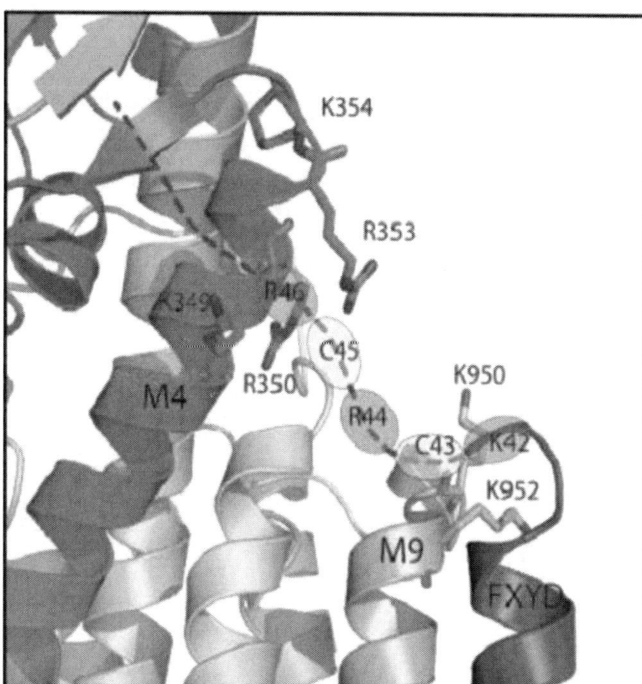

Figure 4. Cysteines at interface between cytoplasm and membrane extrapolated from the 3D structure of shark rectal gland Na^+-K^+ ATPase (Shinoda et al. 2009). The last amino acid residue of FXYD10 (non-mammalian homologue of FXYD1) detected in the crystal structure is K42 just before C1 equivalent (C43). The trajectory of the cytoplasmic domain (dotted line) was derived as described [31].

Observational Studies on FXYD3 Protein and Cancer Treatment Response

Up-regulation of cytosolic antioxidant defenses reduces the therapeutic response to radiotherapy in breast cancer, consistent with the role radiation-induced synthesis of reactive oxygen species has in the cytotoxic effects of the treatment [41]. Due to the role FXYD3 has in reversing oxidative Na^+-K^+ pump inhibition, FXYD3 might provide a "boundary defense." Consistent with such protection, high levels of FXYD3 expression reduced radiotherapy-induced necrosis in human colorectal cancers [42]. Expression levels had no effect on prognosis for patients randomized not to receive radiotherapy, but a high level was a strong predictor of poor survival for patients randomized to have radiotherapy (Figure 5), even after adjusting for other possible predictors [42]. FXYD3 may also protect against chemotherapy-induced ROS as suggested by resistance to 5-fluorouracil with increased FXYD3 expression in a colorectal cancer cell line [43].

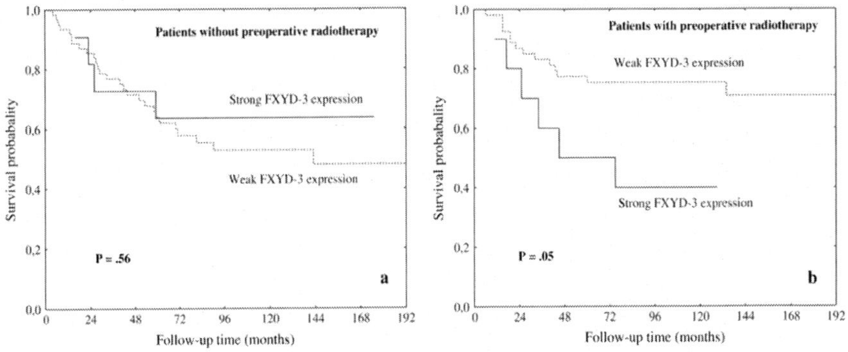

Figure 5. Outcome for patients with colorectal cancer randomly assigned to pre-operative radiotherapy. Strong FXYD3 expression had no effect on outcome without radiotherapy but predicted a poor outcome for those receiving radiotherapy. Adapted from Loftas et al. [42].

FXYD3 as a Promising Biomarker in Cancer

FXYD3, also known as Mat-8 (mammary tumor, 8 kDa), was originally cloned from murine mammary breast tumors induced by the neu and ras oncogenes [44]. Although it is a member of the FXYD family, it differs from most of the other members. It has a signal peptide that is uncleaved and completely different due to being the only FXYD protein with two transmembrane domains. The other members have only one transmembrane domain [16, 45]. It is reported that FXYD3 is expressed not only in normal tissues but also in tumors. In normal tissues FXYD3 is predominantly expressed in the stomach and colon. In tumors, it has been found in breast cancer, colon cancer, bladder cancer, benign and malignant prostate tumors, as well as pancreatic ductal adenocarcinoma [46]. Moreover, in some types of cancers such as breast [44], pancreatic [5], and prostate cancer [4], FXYD3 is overexpressed. The predominant expression of FXYD3 in cancer has led to its use as a biomarker of malignancy.

FXYD3 AS A THERAPEUTIC TARGET IN CANCER

Grzmil et al. [4] showed that small interfering RNA (siRNA)-mediated inhibition of FXYD-3 expression promotes a decreased proliferation in prostate cancer cell line. Downregulation of FXYD3 in pancreatic cancer cell lines alters its growth behavior both *in vitro* and *in vivo*, suggesting that FXYD3 promotes the proliferation of pancreatic cancer cells [5]. Recent developments in our understanding of the role FXYD proteins have in cellular function have suggested that FXYD3 up-regulation could also have a functional role that might be targeted therapeutically [6].

In support of the importance of Na^+-K^+ pump function for survival inhibition of pump activity with cardiotonic steroids, CTS induces apoptosis in a variety of cancer cells [47]. FXYD3-dependent protection of cancer cells against oxidative stress would be important from a therapeutic perspective because cytotoxic effects of treatment with many antitumor agents and radiotherapy are largely due to the increase in oxidative stress the treatments induce [48].

We have examined the effect of siRNA-induced knockdown of FXYD3 on the Na^+-K^+ pump of MCF-7 and MDA-MB 468 human breast cancer cells and on the sensitivity of the cells to Doxorubicin (Dox) and γ-radiation. MCF-7 cells belong to the luminal A group of breast cancers that are endocrine responsive and often chemotherapy responsive while MDA-MB 468 belong to the basal breast cancer group that are endocrine nonresponsive and often chemotherapy responsive [49].

FXYD3 in MCF-7 human breast cancer cells co-localizes with the membrane Na^+-K^+ pump [3]. Such co-localization is a cardinal feature of the family of FXYD membrane proteins [2, 16]. Exposure of human breast cancer cells to FXYD3 increased glutathionylation of the β1 Na^+-K^+ pump subunit consistent with the role FXYD3 has in promoting the reversal of glutathionylation of the β1 subunit [32]. Since glutathionylation of the β1 pump subunit causes pump inhibition [26] reversal of glutathionylation also reverses pump inhibition [32] and the decrease in Na^+/K^+-ATPase activity with suppression of FXYD3 is expected from the increase in β1 subunit glutathionylation.

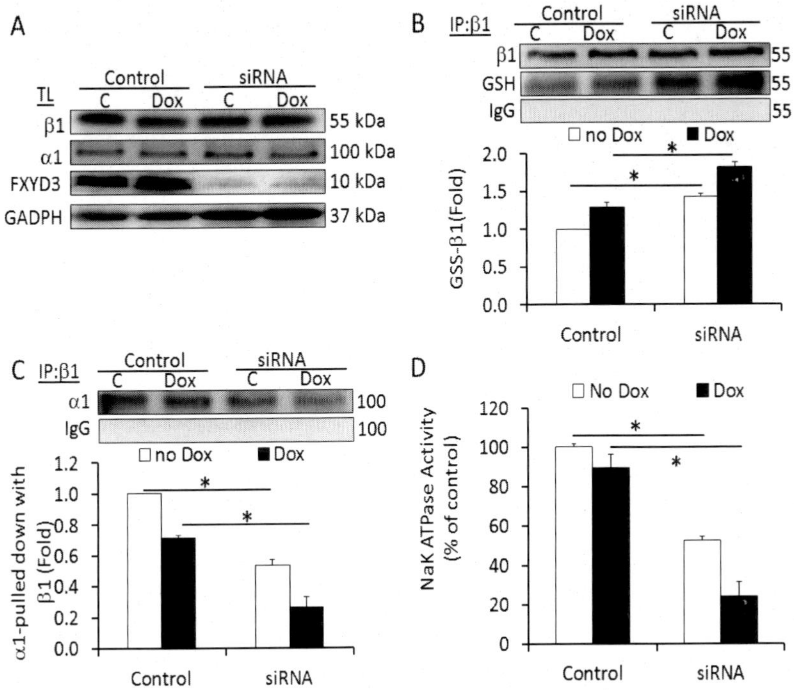

Figure 6. Effects of FXYD3 siRNA and Dox on β1 subunit glutathionylation, α1/β1 subunit interaction and Na^+/K^+ ATPase activity in MCF-7 cells. (A) Expression of α1-, β1 subunit and FXYD3 protein. TL: total lysate; GADPH: internal loading control; siRNA: siRNA transfection for 48 hrs; Control: No siRNA transfection; C: No Dox treatment; Dox: Dox treatment for 24 hrs. (B) β1 Na^+-K^+ subunit glutathionylation as detected in an immunoprecipitate with β1 subunit antibody and immunoblotted with a GSH antibody. The β1 subunit was used as a loading control. IgG is negative control for the co-immunoprecipitation. (C) Co-immunoprecipitation of α1 and β1 pump subunits. IP: immunoprecipitate. (D) Ouabain-sensitive ATPase activity. Adapted from Liu et al. [6].

Since Dox also increases oxidative stress these effects in cardiac myocytes are in good agreement with the effect of a reduction in FXYD3 expression in MCF-7 cells to amplify Dox -induced inhibition of Na^+/K^+-ATPase in this study (Figure 6D). We conclude protection of the β1 Na^+-K^+ pump subunit against glutathionylation and hence pump function against inhibition induced by oxidative stress is a feature of FXYD proteins that extends to cancer cells.

FXYD proteins stabilize the interaction of α- with β subunits in the Na^+-K^+ pump molecular complex [50] and the decrease in α1/β1 subunit coimmunoprecipitation with the siRNA-induced down-regulation of FXYD3 we report here (Figure 6C) is as expected. In contrast, a primary effect of FXYD3 to stabilize α/β subunit interaction cannot account for the decrease in coimmunoprecipitation of α1/β1 subunits we found in lysate of MCF-7 cells that had been exposed to Dox but not to FXYD3 siRNA because Dox alone up-regulated FXYD3 expression (Figure 6A). However, exposure to the drug also increased glutathionylation of the β1 subunit and the three dimensional structure of the Na^+-K^+ pump molecular complex indicates the glutathionylation is expected to reduce the strength of association between α1 and β1 subunits [34]. Any effect of increased expression of FXYD3 with Dox exposure to strengthening α/β subunit interaction would therefore be opposed by β1 subunit glutathionylation weakening the interaction. Within this scheme, supressing FXYD3 synthesis with siRNA and increasing β1 subunit glutathionylation with exposure to Dox should have additive effects on α1/β1 subunit coimmunoprecipitation.

While exposure of MCF-7 cells to FXYD3 siRNA alone induced a ~50% decrease in Na^+/K^+-ATPase activity it had had no significant effect on indices of cell viability or apoptosis (Figure 6D). This is consistent with the absence of an effect on viability of MCF-7 cells of cardiac glycosides in concentrations that inhibits Na^+/K^+-ATPase by 50% [51]. Effects of cardiac glycosides on proliferation of different malignant and non-malignant cells lines decrease with an increase in the level of expression of Na^+-K^+ pumps and expression of pumps is high in MCF-7 cells [52]. Overexpression of Na^+-K^+ pumps may therefore have left enough pump capacity for a 50% decrease not to affect viability or apoptosis. However, a further decrease in Na^+/K^+-ATPase activity in cells exposed to FXYD3 siRNA and Dox (Figure 7A & 7B) or γ-radiation (Figure 7 C & 7D) was associated with a large increase in the effects on these indices. These results are reminiscent of the effect of glycosides to enhance cytotoxic effects of radiation [47].

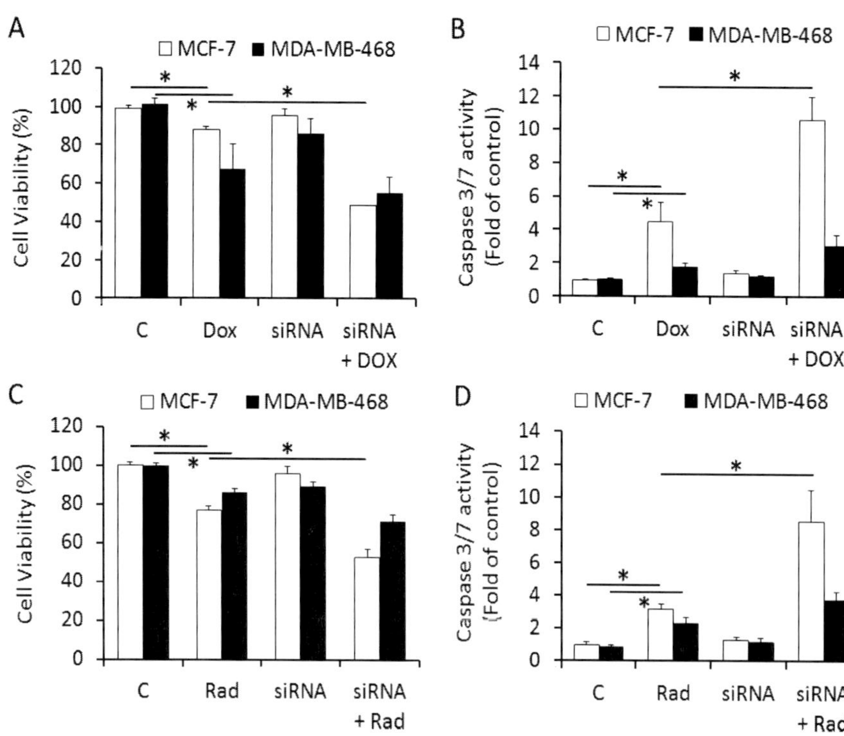

Figure 7. Effects of FXYD3 siRNA with Doxorubicin or γ-radiation on cell survival. MCF-7 or MDA-MB 468 cells were treated or not treated with 1 μM doxorubicin (Dox) for 24 hr after siRNA transfection (48 hrs) and cell viability (A) and caspase-3 activity (B) measured. MCF-7 or MDA-MB 468 cells were treated or not treated with γ-radiation (Rad) for 24 hr after siRNA transfection (48 hrs). Cell viability (C) and caspase-3 activity (D) were also measured. Adapted from Liu et al. [6].

As for MCF-7 cells, exposure of MDA-MB 486 cells to FXYD3 siRNA decreased Na^+/K^+-ATPase activity but did not affect cell viability or apoptosis (Figure 7). A cell-line related to MDA-MB 486, MDA-MB 231, expresses Na^+-K^+ pumps at even higher levels than MCF-7 cells [52], and if MDA-MB 486 cells express similarly high levels of pumps the cells might be able to resist even large decreases in Na^+/K^+-ATPase activity before the decrease *per se* affects viability. In contrast to what was the case for MCF-7 cells, FXYD3 suppression MDA-MB 486 did not augment the effects of Dox or γ-radiation on Na^+/K^+-ATPase activity or cell survival. Expression of FXYD3 at baseline in MDA-MB 486 cells was significantly lower than in MCF-7 cells and our results suggest defences of MDA-MB 486 cells against treatment are relatively independent of FXYD3 levels.

The increase in the sensitivity of MCF-7 cells to Dox or γ-radiation with FXYD3 suppression implies that FXYD3 confers treatment resistance. This is consistent with a study on colorectal cancer cells in which cell lines that were resistant to 5-Fluorouracil (5-FU) had markedly increased levels of FXYD3 mRNA compared with cells that were sensitive [43]. In pancreatic cancer cells antisense FXYD3 transfection had no effect on the sensitivity of cells to 5-FU or Cisplatin and Gemcitabine. However, growth rates of tumors in nude mice transplanted with antisense transfected cells was reduced relative to growth rates in mice transplanted with non-transfected cells [53].

In a clinical study patients with colorectal cancer were assigned to groups receiving or not receiving radiotherapy before tumors were surgically resected [54]. The resected tumors were classified according to strong or weak FXYD3 expression as identified by immunohistochemistry. Expression level had no effect on prognosis for patients not receiving radiotherapy. However, strong expression was a predictor of poor survival for patients randomized to have radiotherapy, even after adjusting for other possible predictors [42].

Exposure of MCF-7 cells to Doxorubicin or γ-radiation up-regulated FXYD3 expression in MCF-7 cells in this study but had no effect on the expression in MDA-MB 468 cells suggesting that treatment resistance is not only predicted by high FXYD3 baseline levels. If *in vitro* drug sensitivity is used to select treatment [55] the response of FXYD3 expression to drug exposure might add predictive accuracy to the clinical response to treatment and improve choice of drugs. Exposure of MCF-7 cells to Tomudex or Oxaliplatin and exposure of colonic cancer cells to 5-FU increases expression FXYD3 mRNA [43]. If also reflected in protein levels the increase might be useful in predicting the development of treatment resistance for drugs and cells other than Doxorubicin and MCF-7.

Radiotherapy kills cancer cells predominantly by generating reactive oxygen species [48] and redox proteins important for glutathione metabolism have been identified as predictors of outcome after radiotherapy for breast cancer [56]. The present study indicates that the role of FXYD3 to facilitate de-glutathionylation of the membrane Na^+-K^+ pump and protect pump function against oxidative inhibition previously shown in cardiac myocytes and *Xenopus* oocytes [32] also applies for MCF-7 breast cancer cells. It also shows that a modest downregulation of FXYD3 enhances the sensitivity of the cancer cells to γ-radiation. Woolston et al. (2011) proposed that selected redox proteins in tumors that regulate glutathione metabolism might be useful for predicting optimal radiotherapy treatment modalities. On the basis of the present study we suggest FXYD3 is a redox protein that might be useful as part of a predictive algorithm in future clinical studies aimed at such optimization. A similar approach might be useful for chemotherapeutics for which oxidative stress is thought to contribute to their cytotoxic effects [6].

FXYD3 as Personalized Drug for Cancer Treatments

Knockdown using siRNA is not easily translated to clinical practice and we have developed an alternative strategy based on the concept that transmembrane mimetic peptides can be used to disrupt membrane protein function. Specifically, in preliminary work we have developed derivatives of FXYD3 that are mutated to eliminate function of the WT protein (unpublished). These analogues induce knockdown of FXYD protein function by displacing WT FXYD proteins in cell membranes. The use of FXYD3 derivatives we propose offers a highly novel and clinically realistic method to "knock-out" protective effects of overexpressed FXYD3 in common cancers, one of which (pancreatic cancer) is particularly resistant to treatment. Further validation of the concept will need *in vivo* studies of efficacy and safety and perhaps further development of a compound that can selectively deliver the FXYD3 derivative to targeted cancers. The development of the FXYD analogues and their adjuvant anticancer activity are major focuses in my Laboratory.

For *in vivo* delivery of FXYD3 derivatives to cancers and insertion into the targeted membranes, the proteins will have to be administered systemically, typically intravenously. In support of the feasibility of this, systemic *in vivo* administration of transmembrane-mimetic peptides of multimeric membrane proteins has previously been shown to markedly reduce the membrane protein function within minutes of peptide administration [57]. We have also found that recombinant WT FXYD3 co-immunoprecipitates with the β_1 Na^+-K^+ pump subunit in the myocardium after its intra-peritoneal injection in mice (unpublished) supporting feasibility of *in vivo* delivery.

Last Word

I sincerely hope that this chapter will attract significant attention by investigators in cancer research and therefore stimulate further work to explore the molecular mechanism(s) by which FXYD3 protein is involved in tumor development and progression. I believe that this experimental evidence will also provide the rationale for designing FXYD3 specific inhibitors (such as mutated FXYD3 derivatives) as novel approaches to achieve better clinical benefit in treating a number of types of human cancer.

References

[1] Zhang, Y., K. Satoh, and M. Li. (2012). "Novel therapeutic modalities and drug delivery in pancreatic cancer - an ongoing search for improved efficacy." *Drugs Context* 2012:212244. doi: 10.7573/dic.212244.

[2] Sweadner, K. J., and E. Rael. (2000). "The FXYD gene family of small ion transport regulators or channels: cDNA sequence, protein signature sequence, and expression." *Genomics* 68 (1):41-56. doi: 10.1006/geno.2000.6274.

[3] Yamamoto, H., K. Okumura, S. Toshima, K. Mukaisho, H. Sugihara, T. Hattori, et al. (2009). "FXYD3 protein involved in tumor cell proliferation is overproduced in human breast cancer tissues." *Biol Pharm Bull* 32 (7):1148-54.

[4] Grzmil, M., S. Voigt, P. Thelen, B. Hemmerlein, K. Helmke, and P. Burfeind. (2004). "Up-regulated expression of the MAT-8 gene in prostate cancer and its siRNA-mediated inhibition of expression induces a decrease in proliferation of human prostate carcinoma cells." *Int J Oncol* 24 (1):97-105.

[5] Kayed, H., J. Kleeff, A. Kolb, K. Ketterer, S. Keleg, K. Felix, et al. (2006). "FXYD3 is overexpressed in pancreatic ductal adenocarcinoma and influences pancreatic cancer cell growth." *Int J Cancer* 118 (1):43-54. doi: 10.1002/ijc.21257.

[6] Liu, C. C., R. Teh, C. A. Mozar, R. C. Baxter, and H. H. Rasmussen. (2016). "Silencing overexpression of FXYD3 protein in breast cancer cells amplifies effects of doxorubicin and gamma-radiation on Na^+/K^+-ATPase and cell survival." *Breast Cancer Res Treat* 155 (2):203-13. doi: 10.1007/s10549-015-3667-x.

[7] Albers, R. W. (1967). "Biochemical aspects of active transport." *Annu Rev Biochem* 36:727-56. doi: 10.1146/annurev.bi.36.070167.003455.

[8] Post, R. L., C. Hegyvary, and S. Kume. (1972). "Activation by adenosine triphosphate in the phosphorylation kinetics of sodium and potassium ion transport adenosine triphosphatase." *J Biol Chem* 247 (20):6530-40.

[9] Ratheal, I. M., G. K. Virgin, H. Yu, B. Roux, C. Gatto, and P. Artigas. (2010). "Selectivity of externally facing ion-binding sites in the Na/K pump to alkali metals and organic cations." *Proc Natl Acad Sci US* 107 (43):18718-23. doi: 10.1073/pnas.1004214107.

[10] Fuller, W., L. B. Tulloch, M. J. Shattock, S. C. Calaghan, J. Howie, and K. J. Wypijewski. (2013). "Regulation of the cardiac sodium pump." *Cell Mol Life Sci* 70 (8):1357-80. doi: 10.1007/s00018-012-1134-y.

[11] Hu, Y. K., and J. H. Kaplan. (2000). "Site-directed chemical labeling of extracellular loops in a membrane protein. The topology of the Na,K-ATPase alpha-subunit." *J Biol Chem* 275 (25):19185-91. doi: 10.1074/jbc.M000641200.

[12] Kaplan, J. H. (2002). "Biochemistry of Na,K-ATPase." *Annu Rev Biochem* 71:511-35. doi: 10.1146/annurev.biochem.71.102201.141218.

[13] Durr, K. L., N. N. Tavraz, R. E. Dempski, E. Bamberg, and T. Friedrich. (2009). "Functional significance of E2 state stabilization by specific alpha/beta-subunit interactions of Na,K- and H,K-ATPase." *J Biol Chem* 284 (6):3842-54. doi: 10.1074/jbc.M808101200.

[14] Therien, A. G., H. X. Pu, S. J. Karlish, and R. Blostein. (2001). "Molecular and functional studies of the gamma subunit of the sodium pump." *J Bioenerg Biomembr* 33 (5):407-14.

[15] Geering, K. (2005). "Function of FXYD proteins, regulators of Na, K-ATPase." *J Bioenerg Biomembr* 37 (6):387-92. doi: 10.1007/s10863-005-9476-x.

[16] Geering, K. (2006). "FXYD proteins: new regulators of Na-K-ATPase." *Am J Physiol Renal Physiol* 290 (2):F241-50. doi: 10.1152/ajprenal.00126.2005.

[17] Liu, C. C., N. A. Fry, E. J. Hamilton, K. K. Chia, A. Garcia, K. Karimi Galougahi, et al. (2013). "Redox-dependent regulation of the Na(+)-K(+) pump: new twists to an old target for treatment of heart failure." *J Mol Cell Cardiol* 61:94-101. doi: 10.1016/j.yjmcc.2013.05.013.

[18] Shinoda, T., H. Ogawa, F. Cornelius, and C. Toyoshima. (2009). "Crystal structure of the sodium-potassium pump at 2.4 A resolution." *Nature* 459 (7245):446-50. doi: 10.1038/nature07939.

[19] McLaughlin, S., and A. Aderem. (1995). "The myristoyl-electrostatic switch: a modulator of reversible protein-membrane interactions." *Trends Biochem Sci* 20 (7):272-6.

[20] White, C. N., C. C. Liu, A. Garcia, E. J. Hamilton, K. K. Chia, G. A. Figtree, et al. (2010). "Activation of cAMP-dependent signaling induces oxidative modification of the cardiac Na+-K+ pump and inhibits its activity." *J Biol Chem* 285 (18):13712-20. doi: 10.1074/jbc.M109.090225.

[21] Rasmussen, H. H., E. J. Hamilton, C. C. Liu, and G. A. Figtree. (2010). "Reversible oxidative modification: implications for cardiovascular physiology and

pathophysiology." *Trends Cardiovasc Med* 20 (3):85-90. doi: 10.1016/j.tcm.2010.06.002.

[22] Sweadner, K. J., and M. S. Feschenko. (2001). "Predicted location and limited accessibility of protein kinase A phosphorylation site on Na-K-ATPase." *Am J Physiol Cell Physiol* 280 (4):C1017-26. doi: 10.1152/ajpcell.2001.280.4.C1017.

[23] Poulsen, H., P. Morth, J. Egebjerg, and P. Nissen. (2010). "Phosphorylation of the Na+,K+-ATPase and the H+,K+-ATPase." *FEBS Lett* 584 (12):2589-95. doi: 10.1016/j.febslet.2010.04.035.

[24] Pimentel, D., D. J. Haeussler, R. Matsui, J. R. Burgoyne, R. A. Cohen, and M. M. Bachschmid. (2012). "Regulation of cell physiology and pathology by protein S-glutathionylation: lessons learned from the cardiovascular system." *Antioxid Redox Signal* 16 (6):524-42. doi: 10.1089/ars.2011.4336.

[25] Gallogly, M. M., D. W. Starke, and J. J. Mieyal. (2009). "Mechanistic and kinetic details of catalysis of thiol-disulfide exchange by glutaredoxins and potential mechanisms of regulation." *Antioxid Redox Signal* 11 (5):1059-81. doi: 10.1089/ARS.2008.2291.

[26] Figtree, G. A., C. C. Liu, S. Bibert, E. J. Hamilton, A. Garcia, C. N. White, et al. (2009). "Reversible oxidative modification: a key mechanism of Na+-K+ pump regulation." *Circ Res* 105 (2):185-93. doi: 10.1161/CIRCRESAHA.109.199547.

[27] Rajagopalan, S., S. Kurz, T. Munzel, M. Tarpey, B. A. Freeman, K. K. Griendling, et al. (1996). "Angiotensin II-mediated hypertension in the rat increases vascular superoxide production via membrane NADH/NADPH oxidase activation. Contribution to alterations of vasomotor tone." *J Clin Invest* 97 (8):1916-23. doi: 10.1172/JCI118623.

[28] White, C. N., G. A. Figtree, C. C. Liu, A. Garcia, E. J. Hamilton, K. K. Chia, et al. (2009). "Angiotensin II inhibits the Na+-K+ pump via PKC-dependent activation of NADPH oxidase." *Am J Physiol Cell Physiol* 296 (4):C693-700. doi: 10.1152/ajpcell.00648.2008.

[29] Liu, C. C., A. Garcia, Y. A. Mahmmoud, E. J. Hamilton, K. K. Galougahi, N. A. Fry, et al. (2012). "Susceptibility of beta1 Na+-K+ pump subunit to glutathionylation and oxidative inhibition depends on conformational state of pump." *J Biol Chem* 287 (15):12353-64. doi: 10.1074/jbc.M112.340893.

[30] Felippe Goncalves-de-Albuquerque, C., A. Ribeiro Silva, C. Ignacio da Silva, H. Caire Castro-Faria-Neto, and P. Burth. (2017). "Na/K Pump and Beyond: Na/K-ATPase as a Modulator of Apoptosis and Autophagy." *Molecules* 22 (4). doi: 10.3390/molecules22040578.

[31] Babula, P., M. Masarik, V. Adam, I. Provaznik, and R. Kizek. (2013). "From Na+/K+-ATPase and cardiac glycosides to cytotoxicity and cancer treatment." *Anticancer Agents Med Chem* 13 (7):1069-87.

[32] Bibert, S., C. C. Liu, G. A. Figtree, A. Garcia, E. J. Hamilton, F. M. Marassi, et al. (2011). "FXYD proteins reverse inhibition of the Na+-K+ pump mediated by glutathionylation of its beta1 subunit." *J Biol Chem* 286 (21):18562-72. doi: 10.1074/jbc.M110.184101.

[33] Arimochi, J., A. Ohashi-Kobayashi, and M. Maeda. (2007). "Interaction of Mat-8 (FXYD-3) with Na+/K+-ATPase in colorectal cancer cells." *Biol Pharm Bull* 30 (4):648-54. doi: JST.JSTAGE/bpb/30.648 [pii].

[34] Liu, C. C., K. Karimi Galougahi, R. M. Weisbrod, T. Hansen, R. Ravaie, A. Nunez, et al. (2013). "Oxidative inhibition of the vascular Na+-K+ pump via NADPH oxidase-dependent beta1-subunit glutathionylation: implications for angiotensin II-induced vascular dysfunction." *Free Radic Biol Med* 65:563-72. doi: 10.1016/j.freeradbiomed.2013.06.040.

[35] Geering, K. (2008). "Functional roles of Na,K-ATPase subunits." *Curr Opin Nephrol Hypertens* 17 (5):526-32. doi: 10.1097/MNH.0b013e3283036cbf00041552-200809000-00014 [pii].

[36] Karimi Galougahi, K., C. C. Liu, A. Garcia, N. A. Fry, E. J. Hamilton, G. A. Figtree, et al. (2015). "beta3-Adrenoceptor activation relieves oxidative inhibition of the cardiac Na+-K+ pump in hyperglycemia induced by insulin receptor blockade." *Am J Physiol Cell Physiol* 309 (5):C286-95. doi: 10.1152/ajpcell.00071.2015.

[37] Pastore, A., and F. Piemonte. (2012). "S-Glutathionylation signaling in cell biology: progress and prospects." *Eur J Pharm Sci* 46 (5):279-92. doi: 10.1016/j.ejps.2012.03.010.

[38] Thogersen, L., and P. Nissen. (2012). "Flexible P-type ATPases interacting with the membrane." *Curr Opin Struct Biol* 22 (4):491-9. doi: 10.1016/j.sbi.2012.05.009.

[39] Fuller, W., L. B. Tulloch, M. J. Shattock, S. C. Calaghan, J. Howie, and K. J. Wypijewski. (2012). "Regulation of the cardiac sodium pump." *Cell Mol Life Sci.* doi: 10.1007/s00018-012-1134-y.

[40] Burgoyne, J. R., H. Mongue-Din, P. Eaton, and A. M. Shah. (2012). "Redox signaling in cardiac physiology and pathology." *Circ Res* 111 (8):1091-106. doi: 10.1161/CIRCRESAHA.111.255216.

[41] Woolston, C. M., A. Al-Attar, S. J. Storr, I. O. Ellis, D. A. Morgan, and S. G. Martin. (2011). "Redox protein expression predicts radiotherapeutic response in early-stage invasive breast cancer patients." *International journal of radiation oncology, biology, physics* 79 (5):1532-40. doi: 10.1016/j.ijrobp.2010.11.002.

[42] Loftas, P., S. Onnesjo, E. Widegren, G. Adell, H. Kayed, J. Kleeff, et al. (2009). "Expression of FXYD-3 is an independent prognostic factor in rectal cancer patients with preoperative radiotherapy." *International journal of radiation oncology, biology, physics* 75 (1):137-42. doi: 10.1016/j.ijrobp.2008.10.076.

[43] Maxwell, P. J., D. B. Longley, T. Latif, J. Boyer, W. Allen, M. Lynch, et al. (2003). "Identification of 5-fluorouracil-inducible target genes using cDNA microarray profiling." *Cancer research* 63 (15):4602-6.

[44] Morrison, B. W., J. R. Moorman, G. C. Kowdley, Y. M. Kobayashi, L. R. Jones, and P. Leder. (1995). "Mat-8, a novel phospholemman-like protein expressed in human breast tumors, induces a chloride conductance in Xenopus oocytes." *J Biol Chem* 270 (5):2176-82.

[45] Bibert, S., S. Roy, D. Schaer, E. Felley-Bosco, and K. Geering. (2006). "Structural and functional properties of two human FXYD3 (Mat-8) isoforms." *J Biol Chem* 281 (51):39142-51. doi: 10.1074/jbc.M605221200.

[46] Zhu, Z. L., Z. R. Zhao, Y. Zhang, Y. H. Yang, Z. M. Wang, D. S. Cui, et al. (2010). "Expression and significance of FXYD-3 protein in gastric adenocarcinoma." *Dis Markers* 28 (2):63-9. doi: 10.3233/DMA-2010-0669.

[47] Mijatovic, T., F. Dufrasne, and R. Kiss. (2012). "Cardiotonic steroids-mediated targeting of the Na^+/K^+-ATPase to combat chemoresistant cancers." *Curr Med Chem* 19 (5):627-46. doi: BSP/CMC/E-Pub/2012/052 [pii].

[48] Pani, G., E. Giannoni, T. Galeotti, and P. Chiarugi. (2009). "Redox-based escape mechanism from death: the cancer lesson." *Antioxid Redox Signal* 11 (11):2791-806. doi: 10.1089/ars.2009.2739.

[49] Holliday, D. L., and V. Speirs. (2011). "Choosing the right cell line for breast cancer research." *Breast Cancer Res* 13 (4):215. doi: 10.1186/bcr2889.

[50] Mishra, N. K., Y. Peleg, E. Cirri, T. Belogus, Y. Lifshitz, D. R. Voelker, et al. (2011). "FXYD proteins stabilize Na,K-ATPase: amplification of specific phosphatidylserine-protein interactions." *J Biol Chem* 286 (11):9699-712. doi: 10.1074/jbc.M110.184234.

[51] Clifford, R. J., and J. H. Kaplan. (2013). "Human breast tumor cells are more resistant to cardiac glycoside toxicity than non-tumorigenic breast cells." *PLoS One* 8 (12):e84306. doi: 10.1371/journal.pone.0084306.

[52] Cherniavsky Lev, M., S. J. Karlish, and H. Garty. (2015). "Cardiac glycosides induced toxicity in human cells expressing alpha1-, alpha2-, or alpha3-isoforms of Na-K-ATPase." *Am J Physiol Cell Physiol* 309 (2):C126-35. doi: 10.1152/ajpcell.00089.2015.

[53] Kayed, H., J. Kleeff, S. Keleg, X. Jiang, R. Penzel, T. Giese, et al. (2006). "Correlation of glypican-1 expression with TGF-beta, BMP, and activin receptors in pancreatic ductal adenocarcinoma." *Int J Oncol* 29 (5):1139-48.

[54] Trial, Swedish Rectal Cancer. (1997). "Improved survival with preoperative radiotherapy in resectable rectal cancer." *N Engl J Med* 336 (14):980-7. doi: 10.1056/NEJM199704033361402.

[55] (2013). "Dishing out cancer treatment." *Nat Biotechnol* 31 (2):85. doi: 10.1038/nbt.2516.

[56] Woolston, C. M., L. Zhang, S. J. Storr, A. Al-Attar, M. Shehata, I. O. Ellis, et al. (2012). "The prognostic and predictive power of redox protein expression for anthracycline-based chemotherapy response in locally advanced breast cancer." *Mod Pathol* 25 (8):1106-16. doi: 10.1038/modpathol.2012.60.

[57] George, S. R., G. Y. Ng, S. P. Lee, T. Fan, G. Varghese, C. Wang, et al. (2003). "Blockade of G protein-coupled receptors and the dopamine transporter by a transmembrane domain peptide: novel strategy for functional inhibition of membrane proteins in vivo." *J Pharmacol Exp Ther* 307 (2):481-9. doi: 10.1124/jpet.103.053843.

In: Research on New Generation …
Editor: Yildiz Dincer
ISBN: 978-1-53614-367-6

Chapter 3

CIRCULATING CELL-FREE DNA

***Rahiye Dilhan Kuru**[*]*
Istanbul University, Cerrahpasa Medical Faculty,
Department of Medical Biology, İstanbul/Turkey

ABSTRACT

The analysis of cell-free circulating DNA (cfDNA) from blood is called liquid biopsy. cfDNA in blood constitutes a mixture of DNA released from cells from different tissues of the body. Liquid biopsy allows for detection of DNA in different aspects such as quantity, chromosomal alterations, sequence mutations and epigenetic changes. Circulating tumor DNA (ctDNA) can be obtained from the plasma or serum of cancer patients and provides a suitability for non-invasive sampling of tumor DNA. The ctDNA is a very promising tool and might be beneficial for early detection of cancer, identification of minimal residual disease, guidance for treatment and monitoring treatment response and tumor evolution. Invasive tissue biopsies are not suitable for continuous monitoring of tumoral changes during course of the disease. When analysis of ctDNAs in cancer patients is compared to the conventional physical and biochemical methods, blood-based ctDNA detection offers a non-invasive and easily applicable tool for cancer diagnosis and follow up, identification of molecular targets for cancer therapy, monitoring of tumor molecular profile in real time, detection of emerging molecular aberrations associated with resistance to a particular therapy and diagnosis of metastasis. This chapter focuses on the origins, structures and the clinical application of cfDNA/ctDNA.

Keywords: cfDNA, ctDNA, liquid biopsy, solid tumors

---

[*] Corresponding Author Email: Email:dilhank@istanbul.edu.tr.

Introduction

The availability of easy diagnostic tools for identifying subsets of patients with solid tumors sensitive to targeted therapies, real-time molecular monitoring of treatment, the detection of recurrence, and tracking resistance are necessary for the successful application of such personalized approaches to chemotherapy [1, 2]. Biopsy materials (fresh and archived) are still commonly used for solid tumor genomic profiling. However, there are many limitations for biopsies such as limited availability, repeatability, and high failure rates [3]. Therefore, less invasive, blood-based tumor genomic profiling methods using either circulating tumor cells (CTCs) or circulating tumor DNA (ctDNA) are being developed to overcome these limitations [2].

Tissue biopsy is a common standard procedure in medicine for diagnosis and monitoring of the patients. In recent years, researchers have reported the utility of CTCs as candidate tumor biomarkers in cancer cases especially in breast and colorectal cancer and refered as liquid biopsy [1, 4, 5]. Also, circulating subcellular structures, essentially exosomes and other biomolecules, such as microRNA (miRNA) and cell-free DNA (cfDNA) are involved in terms of liquid biopsy especially in the field of cancer research. All of these are biomarkers obtained from biological fluids and they take the characteristic of the tissue which they originate from. Blood samples are preferred in most of the studies, but other fluids such as urine, saliva, ascites fluids and pleural effusions are also being used. Some collaborations have already begun to use liquid biopsy in diagnosis and treatment protocols. Also, different commercial systems have got government approval for clinical use of these biomarkers in cancer patients [5].

History

The presence of DNA in human plasma of healthy and ill people was first described by Mandel and Métais in 1948 [6, 7]. Stroun and Anker demonstrated the spontaneous release of nucleic acids by living frog auricles in 1963-1970 [6, 8-10]. After a short time, increased levels of cfDNA in the serum of patients suffering from systemic lupus erythematosus (SLE) or rheumatoid arthritis was demonstrated by Koffler et al. in 1973. Firstly, prognostic value of the cfDNA has been demonstrated in patients with rheumatoid arthritis by Leon et al. in 1977 [4, 11-13]. At the same time, Anker et al. reported that cfDNA was involved in humoral immune response and they described that human blood leokocytes spontaneously release DNA while in 1989 Stroun et al. suggested that cfDNA in cancer patients was derived from cancer cells [14-16]. In 1994, Vasioukhin et al. and Sorenson et al., in collaboration with P. Anker, detected *RAS* mutations (*NRAS*; myelo dysplastic syndrome (MDS), and *KRAS* (pancreatic cancer)) using cfDNA analysis in tumor cells [4, 17, 18]. Lo et al. showed that DNA of fetal origin circulated in blood of pregnant women in 1997 [10]. So, cfDNA gained attention in another clinical application which is prenatal diagnosis. On the other hand, cfRNA, in contrast to cfDNA, was not discovered till 1999. Lo et al. first identified (viral) cfRNA in blood of patients with nasopharyngeal carcinoma [6, 12]. cfRNA analysis and first studies were performed on gene aberrance in plasma of patients with solid tumors [20-22]. In 2001, established circulating nucleosome in serum by Holdenricder et al. [23]. Fujiwara et al. demonstrated the presence of aberrant methylations in promoters of five tumor suppressor genes in serum DNAs of patients with lung cancers in 2005 [24]. Larrea et al. reported the presence of miRNAs in blood of lymphoma patients in 2007 [25].

Large cohorts of studies have focused on detection of ctDNAs in multiple types of tumors such as cancers of the breast, colorectal region, prostate, lungs and pancreas in the last decade [26-30]. So, it was suggested that miRNAs could be used as non-invasive biomarkers of cancer in following years [12]. Murtaza et al. performed first exome sequencing analysis of cfDNA in 2012 [31]. In addition to this study, several groups suggested that ctDNA analysis in blood of cancer patients could be useful as a novel technique for monitoring response of targeted therapy and prognosis [26, 28, 32-37]. Chan et al. performed whole-methylome analysis of cfDNA in 2013 [38, 39]. First clinical validation of cfDNA analysis was achieved by Thiery et al. in 2014 [29]. In 2015, Mitra et al. reported that cfDNA induces apoptosis and DNA damage into recipient cells [40]. In 2016, Ulz et al. [41] showed that gene expression could be inferred from cfDNA, and Snyder et al. [42] presented that nucleosome phasing is reflected in cfDNA. Table 1 shows a timeline of the main discoveries on cfDNA.

Biology of Cell-Free DNA

Origins of Cell-Free DNA

cfDNA is released into circulation by various pathologic and normal physiologic mechanisms such as dying cells during cellular turnover or apoptotic and necrotic cells. Researcher suggest apoptosis as the predominant source of cfDNA in the circulation. Under normal physiological conditions, apoptotic and necrotic cells are eliminated by infiltrating phagocytes [43, 44]. cfDNA often accumulates at higher levels in arthritis, hepatitis and cancer patients, than in normal subjects [4].

In cancer patients, cfDNA is derived from tumor mass and is termed as ctDNA. Contrary to apoptosis, necrosis generates larger DNA fragments, due to an incomplete and random digestion of genomic DNA [43]. Biophysical particularities of ctDNA shows similar features to cancer cell DNA. Especially, strand stability following *in vitro* carcinogen treatment is decreased in cfDNA of cancer patients when compared to normal cfDNA. On the contrary, in healthy people, cfDNA shows the same strand stability as normal cellular DNA. The percentage of ctDNA varies from one patient to another [7]. cfDNA generates a mixture of DNA released from cells from different tissues of the body. For example, placenta the origin of cell-free fetal DNA (cffDNA) detectable in maternal circulation.

Table 1. Timeline of the main discoveries on cfDNA and cfDNA analysis applications [6, 26]

Timeline discoveries on circulating cell-free DNA and applications		
Period	**Researchers**	**Discovery**
1948	*Mandel &Métais*	"cell-free DNA molecules in the human blood compartment"
1963-1970	*Stroun, Anker & Gahan*	"Active cellular release of cirDNA as a Messenger"
1972-1984	*Anker* et al.	"cirDNA involved in humoral immune response"
	Kofler et al.	"increased levels of cfDNA in SLE or rheumatoid arthritis"
1977	*Leon* et al.	"prognostic value of the cfDNA in rheumatoid arthritis"
1989	*Stroun &Anker*	"cirDNA dreives from the tumor in cancer patients"
1994	*Anker* et al.	"*RAS* mutations detection by cirDNA analysis"
	Vasioukhin et al.	"Detected the mutated *RAS* gene fragments in the blood of cancer patients "
1997	*Lo* et al.	"Foetal DNA circulates in the blood of pregnant women"
	Kopreski Yamada Sanchez-Cespedes	"First studies on gene aberrance in the plasma of colorectal cancer, pancreatic, NSCLC patients, etc."
1999	*Kopreski* et al.	" cirRNA analysis"
2001	*Holdenrieder* et al.	"circulating nucleosomes"
2005	*Diehl* et al.	" First clinical study on the detection of point mutations by cirDNA analysis"
2008	*Diehl* et al.	"Study of the tumor dynamics by cirDNA analysis"
2010	*Thierry* et al.	"cirDNA deriving from the tumor is highly fragmented"
	Zhang et al.	"cir mitokondrial DNA is a DAMP"
2012	*Bardelli* et al. *Diaz* et al.	"Tracking tumor resistance by cirDNA analysis"
	Leary et al.	"Wirst whole-genome analyses of cfDNA"
	Murtaza et al.	"First exome sequencing analyses of cfDNA"
	Shaw et al.	"Landmark study showing that cfDNA analysis may detect minimal residual disease"
2013	*Dawson* et al.	"cirDNA analysis to follow-up cancer"
	Misale et al.	"Resistance mechanisms identified in plasma DNA"
	Chan et al.	"Whole-methylome analyses of cfDNA"
2014	*Thierry* et al.	"Firs clinical validation of cirDNA analysis"
2015	*Mittra* et al.	"cirDNA induces apoptosis and DNA dmage into recipient cells"
	Cools-Lartigue et al.	"Involvement of NETs in oncogenesis"
2016	*Ulz* et al.	"Gene expression can be inferred from cfDNA"
	Snyder et al.	"Nucleosome phasing is reflected in cfDNA"

SLE: systemic lupus erythematosus; DAMP: damage associated molecular pattern; NSCLC: non small cell lung cancer; NETs: Neutrophil extracellular traps

According to previous studies in healthy individuals, cfDNA is mostly derived from apoptosis of normal cells of the hematopoietic lineage (especially lymphoid and myeloid cells) and other solid tissues contribute only to a small part of cfDNA. Although in progressive cancer patients, a large part of the circulating DNA fragments could be associated with the primary tumor tissue [4]. Early studies have shown that ctDNA is single- or double-stranded DNA and possessed many cancer-associated single-nucleotide mutations, methylation changes [45].

Mechanisms of Release and Clearance of Cell-Free DNA in the Blood

cfDNA is released into blood circulation by various pathological and physiological mechanisms. There are two biological mechanisms for release of free DNA into the bloodstream. These mechanisms are apoptosis/necrosis and lysis of dying cells during cellular turnover in the bloodstream. Under normal physiological conditions, apoptotic and necrotic cells are cleared by infiltrating phagocytes and cfDNA is sensitive to plasma nucleases such as DNAse1 and is discarded from the bloodstream rapidly. So, cfDNA levels are relatively low. However, this mechanism does not similarly function in the tumor mass. High levels of ctDNA established in cancer patients due to DNase activity is lower in blood plasma of these patients than in healthy ones [4]. In addition, different study groups have demonstrated that, by using bacterial DNase, the tumor DNA could be resistant to DNase digestion [46, 47].

In healthy people, cfDNA concentration is between 0 and 100 ng/ml in blood with an average of 30 ng/ml. Usually, cfDNA in plasma is double-stranded and in the form of a nucleoprotein complex. It consists of DNA molecules sized between 0.18-21 kilobases (kb). A vast majority of cfDNA fragments released from apoptotic cells are measured between 180 and 200 base pairs (bp) [48].

In solid tumors, ctDNA can be also released by necrosis, autophagy, and other physiological conditions stimulated by micro environmental stress and treatment pressure and due to an incomplete and random digestion of genomic DNA, necrosis generates larger DNA fragments. Generally, levels of ctDNA is higher than healthy individuals in cancer patients but the levels vary widely. The concentration of ctDNA in plasma or serum is found at concentrations between 0 and over 1000 ng/ml of blood, with an average of 180 ng/ml in cancer patients. The instability of ctDNA levels in cancer patients is likely to be associated with tumor burden, stage, vascularity, cellular turnover, and response to the therapy [43, 44]. Moreover, circadian rhythms, inflammation or particular therapies most likely affect ctDNA release and clearance [4]. Diehl et al. have suggested that, up to 3.3% of the tumor DNA enters the circulation everyday for a patient with a tumor load of 100 g in size [27]. However, spontaneously newly synthesized DNA also is released by living cells. For example, stimulation of lymphocytes results in the release of large amounts of cfDNA into normal physiological process [43, 44].

The clearance and stability of cfDNA has not yet been clarified. Experimental studies on animal models show that cfDNA appears to be rapidly cleared by the liver, second extent by the kidneys and the spleen [48]. Various studies with healthy subjects performing extensive exercise have reported data about the stability and half-life of cfDNA. Available data are obtained from fetal DNA studies, as cfDNA is cleared rapidly from the maternal plasma after delivery. The half-life of cell-free fetal DNA (cffDNA) was previously estimated to be 16 minute.

In addition, cfDNA is not detectable 1–2 days after delivery [4, 43, 49]. Yu et al. found that the clearance of cffDNA occurred in two phases with different kinetics as rapid and slow phase by using paired-end massive parallel sequencing (MPS). Firstly, rapid phase has a mean half-life of approximately 1h, while slow phase has a mean half-life of approximately 13h [50]. Recently, a few study groups (Lo et al., To et al., and Chan et al.) reported the clearance mechanism of ctDNA from plasma. Their studies about the kinetics and clearance of circulating Epstein–Barr virus (EBV) DNA may indicate equivalent mechanisms [51-53]. Secondly, the most extensively analyzed source of cfDNA is urine. Studies performed in mice and humans have shown that DNA molecules are passed through from the kidney barrier in rodents and humans and could be analyzed by standard genetic methods [54].

In addition, urinary DNA inspection showed two different DNA fractions in urine as the first fraction consists of DNA molecules which mainly originate from the cells found in the urinary tract and larger than one kilobase, and second fraction consists of DNA molecules found in urine supernatant after centrifugation which at least partially enters into the urine from circulation which are shorter than the DNA molecules found in first fraction (sized 150-250 bp) [55]. These findings have led to new methods for cfDNA analysis, because the analysis of cfDNA in urinary has numerous advantages over cfDNA analysis in plasma such as being less infectious and obtaining larger sample pool [48].

Clinical Application of cfDNA/ctDNA

In Autoimmune Diseases

Since a long time, research about cfDNA has focused on analysing the level of cfDNA in autoimmune diseases like rheumatoid arthritis systemic, lupus erythematosus, systemic sclerosis and primary Sjogren's syndrome [4, 11, 12, 56].

In Transplantation

Microchimerism is the presence of small quantitities of foreign cells or DNA either in the tissue or circulatory system. Microchimerism occurs in two-ways: First, transplacental traffic between the mother and fetus. Second, transfusion or transplantation. Free DNA microchimerism in plasma and urine has been defined in female patients with a transplanted organ from a male donor in 1998 [57, 58].

In Trauma and Sepsis

After the severe trauma, a systemic anti inflammatory response appear which can lead to the organ failure and increased cell death. These conditions are caused due to the increase of cfDNA level in circulation. Lo et al. and Chiu et al. have reported that the level of cfDNA will change in patients with trauma and burns and augmentation depends on the severity of the injury. cfDNA in those patients can be used as a prognostic marker for the complication development and the recovery process [59, 60]. Also, different researchers have reported that changed levels of cfDNA concentration detected from patients in intensive care units can be used as a prognostic marker for development of sepsis, organ failure, disease severity and mortality [61-63].

In Non-Invasive Prenatal Diagnosis

In 1997, Lo et al. showed DNA of fetal origin circulated in maternal blood plasma and serum and this discovery led to the possibilities for development of non-invasive prenatal diagnostic methods (NIPT) [19]. Some advantages of this method are, rapid clearance after termination of pregnancy, detectability from 5th week of gestation and easy reasonable (compared to invasive sampling), but also some disadvantages can be counted as being practically inseparable from maternal cfDNA, percentage of cffDNA in total free DNA amounts to only 5 - 10%, depending on week of gestation. Present fetal DNA studies are supported on three basic experimental approaches: enriching free fetal DNA fraction in plasma; detecting fetal-specific epigenetic markers or placenta-specific mRNA markers; developing precise digital PCR and second-generation genetic sequencing [48, 64].

In Cancer

In the last 25 years studies have suggested that free DNA concentration changes are effective in various types of cancers such as pancreatic [65, 66], colorectal [20, 27, 67-69], liver [70, 71], lung [72, 73], head and neck (74-77), nasopharynx [78], ovary [79], breast [80, 81], prostate [82, 83], testes [84], leukemia [17, 85], lymphoma [86, 87], etc. These studies on various cancer types drew attention to increased cfDNA levels in cancer patients compared to healthy people. Researchers found that ctDNA existed at significantly different levels among the patients with different cancer stages by testing blood samples from various cancer patients. Especially, patients with advanced-stages of gastro-esophageal, pancreatic, breast and colorectal cancers had higher levels of ctDNA than patients in early stages of those diseases [45, 88, 89]. Tissue biopsy is still the gold standard for tumor diagnosis but there are risks associated with invasive sampling in some fragile organs such as lung and sensitivity is suboptimal, frequently resulting in inability to detect early-stage tumors. Otherwise, results of biopsies are often unable to determine tumor progression accurately, because tumor tissues are heterogeneous structures [45, 90]. ctDNA level and tumor specific genome changes could be useful markers for monitoring the treatment and relapse. In addition, researchers reported that ctDNA could be used to monitor relapse status after 10-months on detection of relapse when compared with the conventional follow-up, in comparison to relapsing and non-relapsing patients.

In recent years, researchers have reported effective applications of ctDNA in clinical practice. These studies indicate that liquid biopsy containing ctDNA has more powerful sensitivity and provides better clinical correlations than protein biomarkers and CTC. Researchers have suggested that the screening of genetic lesions using ctDNA is highly sensitive and specific. ctDNA analysis may be suitable for determining the tumor progression, prognosis and can assist in targeted therapy [91-96]. So, ctDNA analysis might be used for early detection of cancer, evaluation of tumor progression and prognosis. Therapy resistance is the major cause of cancer treatment failure. However, so far appropriate methods of investigating tumor resistance have not yet been established [45]. The study groups in the field of personalized targeted therapies have reported that, ctDNA is useful for detection of resistant tumor cells. For example, the mutant allele frequency has significantly increased in a non small cell lung cancer (NSCLC) patient following the treatment. A resistant conferring mutation in *EGFR* liquid biopsy has been shown to be possible by repeated sampling of plasma ctDNA for patients with resistance to cancer therapy [1, 97, 98]. Although *ERBB2* over expression in breast cancer, *ABL1* rearrangements in chronic myeloid leukemia, *KIT* mutations in gastrointestinal stroma tumors, *KRAS* gene mutations in colorectal cancer, and *V600E BRAF* mutation in melanoma patients have been identified as targets in late invasive and metastatic cancers, thereby limiting the success of the treatment. Use of liquid biopsy may be beneficial for these cancer types. However, authors suggested that ctDNA biomarkers in oral cavity and pharynx, digestive system, respiratory system, skin (excluding basal and squamous), endocrine system, breast, and different other tumor types may be deficient and inadequate for clinical practices [99, 100]. Pancreatic ductal adenocarcinoma (PDAC) is a rare and very aggressive tumor representing around 80% of all pancreatic cancers. Early diagnosis of PDAC can have a dramatic impact on survival, but current screening methods are not effective. Non-invasive "liquid biopsy" approaches using ctDNA as new effective and reliable biomarkers could improve early detection and diagnosis of PDAC, monitoring and treatment decisions. One of the most common genetic alterations in pancreatic adenocarcinoma is activating mutations of *KRAS* which are located in recurrent hotspots (e.g., codon 12 and 13) and involve single nucleotide variations. These mutations are therefore ideal targets to detect and quantify the presence of ctDNA in body samples [101].

Earlier detection of metastatic breast cancer might be clinically beneficial, because of identification of recurrent disease at the earliest moment will allow for initiation of therapies [102]. Dawson et al. suggested that increasing ctDNA levels were significantly associated with poor overall survival in breast cancer [34]. Breast cancer genome dynamically changes with mutations, somatic copy number alterations (SCNAs), rearrangements and promoter methylation and it may evolve during cancer progression and *throughout the treatment course, therefore* a continuous monitoring via ctDNA is especially important [4].

Over the last few years, several studies using ctDNA have been performed in haematological malignancies, and these studies have explained the role of ctDNA for risk stratification and monitoring tumor burden and response to treatment in both lymphoid and myeloid malignancies. For example, Patient-specific IgH rearrangements in non-Hodgkin's lymphoma are easily detectable by ctDNA analysis. ctDNA is also a powerful tool for monitoring chronic lymphocytic leukemia (CLL), MDS and multiple myeloma (MM). Furthermore, clonal dynamics could be monitored with ctDNA analysis in CLL, so genomic changes associated with Richter's syndrome can be identified. In MM, genomic changes in the bone marrow can be detected by ctDNA analysis. Also, these techniques can be used as monitoring strategies to limit invasive bone marrow biopsies [103].

In summary, numerous studies based on ctDNA analysis indicate that ctDNA detection can quickly predict the result of therapy, and provide an important reference for clinicians in determining the next treatment protocols [28, 88, 104, 105].

Methods for Detecting ctDNA

At first, ctDNA was primarily analyzed by Sanger sequencing and allele-specific PCR. With the advent of new technologies, mainly digital PCR and MPS, the sensitivity and specificity of ctDNA-based assays have developed significantly [106].

In the present day, many methods such as microarray-based comparative genome hybridization (CGH), single nucleotide polymorphism (SNP) analysis, and next-generation sequencing (NGS) are being used using for detecting genetic mutations in cancers and these methods are notably sensitive and accurate for detection of tumor genomes. Furthermore, array-based CGH and SNP arrays allow the identification of genetic variations. In addition, genome-wide association studies (GWAS) could detect tumor-specific DNAs such as blood-based diagnostic tests in cancer patients. Firstly, targeted plasma re-sequencing (TAm-Seq) could identify TP53 mutations by tracking ctDNAs in ovarian cancer patients in 2012. Alterations of ctDNA levels could be determined through MPS technique, including the personalized analysis of rearranged ends (PARE) and shotgun MPS in patients before and after surgery [26]. Whole-exome sequencing (WES) and whole-genome sequencing (WGS) of ctDNA have also been applied successfully in genetic profiling of advanced cancer stages these assays are usually more sensitive in detecting low level ctDNA [106, 107].

Conclusion

In the past decade, cfDNA/ctDNA, isolated from body fluids such as plasma, serum, urine, saliva, ascites fluids and pleural effusions have been shown as important tools for clinical diagnostics. Especially, ctDNA is released by the tumor cells into the blood and thus including the mutations of the original tumor. Several studies on ctDNA have been performed by new applications of NGS technology. Sensitive and specific ctDNA methods are beneficial for the early detection of recurrence and cancer screening at asymptomatic stages. Furthermore, cfDNA methods can be applied in non-cancer conditions, too. For example, NIPT using cffDNA have some advantages such as rapid clearance after termination of pregnancy, detectability from 5th week of gestation. ctDNA will provide genomic information when biopsy is not performed recurrently, as well as enabling longitudinal monitoring of the prognosis in the metastatic tumors and follow up on the evolving genome for timely personalized targeted therapies.

References

[1] Oellerich, M., Schütz, E., Beck, J., Kanzow, P., Plowman, P. N., Weiss, G. J. & Walson, P. D. (2017). Using circulating cell-free DNA to monitor personalized cancer therapy. *Critical Reviews in Clinical Laboratory Sciences*, Vol. *54*, No. 3, 205–218.

[2] Oxnard, G. R., Paweletz, C. P., Kuang, Y., Mach, S. L., O'Connell, A., Messineo, M. M., et al. (2014). Noninvasive detection of response and resistance in EGFR-mutant lung cancer using quantitative next-generation genotyping of cell-free plasma DNA. *Clin Cancer Res*, 20, 1698–1705.

[3] Mao, C., Yuan, J. Q., Yang, Z. Y., Fu, X. H., Wu, X. Y. & Tang, J. L. (2015). Blood as a substitute for tumor tissue in detecting EGFR mutations for guiding EGFR TKIs treatment of nonsmall cell lung cancer: a systematic review and meta-analysis. *Medicine (Baltimore).*, *94*, e775.

[4] Perakis, S., Auer, M., Belic, J. & Heitzer, E. (2017). Advances in Circulating Tumor DNA Analysis. *Adv Clin Chem*, *80*, 73-153. (Epub).

[5] Macías, M., Alegre, E., Díaz-Lagares, A., Patiño, A., Pérez-Gracia, J. L., Sanmamed, M., et al. (2018). Liquid Biopsy: From Basic Research to Clinical Practice. *Advances in Clinical Chemistry*, Volume *83*, 73-119, https://doi:10.1016/bs.acc.2017.10.003 .

[6] Thierry, A. R., El Messaoudi, S., Gahan, P. B., Anker, P. & Stroun, M. (2016). Origins, structures, and functions of circulating DNA in oncology. *Cancer Metastasis Rev*, *35*, 347–376.

[7] Mandel, P. & Metais, P. (1948). Les acides nucleiques du plasma sanguin chez l'homme. [Nucleic acids of blood plasma in humans.] *CR Acad Sci* Paris, *142*, 241–3.

[8] Stroun, M. (1962). Modifications des caractères à la suite de greffes intervariétales chez le Solanum melongena. [Character changes following intervarietal grafts in Solanum melongena.] *Comptes Rendues á l'Academie de Science Paris*, *255*, 361–363.

[9] Gahan, P. B., Chayen, J. & Silcox, A. A. (1962). Cytoplasmic localization of deoxyribonucleic acid in Allium cepa. *Nature*, *195*, 1115–1116.

[10] Gahan, P. B. & Chayen, J. (1965). Cytoplasmic deoxyribonucleic acid. *International Review of Cytology*, *18*, 223–24.

[11] Rapisuwon, S., Vietsch, E. E. & Wellstein, A. (2016). Circulating biomarkers to monitor cancer progression and treatment. *Computational and Structural Biotechnology Journal*, *14*, 211–222.

[12] Larrea, E., Sole, C., Manterola, L., Goicoechea, I., Armesto, M., Arestin, M., et al. (2016). New Concepts in Cancer Biomarkers: Circulating miRNAs in Liquid Biopsies. *Int. J. Mol. Sci*, *17*, 627.

[13] Leon, S. A., Shapiro, B., Sklaroff, D. M. & Yaros, M. J. (1977). Free DNA in the serum of cancer patients and the effect of therapy. *Cancer Res*, *37*, 646–50.

[14] Stroun, M., Anker, P., Maurice, P. & Gahan, P. B. (1977). Circulating nucleic acids in higher organisms. *International Review of Cytology*, *51*, 1–48.

[15] Stroun, M. & Anker, P. (1972). Nucleic acids spontaneously released by living frog auricles. *The Biochemical Journal*, *128*(3), 100P–101P.

[16] Stroun, M., Anker, P., Maurice, P., Lyautey, J., Lederrey, C. & Beljanski, M. (1989). Neoplastic characteristics of the DNA found in the plasma of cancer patients. *Oncology*, *46*, 318–22.

[17] Vasioukhin, V., Anker, P., Maurice, P., Lyautey, J., Lederrey, C. & Stroun, M. (1994). Point mutations of the N-ras gene in the blood plasma DNA of patients with myelodysplastic syndrome or acute myelogenous leukaemia. *Br J Haematol*, *86*, 774-9.

[18] Sorenson, G. D., Pribish, D. M., Valone, F. H., Memoli, V. A., Bzik, D. J. & Yao, S. L. (1994). Soluble normal and mutated DNA sequences from single-copy genes in human blood, Cancer Epidemiol. *Biomarkers Prev*, *3* (1), 67–71.

[19] Lo, Y. M. D., Corbetta, N., Chamberlain, P. F., Rai, V., Sargent, I. L., Redman, C. W. G., et al. (1997). Presence of fetal DNA in maternal plasma and serum. *Lancet*, *350*, 485–7.

[20] Kopreski, M. S., Benko, F. A., Kwee, C., Leitzel, K. E., Eskander, E., Lipton, A., et al. (1997). Detection of mutant K-ras DNA in plasma or serum of patients with colorectal cancer. *Br J Cancer*, *76*, 1293-9.

[21] Yamada, T., Nakamori, S., Ohzato, H., Oshima, S., Aoki, T., Higaki, N., et al. (1998). Detection of K-ras gene mutations in plasma DNA of patients with pancreatic adenocarcinoma: correlation with clinicopathological features. *Clin Cancer Res*, *4*, 1527–32.

[22] Sanchez-Cespedes, M., Monzo, M., Rosell, R., Pifarre, A., Calvo, R., Lopez-Cabrerizo, M. P., et al. (1998). Detection of chromosome 3p alterations in serum DNA of non-small-cell lung cancer patients. *Ann Oncol*, *9*, 113–6.

[23] Holdenrieder, S., Stieber, P., Bodenmüller, H., Busch, M., Von Pawel, J., Schalhorn, A. & Seidel, D. (2001). Circulating Nucleosomes in Serum. *Annals of the New York Academy of Sciences*, *945*(1), 93–102. doi:10.1111/j.1749-6632.2001. tb03869.

[24] Fujiwara, K., Fujimoto, N., Tabata, M., Nishii, K., Matsuo, K., Hotta, K., et al. (2005). Identification of epigenetic aberrant promoter methylation in serum DNA is useful for early detection of lung cancer. *Clin Cancer Res*, *11*, 1219–25.

[25] Lawrie, C. H. (2007). MicroRNA expression in lymphoma. *Expert Opin. Biol. Ther*, *7*, 1363–1374.

[26] Han, X., Wang, J. & Sun, Y. (2017). Circulating Tumor DNA as Biomarkers for Cancer Detection. *Genomics Proteomics Bioinformatics*, *15*, 59–72.

[27] Diehl, F., Li, M., Dressman, D., He, Y., Shen, D., Szabo, S., et al. (2005). Detection and quantification of mutations in the plasma of patients with colorectal tumors. *Proc Natl Acad Sci USA*, *102*, 16368-73.

[28] Diehl, F., Schmidt, K., Choti, M. A., Romans, K., Goodman, S., Li, M., et al. (2008). Circulating mutant DNA to assess tumor dynamics. *Nat Med*, *14*, 985-90.

[29] Thierry, A. R., Mouliere, F., El Messaoudi, S., Mollevi, C., Lopez- Crapez, E., Rolet, F. & Ychou, M. (2014). Clinical validation of the detection of KRAS and BRAF mutations from circulating tumor DNA. *Nature Medicine*, *20*(4), 430–435. doi:10.1038/nm.3511.

[30] Zhang, Q., Itagaki, K. & Hauser, C. J. (2010). Mitochondrial DNA is released by shock and activates neutrophils via p38 map kinase. *Shock* (Augusta, Ga), *34*(1), 55–59. doi:10.1097/SHK.

[31] Murtaza, M., Dawson, S. J., Tsui, D. W., Gale, D., Forshew, T., Piskorz, A. M., et al. (2013). Non-invasive analysis of acquired resistance to cancer therapy by sequencing of plasma DNA. *Nature*, *497*, 108–112.

[32] Bettegowda, C., Sausen, M., Leary, R. J., Kinde, I., Wang, Y., Agrawal, N. & Diaz, L. A. (2014). Detection of Circulating Tumor DNA in Early and Late-Stage Human Malignancies. *Science Translational Medicine*, *6*(224), 224ra24–224ra24. doi:10. 1126/scitranslmed.3007094.

[33] Misale, S., Yaeger, R., Hobor, S., Scala, E., Janakiraman, M., Liska, D. & Bardelli, A. (2012). Emergence of KRAS mutations and acquired resistance to anti-EGFR therapy in colorectal cancer. *Nature*, *486*(7404), 532–536. doi:10.1038/nature 11156.

[34] Dawson, S. J., Tsui, D. W. Y., Murtaza, M., Biggs, H Rueda O. M., Chin, S. F. & Rosenfeld, N. (2013). Analysis of Circulating Tumor DNA to Monitor Metastatic Breast Cancer. *New England Journal of Medicine*, *368*(13), 1199–1209. doi:10. 1056/NEJMoa1213261.

[35] Shaw, J. A., Page, K., Blighe, K., Hava, N., Guttery, D., Ward, B., et al. (2012). Genomic analysis of circulating cell-free DNA infers breast cancer dormancy. *Genome Res*, *22*, 220–31.

[36] Misale, S. (2012). Emergence of KRAS mutations and acquired resistance to anti-EGFR therapy in colorectal cancer, *Nature*, *486*, 532–536.

[37] Leary, R. J., Sausen, M., Kinde, I., Papadopoulos, N., Carpten, J. D., Craig, D., et al. (2012). Detection of chromosomal alterations in the circulation of cancer patients with whole-genome sequencing, *Sci. Transl. Med*, *4* (162), 162ra154.

[38] Chan, K. C. (2013). Cancer genome scanning in plasma: detection of tumor-associated copy number aberrations, single-nucleotide variants, and tumoral heterogeneity by massively parallel sequencing, *Clin. Chem*, *59*, 211–224.

[39] Chan, K. C., Jiang, P., Chan, C. W., Sun, K, Wong, J., Hui, E. P., et al. (2013). Noninvasive detection of cancer-associated genome-wide hypomethylation and copy number aberrations by plasma DNA bisulfite sequencing, *Proc. Natl. Acad. Sci. USA*, *110* (47), 18761–18768.

[40] Mittra, I. (2015). Circulating nucleic acids: a new class of physiological mobile genetic elements. *F1000Research*, *4*, 924. doi:10.12688/f1000research.7095.1.

[41] Ulz, P., Thallinger, G. G., Auer, M., Graf, R., Kashofer, K., Jahn, S. W., et al. (2016). Inferring expressed genes by whole-genome sequencing of plasma DNA. *Nat. Genet*, *48* (10), 1273–1278. http://dx.doi.org/10.1038/ng.3648. Epub 2016 Aug 29.

[42] Snyder, M. W., Kircher, M., Hill, A. J., Daza, R. M., Shendure, J. Cell-free DNA comprises an *in vivo* nucleosome footprint that informs its tissues-of-origin, *Cell*, *164* (1–2), (2016), 57–68.

[43] Qin, Z., Ljubimov, V. A., Zhou, C., Tong, Y. & Liang, J. (2016). Cell-free circulating tumor DNA in cancer. *Chin J Cancer*, *35*, 36, pp. 1-9.

[44] Gormally, E., Caboux, E., Vineis, P. & Hainaut, P. (2007). Circulating free DNA in plasma or serum as biomarker of carcinogenesis: Practical aspects and biological significance. *Mutation Research*, *635*, 105–1170.

[45] Cheng, F., Su, L. & Qian, C. (2016). Circulating tumor DNA: a promising biomarker in the liquid biopsy of cancer *Oncotarget*, Vol. *7*, No.30 48832-41.

[46] Cox, R. A. & Gokcen, M. (1976). Comparison of serum DNA, native DNA-binding and deoxyribonuclease levels in ten animal species and man, *Life Sci*, *19* (10), 1609–1614.

[47] Reitz, M., Lober, G., Kleemann, P. & Dick, W. (1995). Secretion of neutral and acid DNases in cultivated human lymphocytes after incubation with DNA; possible consequences for inhalation anesthesia, Zeitschrift fur Naturforschung C: *J. Biosci*, *50* (5–6), 419–424.

[48] Wagner, J. (2012). Free DNA – New potential analyte in clinical laboratory diagnostics? *Biochemia Medica*, *22*(1), 24–38.

[49] Beiter, T., Fragasso, A., Hudemann, J., Niess, A. M. & Simon, P. (2011). Short-term treadmill running as a model for studying cell-free DNA kinetics *in vivo*, *Clin. Chem*, *57* (4), 633–636.

[50] Yu, S. C. Y., Lee, S. W. Y., Jiang, P. Y. T., Leung, Chan K. C. A., Chiu, R. W. K. & Lo, Y. M. D. (2013). High-Resolution Profiling of Fetal DNA Clearance from Maternal Plasma by Massively Parallel Sequencing. *Clinical Chemistry*, *59*, 8, 1228–1237.

[51] Lo, Y. M. D., Leung, S. F., Chan, L. Y. S., Chan Lo, K. W., Johnson, P. J. & Huang, D. P. (2000). *Kinetics of Plasma Epstein-Barr Virus DNA during Radiation Therapy for Nasopharyngeal Carcinoma Cancer Research*, *60*, 2351–2355.

[52] To, E. W., Chan, K. C., Leung, S. F., Chan, L. Y., To, K. F., Chan, A. T., et al. (2003). Rapid clearance of plasma Epstein-Barr virus DNA after surgical treatment of nasopharyngeal carcinoma. *Clin Cancer Res*, *9*(9), 3254–9.

[53] Chan, K. C. (2014). Plasma Epstein-Barr virus DNA as a biomarker for nasopharyngeal carcinoma. *Chin J Cancer*, *33*(12), 598–603. doi:10.5732/cjc.014.10192.

[54] Botezatu, I., Serdyuk, O., Potapova, G., Shelepov, V., Alechina, R., Molyaka, Y., et al. (2000). Genetic Analysis of DNA Excreted in Urine: A New Approach for Detecting Specific Genomic DNA Sequences from Cells Dying in an Organism. *Clin Chem*, *46*, 1078-84.

[55] Su, Y. H., Wang, M., Brenner, D. E., Ng, A., Melkonyan, H., Umansky, S., et al. (2004). Human Urine Contains Small, 150 to 250 Nucleotide-Sized, Soluble DNA Derived from the Circulation and May Be Useful in the Detection of Colorectal Cancer. *J Mol. Diagn*, *6*, 101-7.

[56] Barteloni, E., Ludovini, V., Alunno, A., Pistola, L., Bistoni, O., Crino, L., et al. (2011). Increased levels of circulating DNA in patients with systemic autoimmune diseases: A possible marker of disease activity in Sjogren's syndrome. *Lupus*, *20*, 928-35.

[57] Zhang, J., Tong, K. L., Li, P. K. T., Chan, A. Y. W., Yeung, C. K., Pang, C. C. P., et al. (1999). Presence of Donor- and Recipient-derived DNA in Cell-free Urine Samples of Renal Transplantation Recipients: Urinary DNA Chimerism. *Clin Chem*, *45*, 1741-6.

[58] Lo, Y. M., Tein, M. S., Pang, C. C., Yeung, C. K., Tong, K. L. & Hjelm, N. M. (1998). Presence of donor-specific DNA in plasma of kidney and liver-transplant recipients. *Lancet*, *351*, 1329-30.

[59] Lo, Y. M. D., Rainer, T. H., Chan, L. Y. S., Hjelm, N. M. & Cocks, R. A. (2000). Plasma DNA as a Prognostic Marker in Trauma Patients. *Clin Chem*, *46*, 319-23.

[60] Chiu, T. W., Young, R., Chan, L. Y., Burd, A. & Lo, D. Y. (2006). Plasma cell-free DNA as an indicator of severity of injury in burn patients. *Clin Chem Lab Med*, 44, 13-7.

[61] Rhodes, A., Wort, S. J., Thomas, H., Collinson, P. & Bennett, E. D. (2006). Plasma DNA concentration as a predictor of mortality and sepsis in critically ill patients. *Crit Care*, *10*, R60.

[62] Saukkonen, K., Lakkisto, P., Varpula, M., Varpula, T., Voipio-Pulkki, L. M., Pettila, V. & Pulkki, K. (2007). Association of cell-free plasma DNA with hospital mortality and organ dysfunction in intensive care unit patients. *Intensive Care Med*, *33*, 1624-7.

[63] Saukkonen, K., Lakkisto, P., Pettila, V., Varpula, M., Karlsson, S., Ruokonen, E. & Pulkki, K. (2008). Cell-Free Plasma DNA as a Predictor of Outcome in Severe Sepsis and Septic Shock. *Clin Chem*, *54*, 1000-7.

[64] Gregg, A. R., Skotko, B. G., Benkendorf, J. L., Monaghan, K. G., Bajaj, K., Best, R. G., et al., (2016). Noninvasive prenatal screening for fetal aneuploidy, 2016

update: a position statement of the American College of Medical Genetics and Genomics, *Genet. Med*, *18*, 1056–1065.

[65] Castells, A., Puig, P., Mora, J., Boadas, J., Boix, L., Urgell, E., et al. (1999). K-ras Mutations in DNA Extracted From the Plasma of Patients With Pancreatic Carcinoma: Diagnostic Utility and Prognostic Significance. *J Clin Oncol*, *17*, 578-84.

[66] Mulcahy, H. & Farthing, M. J. (1999). Diagnosis of pancreatico-biliarymalignancy: detection of gene mutations in plasma and stool. *Ann Oncol*, *10*, 114-7.

[67] Diehl, F., Schmidt, K., Durkee, K. H., Moore, K. J., Goodman, S. N., Shuber, A. P., et al. (2008). Analysis of Mutations in DNA Isolated From Plasma and Stool of Colorectal Cancer Patients. *Gastroenterology*, *135*, 489-98.

[68] Hibi, K., Robinson, C. R., Booker, S., Wu, L., et al. (1998). Molecular Detection of Genetic Alterations in the Serum of Colorectal Cancer Patients. *Cancer Res*, *58*, 1405-7.

[69] Zitt, M., Muller, H. M., Rochel, M., Schwendinger, V., et al. (2008). Circulating cell-free DNA in plasma of locally advanced rectal cancer patients undergoing preoperative chemoradiation: a potential diagnostic tool for therapy monitoring. *Dis Markers*, *25*, 159-65.

[70] Wong, I. H. N., Lo, Y. M. D., Zhang, J., Liew, C. T., Ng, M. H. L., Wong, N., et al. (1999). Detection of Aberrant p16 Methylation in the Plasma and Serum of Liver Cancer Patients. *Cancer Res*, *59*, 71-3.

[71] Chan, K. C., Lai, P. B., Mok, T. S., Chan, H. L., Ding, C., Yeung, S. W. & Lo, Y. M. (2008). Quantitative analysis of circulating methylated DNA as a biomarker for hepatocellular carcinoma. *Clin Chem*, *54*, 1528-36.

[72] Yoon, K. A., Park, S., Lee, S. H., Kim, J. H. & Lee, J. S. (2009). Comparison of circulating plasma DNA levels between lung cancer patients and healthy controls. *J Mol Diagn*, *11*, 182-5.

[73] Chen, X. Q., Stroun, M., Magnenat, J. L., Nicod, L. P., Kurt, A. M., Lyautey, J., et al. (1996). Microsatellite alterations in plasma DNA of small cell lung cancer patients. *Nat Med*, *2*, 1033-5.

[74] Jiang, W. W., Rosenbaum, E., Mambo, E., Zahurak, M., Masayesva, B., Lopes Carvalho, A., et al. (2006). Decreased Mitochondrial DNA Content in Posttreatment Salivary Rinses from Head and Neck Cancer Patients. *Clin Cancer Res*, *12*, 1564-9.

[75] Jiang, W. W., Masayesva, B., Zahurak, M., Carvalho, A. L., Rosenbaum, E., Mambo, E., et al. (2005). Increased Mitochondrial DNA Content in Saliva Associated with Head and Neck Cancer. *Clin Cancer Res*, *11*, 2486-91.

[76] Nawroz, H., Koch, W., Anker, P., Stroun, M. & Sidransky, D. (1996). Microsatellite alterations in serum DNA of head and neck cancer patients. *Nat Med*, *2*, 1035-7.

[77] Capone, R. B., Pai, S. I., Koch, W. M., Gillison, M. L., Danish, H. N., Westra, W. H., et al. (2000). Detection and Quantitation of Human Papillomavirus (HPV) DNA in the Sera of Patients with HPV associated Head and Neck Squamous Cell Carcinoma. *Clin Cancer Res*, *6*, 4171-5.

[78] Mutirangura, A., Pornthanakasem, W., Theamboonlers, A., Sriuranpong, V., Lertsanguansinchi, P., Yenrudi, S., et al. (1998). Epstein-Barr Viral DNA in Serum of Patients with Nasopharyngeal Carcinoma. *Clin Cancer Res*, *4*, 665-9.

[79] Hickey, K. P., Boyle, K. P., Jepps, H. M., Andrew, A. C., Buxton, E. J. & Burns, P. A. (1999). Molecular detection of tumour DNA in serum and peritoneal fluid from ovarian cancer patients. *Br J Cancer*, *80*, 1803–8.

[80] Silva, J. M., Dominguez, G., Garcia, J. M., Gonzalez, R., Villanueva, M. J., Navarro, F., et al. (1999). Presence of Tumor DNA in Plasma of Breast Cancer Patients: Clinicopathological Correlations. *Cancer Res*, *59*, 3251-6.

[81] Kohler, C., Radpour, R., Barekati, Z., Asadollahi, R., Bitzer, J., Wight, E., et al. (2009) Levels of plasma circulating cell free nuclear and mitochondrial DNA as potential biomarkers for breast tumors. *Mol Cancer*, *8*, 105.

[82] Chun, F. K. H., Muller, I., Lange, I., Friedrich, M. G., Erbersdobler, A., Karakiewicz, P. I., et al. (2006). Circulating tumour-associated plasma DNA represents an independent and informative predictor of prostate cancer. *BJU Int*, *98*, 544-8.

[83] Mehra, N., Penning, M., Maas, J., Daal van, N., Giles, R. H. & Voest, E. E. (2007). Circulating Mitochondrial Nucleic Acids Have Prognostic Value for Survival in Patients with Advanced Prostate Cancer. *Clin Cancer Res*, *13*, 421-6.

[84] Ellinger, J., Albers, P., Muller, S. C., Ruecker von, A. & Bastian, P. J. (2009). Circulating mitochondrial DNA in the serum of patients with testicular germ cell cancer as a novel noninvasive diagnostic biomarker. *BJU Int*, *104*, 48-52.

[85] Rogers, A., Joe, Y., Manshouri, T., Dey, A., Jilani, I., Giles, F., et al. (2004). Relative increase in leukemia-specific DNA in peripheral blood plasma from patients with acute myeloid leukemia and myelodysplasia. *Blood*, *103*, 2799-801.

[86] Deligezer, U., Yaman, F., Erten, N. & Dalay, N. (2003). Frequent copresence of methylated DNA and fragmented nucleosomal DNA in plasma of lymphoma patients. *Clin Chim Acta*, *335*, 89-94.

[87] Hohaus, S., Giachelia, M., Massini, G., Mansueto, G., Vannata, B., Bozzoli, V., et al. (2009). Cell-free circulating DNA in Hodgkin's and non-Hodgkin's lymphomas. *Ann Oncol*, *20*, 1408-13.

[88] Bettegowda, C., Sausen, M., Leary, R. J., Kinde, I., Wang, Y., Agrawal, N., et al. (2014) Detection of circulating tumor DNA in early- and late-stage human malignancies. *Sci. Transl. Med*, *6*, 224ra24.

[89] Fleischhacker, M. & Schmidt, B. (2007). Circulating nucleic acids (CNAs) and cancer a survey. *Biochim Biophys Acta.*, *1775*, 181-232.

[90] Yong, E. (2014). Cancer biomarkers: Written in blood. *Nature*, *511*, 524-6.

[91] Tachibana, M., Takemoto, Y., Nakashima, Y., Kinugasa, S., Kotoh, T., Dhar, D. K., et al. (1998). Serum carcinoembryonic antigen as a prognostic factor in resectable gastric cancer. *J Am Coll Surg*, *187*, 64-8.

[92] Ahrens, M. J., Bertin, P. A., Vonesh, E. F., Meade, T. J., Catalona, W. J. & Georganopoulou, D. (2013). PSA enzymatic activity: a new biomarker for assessing prostate cancer aggressiveness. *Prostate.*, *73*, 1731-7.

[93] Dawson, S. J., Tsui, D. W., Murtaza, M., Biggs, H., Rueda, O. M., Chin, S. F., et al. (2013). Analysis of circulating tumor DNA to monitor metastatic breast cancer. *N Engl J Med*, *368*, 1199-209.

[94] He, C. Z., Zhang, K. H., Li, Q., Liu, X. H., Hong, Y. & Lv, N. H. (2013). Combined use of AFP, CEA, CA125 and CAl9-9 improves the sensitivity for the diagnosis of gastric cancer. *BMC Gastroenterol*, *13*, 87.

[95] Duffy, M. J. (2001). Screening for prostate cancer: Carcinoembryonic antigen as a marker for colorectal cancer: is it clinically useful? *Clin Chem*, *47*, 624-30.

[96] Mazzucchelli, R., Colanzi, P., Pomante, R., Muzzonigro, G. & Montironi, R. (2000). Prostate tissue and serum markers. *Adv Clin Pathol*, *4*, 111-120.

[97] Chan, K. C., Hung, E. C., Woo, J. K., Chan, P. K., Leung, S. F., Lai, F. P., et al. (2013). Early detection of nasopharyngeal carcinoma by plasma Epstein-Barr virus DNA analysis in a surveillance program. *Cancer*, *119*, 1838-44.

[98] Garcia-Murillas, I., Schiavon, G., Weigelt, B., Ng, C., Hrebien, S., Cutts, R. J., et al. (2015). *Mutation Mutation tracking in circulating tumor DNA predicts relapse in early breast cancer.*, *26*, 7(302), 302ra133. doi: 10.1126/scitranslmed.aab 0021.

[99] Manish, Kumar1., Yashmin, Choudhury1., Sankar, Kumar Ghosh. & Rosy, Mondal. (2018). *Application and optimization of minimally invasive cell-free DNA techniques in oncogenomics Tumor Biology*, February, 1–12.

[100] Siegel, R. L., Miller, K. D. & Jemal, A. (2016). Cancer statistics. *CA Cancer J Clin*, *66*, 7–30.

[101] Riva, F., Dronov, O. I., Khomenko, D. I., Huguet, F., Louvet, C., Mariani, P., et al. (2016) *Clinical applications of circulating tumor DNA and circulating tumor cells in pancreatic cancer. Molecular oncology*, *10*, 481-493.

[102] Olsson, E., Winter, C., George, A., Chen, Y., Howlin, J., Tang, M. H. E., et al. (2015). Serial monitoring of circulating tumor DNA in patients with primary breast cancer for detection of occult metastatic disease. *EMBO Molecular Medicine*, Vol 7, No 8, 1034-1047.

[103] Caitlin, M., Stewar, T., Prachi, D., Kothari, Mouliere F., Mair, R., et al. (2018). The value of cell-free DNA for molecular pathology. *J Pathol*, *244*, 616–627.

[104] Reinert, T., Schøler, L. V., Thomsen, R., Tobiasen, H., Vang, S., Nordentoft, I., Lamy, P., Kannerup, A. S., Mortensen, F. V., Stribolt, K., Hamilton-Dutoit, S., Nielsen, H. J., Laurberg, S., et al., (2016). Analysis of circulating tumor DNA to monitor disease burden following colorectal cancer surgery. *Gut*, *65*(4), 625-34. doi: 10.1136/gutjnl-2014-308859.

[105] Oxnard, G. R., Thress, K. S., Alden, R. S., Lawrance, R, Paweletz, C. P., Cantarini, M., et al. (2016). Association between plasma genotyping and outcomes of treatment with osimertinib (azd9291) in advanced non-small-cell lung cancer, *J. Clin. Oncol*, *34*, 3375–3382.

[106] Lu, L., Bi, J. & Bao, L. (2018) Genetic profiling of cancer with circulating tumor DNA analysis. *Journal of Genetics and Genomics*, *20*, 45(2), 79-85. doi: 10.1016/j.jgg. 2017.11.006.

[107] Jiang, P. & Lo, Y. M. D. (2016). *The Long and Short of Circulating Cell-Free DNA and the Ins and Outs of Molecular Diagnostics Trends in Genetics*, Vol. *32*, No. 6.

In: Research on New Generation …
Editor: Yildiz Dincer
ISBN: 978-1-53614-367-6

Chapter 4

CIRCULATING FREE DNA IN HEPATOBILIARY AND GASTROINTESTINAL MALIGNANCIES

Nuzhat Husain, MD and Swati Kumari, PhD
Department of Pathology,
Dr. Ram Manohar Lohia Institute of Medical Sciences,
Lucknow, India

ABSTRACT

Liquid Biopsy has moved from bench to bedside as a non-invasive biomarker for early diagnosis of the malignancy, for studying disease evolution and for assessment of treatment response. Circulating free DNA (cfDNA) has been evaluated in serum, plasma, induced sputum, bronchial lavage, milk, urine and stool. Studies using cfDNA have involved a spectrum of estimation including total quantification, cfDNA integrity assessment, cfDNA methylation as well as detection of specific diagnostic and predictive mutations in the circulating free DNA. Increased cfDNA levels has been reported in several deep-seated malignancies including hepatobiliary and gastro-intestinal (GI) malignancies including colonic, gastric, pancreatic, esophageal, hepatocellular & gallbladder cancers. In a first time study reported from our lab on cfDNA levels in Gallbladder cancer (GBC), we observed that mean values were 14-fold higher in cases as compared to healthy controls and 7 fold higher in the cholecystitis group. Diagnostics of cfDNA estimation showed a sensitivity and specificity of 100% in discrimination of cancer vs. healthy controls. CfDNA integrity was determined by amplification of short (ALU115) & long (ALU247) DNA fragments in GBC and showed significant diagnostic discrimination. CfDNA integrity and total cfDNA values showed significant association with histological grade, lymphovascular invasion, T stage, lymph node involvement TNM stage. Similarly, significantly increased cfDNA integrity has been reported in patients of colorectal cancer both in localized lesions and in advanced stage cancer with detection of mutations of KRAS and EGFR gene. In pancreatic cancer, cfDNA analysis may be used as a tool for postoperative follow-up to detect locoregional recurrence or metastatic disease in conjunction with the radiological diagnosis. In hepatocellular carcinoma (HCC) increased cfDNA has been observed with HCV infection and may be helpful in monitoring the evolution of viral infection in chronic liver disease to hepatocellular carcinoma. The association of cfDNA levels with the TNM stage has shown conflicting results while no correlations with age, gender, levels of alpha-fetoprotein (AFP) have been observed in HCC. High cfDNA levels have also been reported in patients with gastric carcinoma when compared to controls and have been shown to be effective in disease monitoring and early diagnosis. Studies have reported significantly higher cfDNA levels in patients with esophageal cancer with return of levels to normal after complete resection of the tumor. Thus cfDNA analysis in esophageal cancer serves as an important tool for early detection and for evaluation of disease progression.

Diagnostic value of cfDNA for individual patients may be limited by presence of elevated levels of cfDNA in benign pathologies which form an important differential diagnosis; For instance radiological studies often fail to differentiate between malignancy and xanthogranulomatous cholecystitis. However, at well determined cut offs cfDNA levels may be interpreted in the context of radiological findings and may help in preoperative exclusion of non-malignant cases resulting in the right surgical planning. In gastrointestinal malignancies, monitoring of KRAS mutations in cfDNA in colonic cancers is reaching the clinics, while in other gastrointestinal malignancies including hepatobiliary tumors specific diagnostic and predictive markers need to be defined. Some candidate genes include EGFR mutation analysis, which is currently used in non-small cell lung carcinoma-adenocarcinoma, along with KRAS, Her2, APC and TP53.

Keywords: circulating free DNA (cfDNA), hepatobiliary, gastro- intestinal, mutation, methylation, quantification

INTRODUCTION

The current diagnostic methods for gastrointestinal malignancies are mainly based on clinicopathological and radiological assessment and often fail to detect the malignancy at an early stage. Screening with established tumor marker such as carcinoembryonic antigen or CA19-9, has limited sensitivity. Consequently, there is a need for a commonly applicable sensitive diagnostic biomarker for malignant tumors [1, 2]. Advanced genetic diagnosis and profiling of tumors have brought forward cancer biomarkers which have been applied for early diagnosis, prediction of treatment response and prognosticating tumor behavior in various tumor categories, cancer stage and metastatic phenotype [3, 4].

Liquid biopsy has moved from bench to bedside as a non-invasive biomarker for early diagnosis of the malignancy, for studying its evolution and for evaluation of treatment response. Liquid biopsy includes circulating free DNA (cfDNA), circulating tumor cells (CTCs), circulating microRNAs, circulating proteins, extracellular vesicles. CfDNA serves as a non-invasive test for molecular diagnosis of pre-cancer & cancer and its quantification may help in optimized medical practice, personalized medicine and improving quality of life [5, 6]. The presence of cfDNA has been reported in serum, plasma, induced sputum, bronchial lavage, milk and stool [7]. A detectable amount of cfDNA has been reported in the body fluid such as urine, synovial fluid, saliva and cerebrospinal fluid [8]. The release of cfDNA in circulation occurs via active release from diseased cells, apoptosis and necrosis and from the interface of the tumor & adjacent non-tumor cells; however, its proportionate release into the circulation is still debatable [5]. Epigenetic changes such as DNA methylation are the most frequent molecular alterations in human neoplasia [9, 10] and have been implicated as molecular biomarkers.

For several malignancies increased level of cfDNA have been reported as compared to healthy individuals, such as in lung [11], ovaries [12], cervix [13], breast [14], colon [15], head and neck [16], stomach [17], bladder [18], prostate [19], testis [2], pancreas [20], glial tissue [21], liver [22] and gallbladder cancer [23]. CfDNA integrity in serum/plasma using quantitative real-time PCR for ALU repeats has been proven to be a reliable biomarker for the detection of cancer. However, there is a need to set a normal ranges of cfDNA for further examination of the possible role of cfDNA as a diagnostic biomarker, as well as for monitoring effect of therapy and in the early detection of recurrence. Quantification of cfDNA levels may represent tumor load and therby estimate its progression. CfDNA contains tumor-associated genetic and epigenetic alterations such as loss of heterozygosity, TP53 gene mutation and oncogenes such as KRAS, BRAF as well as integrity analysis by analyzing the ratio of long to short DNA fragment such as ALU & LINE1, microsatellite alteration and viral DNA (Figure 1) [15, 24].

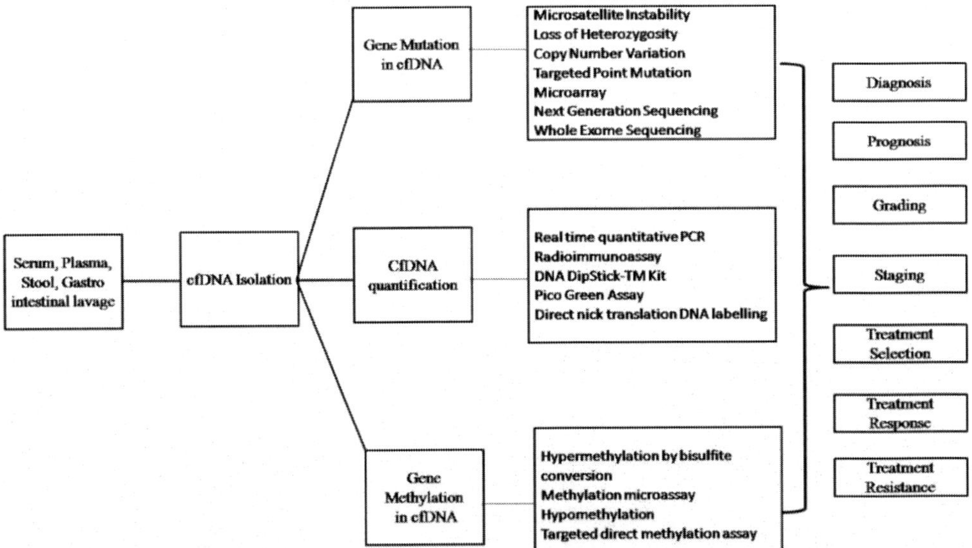

Figure 1. Evolution of cancer and relationship to circulating free DNA levels (cfDNA).

CIRCULATING MARKERS IN GASTROINTESTINAL MALIGNANCIES

The ideal diagnostic marker for cancer is well characterized, a non-invasive test, which is clinically acceptable, with high sensitivity and specificity for all stages including pre-cancerous lesions and correlates with stage and treatment. CfDNA is extracellular, double-stranded, low-molecular-weight DNA molecule that is fragmented into long (up to 21 kilobases) and short (70–200 base pairs) that circulates freely in plasma/serum of humans [25]. The analysis of these long & short fragments has deepened understanding of their origin [26]. CfDNA estimation has been shown to be effective in diagnostics, in monitoring the effectiveness of disease treatment as well as progression of disease in patients who undergo surgery or chemotherapy or radiotherapy. In most of the studies, decrease in cfDNA levels predict a better response of patients to the treatment, whereas constantly high or increasing cfDNA levels indicate a persistence of tumour unresponsive to treatment or a systemic spread of the disease [27, 28]. The ratio of long and short DNA fragments in cancer patients reflects relatively more "apoptotic" short fragments in the circulating DNA as compared to the long necrotic fragments and provides a more sensitive discrimination between normal and cancer patients. Development of malignancy is associated with higher cellular proliferation which initially equilibrated by apoptosis, and later by passive necrosis when tumor differentiates and become invasive. Elevated level of longer DNA fragment (ALU247) and higher DNA Integrity Index values are a potential non-invasive biomarker for detection of malignant tumors. Quantitative amplification of cyclin D1 oncogene in squamous carcinoma of esophagus shows a relative increase in copy number in both tissue and plasma, with prognostic significance [29].

Qualitative studies on cfDNA have demonstrated mutations in various genes including KRAS,p53, microsatellite instability, loss of heterozygosity (LOH), aberrant methylation of several genes and also tumor-specific cell-free mRNA suggesting a tumorigenic origin of cfDNA [30, 31, 32]. Methylation markers can be very useful as ancillary techniques in disease detection or progression. Studies have reported cfDNA methylation through bisulfite modification and methylation- specific PCR of various genes. Alteration in global genome hypomethylation and hypermethylation of tumor suppressor genes at CpG islands in the promoter region are frequently found in several types of tumors. With regards to cancer- specific methylation, certain "prototypical" genes that display aberrant methylation, e.g., p16 in lung and breast cancer and SEPT9 and APC in colon cancer have shown promise as diagnostic and prognostic tools.

However, a maximum diagnostic sensitivity of 80% has been obtained with the concomitant assessment of multiple methylation markers [27, 28].

Whole genome hypomethylation may also contribute to the process of oncogenesis and thus allow the increased replication of cancer- promoting proteins. In colorectal adenocarcinoma circulating lymphocyte DNA shows a significant decrease in genomic methylation status [33]. In patients with gastric and head and neck cancers, line1 hypo-methylation was found to be associated with high-risk [34, 35]. Circulating RNA has also been studied as a diagnostic marker in plasma and or serum of cancer patients. RNA has a high tendency to undergo degradation and is therefore not an attractive molecule [36].

Colorectal Cancer

CRC may develop from premalignant polyps undergoing neoplastic progress to dysplasia, carcinoma in situ and then carcinoma with multiple gene mutations. Mutation of KRAS, BRAF, PTEN, PIK3CA, BAX, TGD-b and TP53 are associated with advanced stage colorectal carcinoma. Colonoscopy is the most common modality used for screening [37].

Much research has been done for the development of blood-based biomarkers in CRC. Circulating cell-free DNA (cfDNA) was one of the earliest targets for researchers studying CRC and has been of diagnostic & prognostic value. Mutation detection in serum and/or plasma of colorectal tumors is a clinically useful biomarker for screening, detecting, and monitoring response to therapy.

Various research groups have shown significantly higher total cfDNA and tumor-specific cfDNA in CRC patients as compared to healthy controls (Table 1) with differentiating between early-stage tumors from benign lesion [38, 39, 40, 41].

In a study by Danese et al. (2010), cfDNA values showed a diagnostic accuracy of 80% in the differentiation of early stage CRC from other benign gastrointestinal disease [41]. In a study, Mead et al. (2011) reported a combination of biomarkers including Line 1, ALU 247, ALU 115, and mitochondrial DNA in plasma of CRC cases and found an AUC of 0.8 to be diagnostic of CRC. A combination of these parameters with CEA measurement increases the AUC to 0.9 [42]. In another study Hao et al. (2014) measured the ALU115 & 247 level and found significantly higher cfDNA integrity (ALU247/115) in CRC and adenomas as compared to healthy controls. Significantly increased cfDNA integrity has been reported for CRC including localized lesions as well as in advanced stage cancer [43]. Gayar et al. (2016) have also reported that serum DNA integrity is significantly higher in CRC as compared to the benign and control groups ($p<0.001$) with sensitivity and specificity of 90.0% & 85% respectively [44].

Mutation of EGFR and KRAS gene has been reported in cfDNA (Table 2). Point mutation of KRAS 2 gene in serum has been reported to affect the management of late-stage colorectal cancer [45, 46]. KRAS gene mutation occurs in 40-50% of CRC cases with the most frequent mutations occurring in codon 12 &13. Lecomte et al. (2002) found detectable amount of cfDNA in 78% CRC cases along with detectable KRAS mutation in 45% cfDNA as compared to 38% in tumor tissue [47]. Thierry et al. (2014) have reported a 96% concordance in KRAS mutation in cfDNA and tumor biopsies [48]. However, few studies have reported no detectable KRAS mutation in cfDNA in gastrointestinal malignancies[49].

Table 1. CfDNA Analysis in Gastrointestinal Malignancies - Techniques and Utility

Cancer	Measurement Technique/ Parameter	No. of Patient	Clinical association	CfDNA level/cfDNA Integrity Mean/ Median (Range) ng/ml	Sensitivity/ Specificity/Positivity	Reference
Colorectal Cancer	Fluorescence measurement	58	Overall survival	105-709	NA	Lecomte et al. 2002 [47]
	Alu115 & Alu247	32	Diagnosis	1.63 ± (0.43) 1.73± (0.45)	Sensitivity 56.0% Specificity 90.0%	Umetani et al. 2006 [15]
	RNAse P gene	67	Diagnosis	4771 (1.75–845.59)	NA	Boni et al. 2007 [39]
	Spectrophot-ometry	70	Diagnosis and Prognosis	437 (191–750)	NA	Frattini et al. 2008 [38]
	Spectrophot-ometry	55	Diagnosis and Overall survival	1157 ± 868	NA	Schwarzenbach et al. 2008 [40]
	GAPDH	118	Diagnosis	9.0 -1334.5	Positivity 80.3%	Danese et al. 2010 [41]
	Dipstick Kit	70	Prognosis	495.7 (100 - 1750),	NA	Frattini et al. 2010 [136]
	Fluorescence measurement	38	Diagnosis and Prognosis	798 ± 409	NA	Czeiger et al. 2011 [137]
	RNAse P gene	223	Early detection of recurrences	47.8 ± 120.6 11.7 (0–1096.6)	NA	Cassinotti et al. 2013 [138]

Cancer	Measurement Technique/ Parameter	No. of Patient	Clinical association	CfDNA level/cfDNA Integrity Mean/Median (Range) ng/ml	Sensitivity/ Specificity/ Positivity	Reference
Colorectal Cancer	Alu115 & Alu247	205	Diagnosis and Prognosis	1046.0 (582.7–1694.0) Integrity 0.62 (0.51–0.65)	Sensitivity 69.2% Specificity 99.0% Sensitivity 73.0% Specificity 97.2%	Hao et al. 2014 [43]
	Alu115 & Alu247	50	Diagnosis	1.54 (0.7-3.1).	Sensitivity 90.0% Specificity 85.0%	Gayar et al. 2016 [44]
Esophageal Cancer	β actin	24	Diagnosis	13.7 (1.3-97.3)	Sensitivity 92.0% Specificity 95.0%	Tomito et al. 2007 [139]
	β actin	44	Prognosis	19	NA	Banki et al. 2007 [83]
	Fluorescence measurement	101	Metastasis and recurrence	67.5 (15-1236)	NA	Tomochika et al. 2010 [82]
	Spectrophotometer	81	Diagnosis and Overall survival	29,907 ± 35,755/14,090 copies/mL	Sensitivity 96.3% Specificity 94.1%	Hsieh et al. 2016 [85]
Gastric Cancer	β actin 102 & β actin 253	53	Diagnostic	5.71 (0.855-303)	Sensitivity 80.0% Specificity 85.0%	Sai et al. 2007 [17]
	Fluorescence measurement	20	Diagnosis	10 ± 4	NA	Kolesnikova et al. 2008 [140]
	Alu81	130	Diagnosis	71.4 ± 55.4 (16.5-220.9)	Sensitivity 75.0% Specificity 63.0%	Park et al. 2012 [88]
Gastric Cancer	Real time PCR	30	Diagnosis	106.88 ± 12.40	Sensitivity 96.6% Specificity 94.1%	Kim et al. 2014 [89]
	TP53	42	Prognosis	43.1 (9.5–1338)	NA	Hamakawa et al. 2015 [141]
	Spectrophot-ometer	277	Prognosis	4280 (44 - 83,991) copies/mL	NA	Fang et al. 2016 [142]
Hepato-Cellular Carcinoma	GSTP1	52	Diagnosis	141.1 ± 161.2	Sensitivity 69.2% Specificity 93.3%	Iizuka et al. 2006 [114]
	GSTP1	96	Predictive marker	115.9 ± 98.3	NA	Tokuhisa et al. 2007 [117]

Table 1 (Continued)

Cancer	Measurement Technique/ Parameter	No. of Patient	Clinical association	CfDNA level/cfDNA Integrity Mean/Median (Range) ng/ml	Sensitivity/ Specificity/ Positivity	Reference
Hepato-Cellular Carcinoma	hTERT	60	Screening and Diagnosis	$4.18\times10^4 \pm 4.94\times10^4$ copies/µl	Sensitivity 64.0% Specificity 90.0%	Yang et al. 2011 [143]
	Fluorescence measurement	72	Diagnosis and Prognosis	173	Sensitivity 90.2% Specificity 90.3%	Huang et al. 2012 [144]
	bDNA-based Alu assay	39	Diagnosis	557.59 (12866.8 – 0)	Sensitivity 56.4% Specificity 95.6%,	Chen et al. 2013 [22]
Hepato-Cellular Carcinoma	hTERT	66	Diagnosis		Sensitivity 91.0% Specificity 43.0%	Piciocchi et al. 2013 [145]
	Alu115 & Alu247	53	Diagnosis and Prognosis	0.545 ± 0.205 (0.203-1.191)	Sensitivity 43.4% Specificity 100%	Huang et al. 2016 [146]
	Fluorescence measurement	24	Diagnosis	64.0 ± 49.2	Sensitivity 62.5% Specificity 93.6%	Yan et al. 2017 [115]
Gall Bladder Cancer	β actin	34	Diagnosis	1245.91 ± 892.46 802 (283.94 - 3848.41)	Sensitivity 100% Specificity 100%	Kumari et al. 2017 [23]
	Alu115	60	Diagnosis	2246.81 1767.86 1790.88 (61.19 - 6352.50)	Sensitivity 71.7% Specificity 66.7%	Kumari et al. 2018 [123]
	Alu247			1186.97 ± 1230.07 673.75 (55.16 - 5445.87)	Sensitivity 80.0% Specificity 86.1%	
	cfDNA Integrity			0.9664 ± 0.7659 0.6790 (0.0650 3.9380)	Sensitivity 78.33% Specificity 80.56%	

Table 2. Gene Mutation Analysis in cfDNA in Gastrointestinal Malignancies

Cancer	Gene	No. of Patient	Clinical association	Concordance CfDNA vs. Tissue	Sensitivity/Specificity/ Positivity	Reference
Colorectal Cancer	KRAS	232	Screening	83%	Sensitivity 83% Specificity 93%	Kopreski et al. 2000 [147]
		35	Diagnosis and Prognosis	74%	NA	Salbe et al.2000 [148]
		58	Overall survival	45% vs. 38%		Lecomte et al. 2002 [47]
		16	Prognosis	NA	NA	Borchers et al. 2002 [149]
		78	Diagnosis & Overall survival	NA	NA	Ryan et al. 2003 [72]
		25	Prognosis	56% vs. 64%	NA	Lindforss et al. 2005 [150]
		31	Prognosis	52% vs. 48%	NA	Lefebure et al. 2010 [67]
		108	Prognosis	78%	NA	Spindler et al. 2012 [76]
		95	Diagnosis	96%		Thierry et al. 2014 [46]
		52	Diagnosis	50% vs. 28.8		Kuo et al. 2014 [151]
		170	Diagnosis	45% vs. 3%	Sensitivity 50% Specificity 90%	Perrone F et al. 2014 [55]
		140	Prognosis	34% vs. 23%		Karen-Lise GarmSpindl er, 2015 [152]
Colorectal Cancer	KRAS	105	Overall survival	NA	NA	Spindler et al. 2014 [77]
		71	Diagnosis	32.39% vs. 50.7%		Wu et al. 2017 [153]
	SEPT9	90	Diagnosis	NA	Sensitivity 75% Specificity 70%	Zitt et al. 2007 [79]
		7941	Screening	NA	Sensitivity 48.2% Specificity 91.5%	Church et al. 2014 [154]
	TP53	46	Diagnosis	NA	NA	Ito et al. 2003 [53]

Table 2 (Continued)

Cancer	Gene	No. of Patient	Clinical association	Concordance CfDNA vs. Tissue	Sensitivity/Specificity/ Positivity	Reference
Colorectal Cancer	RASS-F1A	45	Diagnosis	NA	Sensitivity 28.9%	Wang et al. 2008 [66]
	BRAF	49	Diagnosis		Sensitivity 100% specificity 100%	Thierry et al. 2014 [48]
		105	Overall survival	NA	NA	Spindler et al. 2014 [77]
Esophageal cancer	Multiple gene mutation by NGS	13	Diagnosis and Prognosis	83.3%	NA	Ueda et al. 2016 [155]
Gastric Cancer	Tp53	42	Prognosis	NA	Positivity 24%	Hamakawa et al. 2015 [156]
	HER2 & RPPH1	52	Prognosis and targeted therapy	NA	Sensitivity 53.9% Specificity 96.7%	Shoda et al. 2015 [157]
GIST	C-KIT	38	Diagnosis & Prognosis	NA	Positivity 39.4%	Maier et al. 2013 [95]
Pancreatic cancer	KRAS	24	Diagnosis	54%		Sorenson et al. 1994 [100]
		21	Diagnosis	81%	NA	Mulcahy et al. 1998 [158]
		44	Diagnosis	27%	Sensitivity 27% Specificity 100%	Castells et al. 1999 [159]
		47			Sensitivity 47% Specificity 87%	Maire et al. 2002 [102]
		56	Prognosis	NA	Positivity 36%	Däbritz et al. 2009 [103]
		91	Overall Survival	NA	Positivity 33%	Chen et al. 2010 [104]
		75	Diagnosis	62.6% vs.74.7%	NA	Kinugasa et al. 2015 [106]
		259	Diagnosis and Prognosis	NA	Positivity 32%	Takai et al. 2015 [160]

Cancer	Gene	No. of Patient	Clinical association	Concordance CfDNA vs. Tissue	Sensitivity/Specificity/ Positivity	Reference
Pancreatic Cancer	KRAS	437	Diagnosis	NA	Positivity 21.1%	Calvez-Kelm et al. 2016 [161]
		105	Overall survival	NA	Positivity 31%	Hadano et al. 2016 [162]
		45	Overall Survival	NA	Positivity 24.4%	Nakano et al. 2018 [163]
Hepato-Cellular Carcinoma	Ser-249 TP53	186	Diagnosis	NA	Positivity 39.8%	Kirk et al. 2005 [164]
	CTNN B1	76	Diagnosis	0% vs. 10%	NA	Hosny et al. 2007 [113]

Table 3. CfDNA Methylation in Gastrointestinal Malignancies

Cancer	Gene	No. of Patient	Clinical association	Concordance CfDNA vs. Tissue	Sensitivity/Specificity/ Positivity	Reference
Colorectal Cancer	SEPT9	309	Diagnosis	NA	Sensitivity 48% Specificity 93%	Grutzmann et al. 2008 [59]
		137	Diagnosis	NA	Sensitivity 69% Specificity 86%	Lofton-Day et al.2008 [58]
		97	Diagnosis	NA	Sensitivity 72% Specificity 93%	Tetzner et al. 2009 [56]
		93	Diagnosis	NA	Sensitivity 79.3 Specificity 99.0%	Toth et al. 2012 [165]
		135	Diagnosis	NA	Sensitivity 74.8% Specificity 87.4%	Toth et al. 2014 [57]
	Hmlh1	19	Diagnosis		Sensitivity 33.0% Specificity 100%	Grady et al.2001 [65]
	p16	58	Overall survival	68% vs. 53	NA	Lecomte et al. 2002 [70]
	HLTF, HPP1/ TPEF and hMLH1	104	Overall survival	NA	NA	Wallner et al. 2006 [71]
	RASS-F1A	45	Prognosis	NA	NA	Wang et al.2008 [66]
	MGMT	64	Diagnosis	NA	Sensitivity 86.5% Specificity 90.6%	Lee et al. 2009 [64]
	APC					
	RASSF2 A					
	Wif-1					

Table 3 (Continued)

Cancer	Gene	No. of Patient	Clinical assoc-iation	Concordance CfDNA vs. Tissue	Sensitivity/Specificity/ Positivity	Reference
Colorectal Cancer	RASSF2 A	31	Prognosis	48% vs. 76%	NA	Lefebure et al. 2010 [67]
	NEURO G1	264	Screening	NA	Sensitivity 61% Specificity 91%	Herbst et al. 2011 [166]
Colorectal Cancer	TAC1& SEPT9	36	Overall survival	NA	NA	Tham et al. 2014 [167]
Esophageal Cancer	APC	52	Overall survival	25% vs. 92%	NA	Kawakami et al.2000 [168]
	Whole genome methyl-ation	08	Diagnosis	92%	NA	Zhai et al. 2012 [169]
	MSH2	209	Survival after surgery	36.8% vs.48.3	NA	Ling et al. 2012 [86]
Gastric Cancer	APC	60	Diagnosis & Prognosis	NA	Sensitivity 55% Specificity 86%	Leung et al. 2005 [92]
	E-cadherin					
	GSTP1					
	hMLH1					
	MGMT					
	p15					
	p16					
	SOCS1					
	TIMP3					
	TGF-beta RII					
Gastric Cancer	RASSF1 A	16	Diagnosis	NA	Positivity rate 34%	Wang et al.2008 [66]
	MGMT	20	Diagnosis	NA	Sensitivity 75% Specificity 54%	Kolesnikov a et al.2008 [87]
	p15					
	hMLH1					
	XAF1	202	Prognosis	69.8%vs 83.2%	NA	Ling et al. 2012 [91]
	SOX17	73	Survival	NA	Positivity 58.9%	Balgkourani dou et al. 2013 [170]
	MINT2	92	Prognosis	39.1% vs. 44.6%	NA	Han et al. 2014 [93]

Cancer	Gene	No. of Patient	Clinical association	Concordance CfDNA vs. Tissue	Sensitivity/Specificity/ Positivity	Reference
Gastric Cancer	PYCARD	41	Surgical resection	NA	Sensitivity 97.6% Specificity 66.3%	Shin et al. 2016 [171]
	APAF1					
	MINT1					
	BRCA1					
Pancreatic Cancer	MethDet 56	30	Diagnosis	NA	Sensitivity 81.7% Specificity 78%	Liggett et al. 2010 [172]
	BNC1	143	Diagnosis	NA	Sensitivity 79% Specificity 89%	Yi et al. 2013 [112]
	ADAMTS1				Sensitivity 48% Specificity 92%	
Pancreatic Cancer	CCND2	30	Diagnosis	NA	Sensitivity 76% Specificity 59%	Melnikov et al. 2009 [173]
	SOCS1					
	THBS1					
	PLAU					
	VHL					
Hepato-Cellular Carcinoma	p15	25	Diagnosis and Prognosis	22% vs. 64%	NA	Wong et al. 2000 [174]
	p16			60% vs. 76%		
	p16	29	Detection and monitoring	NA	Positivity 80%	Wong et al,2003 [121]
	p16	46	Diagnosis	NA	Sensitivity 47.8% Specificity 82.6%	Chu et al. 2004 [175]
	RASSF1A	40	Screening and/or Prognosis	42.5% vs. 92.5%	NA	Yeo et al. 2005 [176]
Hepato-Cellular Carcinoma	RUNX3	8	Diagnosis and Prognosis	NA	Positivity 88%	Tan et al. 2007 [177]
	RASSF1A				Positivity 50%	
	p16				Positivity 100%	
	CDH1				Positivity 12%	
	p16	50	Diagnosis	NA	Positivity 44%	Zhang et al. 2007 [178]
	RASSF1A				Positivity 70%	
	p15				Positivity 22%	
	APC	28	Diagnosis	53.3% vs. 64.2%	Sensitivity 82.1% Specificity 90%	Iyer et al. 2010 [179]

Table 3 (Continued)

Cancer	Gene	No. of Patient	Clinical association	Concordance CfDNA vs. Tissue	Sensitivity/Specificity/ Positivity	Reference
Hepato-Cellular Carcinoma	FHIT			67.8% vs. 75%	Sensitivity 85.7% Specificity 86%	
	p15			10.7% vs. 14.2%	Sensitivity 89.2% Specificity 96%	
	p16			46.4% vs. 71.4%	Sensitivity 67.9% Specificity 86%	
Hepato-Cellular Carcinoma	E-cadherin			46.4% vs. 67.8%	Sensitivity 78.5% Specificity 100%	
	MGMT	20	Diagnostic		Positivity 50%	Kolesnikova et al. 2008 [87]
	p15				Positivity 70%	
	hMLH1				Positivity 25%	
Gallbladder Cancer	Whole genome hypomethylation	40	Diagnostic	NA	Sensitivity 55.0% Specificity 50.0%	Kumari et al. 2018 [123]

The APC mutation found in the adenomatous polyposis coli is detected in 60%–70% of CRC, and is the most frequent associated with early carcinogenesis in CRC [50]. Diehl et al. (2005) reported that APC gene mutation in cfDNA correlated significantly with stage [51]. Mutation in TP53 gene is a late event in CRC carcinogenesis and it is mutated in 60-70% of cases [52]. However, TP53 gene mutation is infrequently detected in cfDNA from cases with CRC [53]. In early CRC, BRAF mutation is less common and therefore may have limited diagnostic use. Lilleberg et al. (2004) reported BRAF1 mutation in cfDNA and in matched tumor tissue in CRC [54].

Further, Thierry et al. (2014) reported a 96% concordance in BRAF mutation in cfDNA and tumor tissues [48]. Perrone et al. (2014) demonstrated that cfDNA is a predictive marker for CRC but is not positive in premalignant lesions. They concluded that cfDNA could serve as an adjunct marker to traditional screening modalities for CRC. However, KRAS mutation detection in cfDNA is low as compared to tissue based analysis (3% vs. 45%) [55].

Gene promoter methylation has also been detected in colorectal tissues, stools, and peripheral blood (Table 3). In colorectal cancer detection of methylation of tumor-specific genes may have a potential therapeutic and prediction use. Methylation of SEPT9 is well-studied gene and considered as a clinically useful biomarker for screening of invasive colorectal cancer [56, 57]. Septin 9 methylation is reported in 90% of CRC cases and is the most validated marker of CRC. In a number of studies, SEPT9 methylation in cfDNA showed a sensitivity and specificity in the range of 68%–79.3% & 84.8%–89% for SEPT9 in the diagnosis of CRC as compared to healthy controls [56, 57, 58, 59]. Methylation of SEPT9 in cfDNA is reported in 69%–88% of CRC, 30% of benign adenomas, and 8%–14% of healthy controls [57, 58]. In a study by Ahlquist et al. (2012) reported higher sensitivity & specificity of stool DNA in the diagnosis of CRC as compared to SEPT9 methylation in cfDNA [60]. In a cost-effectiveness analysis by Ladabaum et al. (2013) compared SEPT9 gene methylation with current screening methods (fecal-occult blood testing, fecal immunohistochemical testing, sigmoidoscopy, and colonoscopy) and found it to be inferior to other screening techniques [61]. SEPT9 used as a single test is not a suitable screening marker for CRC, although it may be valuable in combination with other markers. Other genes with abnormal promoter methylation include TMEFF2, NGFR, and p16 [62, 63]. Methylation of HPP1 and/or HLTF in serum cfDNA has been found to be associated with a poor outcome and a relative risk of high mortality. Lee et al. (2009) evaluated the methylation pattern of p14, p16, APC, DAPK, HLTF, hMLH1, MGMT, RARbeta2, RASSF2A, WNT inhibitory factor 1 (WIF1) in tumor tissue and plasma cfDNA from 243 early-stage cases of CRC, 64 patients with colonic adenomas and 276 healthy controls and reported a sensitivity of >40% for each individual gene. However, at a lower cutoff of 1.6 as methylation score, sensitivity of cfDNA methylation detection for the diagnosis of CRC was 85.6% and specificity was 92.1%, with a positive predictive value of 90.6% and a negative predictive value of 88.8% in APC, MGMT, RASSF2A, and WIF-1 modal system [64]. Hypermethylation of WIF-1, RASSF2A, p16, and HMLH1 gene in cfDNA of CRC cases has also been reported in other studies [65- 67]. Microsatellite instability and loss of heterozygosity have been reported in plasma cfDNA of 35% CRC cases [68]. CRC tumors with confirmed microsatellite instability may also have detectable markers of microsatellite instability in cfDNA [69]. However, the high false positive rate of cfDNA detection is likely to limit its utility as a diagnostic marker [68]. There is an urgent need for noninvasive detection of colonic adenomas to conduct early colonoscopy and diagnosis of premalignant change. There are very few studies to support the use of cfDNA in the differentiation of adenomas from healthy controls.

Lefebure et al. (2010) studied the prognosis in CRC patients receiving chemoradiotherapy by analyzing KRAS mutations and RASSF2A hypermethylation in cfDNA which were found to be associated with reduced disease-free survival in metastatic CRC [67]. Lecomte et al. [68] found 100%, 2-year survival in patients who had no detectable cfDNA with KRAS mutation as compared to 48% in CRC patients with detectable KRAS mutation in plasma cfDNA [70]. Further Ito et al. observed that TP53 mutation in cfDNA was associated with advanced clinical stage and liver metastases, but not lymph node metastases, tumor size, or vascular invasion [53]. Hypermethylation of HMLH1, HLTF, HPP1, and APC promoters in cfDNA of CRC cases are also associated with poor overall survival [65, 71]. However, there is still a need to evaluate the prognostic use of gene alteration in cfDNA of cases of CRC.

Detection of KRAS mutation and p16 hypermethylation in cfDNA of CRC patients has been associated with tumor recurrence after resection [38, 72]. Reinert et al. (2016)

found cfDNA to be a useful post-resection marker for determining the completeness of resection, response to adjuvant chemotherapy, recurrence after surgery and development of metastasis [73]. cfDNA based technique may be superior to conventional techniques for detection of tumor recurrence.

CfDNA estimation has also been applied to monitoring treatment response and resistance to chemotherapy. EGFR and KRAS mutation can also be detected in cfDNA of patients who are failing EGFR inhibitor therapy [74, 75]. Analysis for KRAS and BRAF mutation prior to and during therapy with cetuximab and irinotecan predicts response to therapy [76, 77]. Tabernero et al. (2015) analyzed KRAS, BRAF, and PIK3CA mutations in cfDNA of CRC patients on therapy and found that changes in cfDNA occur during chemotherapy and mutation profile may differ from that present at baseline in pre-treatment samples [78]. Zitt et al. (2008) evaluated total cfDNA levels to assess response to neoadjuvant chemoradiation in rectal carcinoma and found a significant reduction in cfDNA level in responders as compared to non responders [79]. Agostini M et al. (2011) reported cfDNA integrity as an independent predictor of response to neoadjuvant chemoradiation. SEPT9 methylation and KRAS mutation in cfDNA of CRC patients apart from being useful diagnostic and prognostic markers are also effective in monitoring treatment response and postsurgical recurrence [80]. In addition, KRAS mutation also predicts response to EGFR based therapy.

Though few studies have specifically evaluated the utility of cfDNA in CRC for distinguishing patients with metastatic disease, pilot data suggests a potential utility of cfDNA for metastatic disease detection in CRC. The time is ripe for further studies validating the diagnostic efficacy and cost- effectiveness of these markers within current management guidelines. However, available data suggest current test sensitivity is inadequate for their use as sole screening markers in CRC compared with current screening strategies. Studies validating of combinations of biomarkers with CEA and further assessment of the impact of cfDNA based biomarkers in screening of CRC and monitoring cases on follow up are urgently needed.

Esophageal Cancer

Most esophageal cancers are squamous cell carcinoma (90%) and adenocarcinoma (10%). In adenocarcinoma, frequent genetic mutations are TP53, CDKN2A, SMAD4, and ARID1A and in squamous cell carcinoma, TP53, retinoblastoma 1 (RB1), CDKN2A, PIK3CA, and NOTCH1 genes are frequently mutated [81]. Lack of biomarkers and preference for non-invasive diagnostic approach for screening and monitoring promotes the use of cfDNA based detection of esophageal cancer. There are very few studies evaluating the usefulness of cfDNA in esophageal cancer (Table 1). Tomochika et al. (2010) have reported significantly higher cfDNA in ESCC cases as compared to controls which was further significantly associated with tumor size, stage of tumor and metastatic disease. In a follwup of cases of ESCC Banki et al. (2007) cfDNA reported that the high cfDNA levels turned to normal after complete resection of the tumor [83, 84]. CfDNA quantification may be a non-invasive tool for diagnosis and prognosis in ESCC cases. cfDNA has also been found to be a marker of relapse of ESCC with high sensitivity and specificity (96.3% and 94.1%). A higher cfDNA level was also associated with tumor relapse and shorter DFS after esophagectomy in ESCC patients [85].

Ling et al. (2012) reported hypermethylation of the MutS homolog 2 (MSH2) promoters in tumor tissue of 48% of esophageal squamous cell carcinoma and hypermethylation of this gene in plasma cfDNA predicted reduced disease-free survival after esophagectomy [86]. Current data is insufficient to define the role of cfDNA in esophageal cancer diagnosis and management and further studies are warranted. Detectable levels of promoter methylation of Death Associated Protein Kinase (DAPK) and adenomatous polyposis coli gene (APC) have been reported in 61% of esophageal cancer patients. In preoperative esophageal cancer methylation of DAPK & APC has been reported to carry a worse prognosis (Table 3).

Gastric Cancer

Higher cfDNA levels have also been reported in patients with gastric carcinoma as compared with controls [87, 88] as summarized in Table 1. Kim et al. have analysed the diagnostic value of cfDNA at a cutoff point of 90 ng/mL, and reported that cfDNA value significantly differentiates cases from control with 96.6% sensitivity and 94.1% specificity [89].

Mutation of the TP53 and HER2/ERBB2/EGFR pathways is frequent in gastric cancer along with mismatch repair genes, mutations in APC, TP53, STK11, CDH1, and CTNN1A [90] summarized in Table 2. Promoter methylation of RUNX3, MGMT, p15, and hMLH1in serum cfDNA of gastric cancer patients may be used for disease monitoring and early diagnosis. In postoperative cases methylation of CEA, P16, E-cadherin, RARbeta and CDH4 genes has also been reported [17, 87]. Studies have also analyzed promoter hypermethylation of XIAP association factor 1 (XAF1), APC, HMLH1, and tissue inhibitor of metalloproteases 3 (TIMP3) genes in cfDNA of gastric cancer cases (Table 3).

Mutation analysis of these genes represents therapeutic targets with available treatments; cfDNA mutation analysis could be useful in monitoring of targeted therapies and in the detection of drug resistance mutations. Ling et al. (2012) have reported XAF1 downregulation in gastric cancer patients and have observed that an AUC of 0.9 is effective in the differentiation of cases from controls and cfDNA levels have been found to be associated with shorter overall survival [91]. Hypermethylation of HMLH1 in 41%,

APC gene in 17%, TIMP3 in 17% and E cadherin in 13% of gastric cancer cases and methylated HMLH1, APC and TIMP3 concentration has been observed to be significantly associated with stage III &IV tumours. A combination of E-cadherin and APC has been associated with shorter overall survival [92]. Promoter methylation of RASSF1A in cfDNA in cases of gastric cancer is higher, as compared to healthy controls, proving its value as a diagnostic marker of gastric cancer [66]. MINT2 methylation has also been reported to be associated with tumor progression, metastatic disease, and shorter overall survival [93].

Gastrointestinal Stromal Tumors (GIST)

The feasibility of liquid biopsy and its application in GIST was reported in the ASCO 2013 meeting for the first time [94, 95]. Several studies have evaluated cfDNA and its genetic and epigenetic alterations as a diagnostic and prognostic marker for GIST (Table 1, 2 & 4); however, studies have been small and require further validation. Maier et al. (2013) reported that cKIT and platelet-derived growth factor receptor, mutation load was higher in cases with active disease and in cases who respond to therapy. Rawnaq et al. (2011) analyzed loss of heterozygosity in cases of GIST and found microsatellite instability in cfDNA in 33% (30 of 92) patients with recurrent disease [96]. Yoo et al. (2014) analyzed cKIT mutations in exon 17, that predict response to tyrosine kinase inhibitor-based chemotherapy in GIST [97]. Demetri et al. (2013) analyzed kinase genotypes in a subgroup of GIST patients in the phase III GRID trial using tissue and plasma samples and found 84% concordance between plasma and tissue for detection of primary *KIT* mutations [94]. A secondary mutation is more easily detected in plasma (47%) as compared to tissue (12%). Secondary C-Kit mutations in cfDNA with a mutant fraction range of 0.010-9.385% have been reported in Imatinib-resistant lesions of GIST [98] and have been summarized in Table 2.

Pancreatic Cancer

Pancreatic Cancer carries a KRAS mutation in more than 90% of cancer cases and this mutation is reported to be an initiating event in carcinogenesis in the pancreas. Other genetic mutations include CDKN2A, TP53, and SMAD family 4 (SMAD4) mutations [99]. There is a lack of effective screening methods for premalignant pancreatic lesions and carcinoma. Surgical resection of the tumor is a promising therapy in patients who are diagnosed at an early stage. However, about 80% of patients suffer from locoregional recurrence even after rigorous surgical techniques. Therefore cfDNA analysis may be used as a tool for follow-up of cases undergoing surgical resection [20] (Table 1). Methylation profiling in genes in cases with inflammatory and pancreatic disease may also be helpful in the development of new diagnostic biomarker. Multiplexed assay of the promoter regions in genes such as CCND2, SOCS1, THBS1, PLAU, and VHL can detect ductal adenocarcinoma of the pancreas with high accuracy (Table 3).

Concordant KRAS mutation in tumor tissue and in cfDNA of pancreatic cancer cases was first studied by Sorenson et al. [100]. Kinugasa et al. (2015) found KRAS mutation in 74.7% of tumor tissue and in 62.6% of cfDNA samples [101]. Maire et al. (2002) analyzed the KRAS mutation in pancreatic cancer and in patients with chronic pancreatitis and found KRAS mutation in 47% of pancreatic cancer cases, 13% of patients with chronic pancreatitis, with a sensitivity & specificity 47% & 87% respectively in the diagnosis of pancreatic cancer cases. However, a combination of KRAS mutation with CA19-9 showed a sensitivity, specificity and negative predictive value of 98.0%, 77.0%, and 96.0% respectively [102]. These studies suggest that KRAS mutation in cfDNA may be used as an adjunct to CA-19.9 testing when CA-19.9 results are equivocal. Däbritz et al. (2009) detected KRAS mutations in 36% of 56 patients with Pancreatic ductal adenocarcinoma (PDA) using real-time PCR [103]. In direct sequencing, Chen et al. (2010) detected KRAS mutations in 33% of 91 patients, and they predicted a worse outcome in patients with unresectable disease [104]. Bettegowda et al. (2014) used digital PCR in a large sample of 155 patients with PDA, detecting cfDNA in more than 75% of patients with advanced disease and in as much as 57% of patients with localized disease [105]. In a large series by Kinugasa et al. (2015), mutation in KRAS in cfDNA was detected in up to 62.6% of patients by ddPCR with a concordance of 77.3% with KRAS mutation status in the solid tumor [106]. While the mutation in KRAS in the primary tumor had no prognostic value, the presence of mutant KRAS in cfDNA was associated with a significantly shorter survival ($P = 0.006$). It seems KRAS-mutant cfDNA in PDA is recurrently associated with a worse prognosis. Due to the high prevalence of KRAS mutations in PDA, it may also serve in early PDA detection and in therapy monitoring [107]. Mutation of KRAS gene appears to be of importance in the diagnosis of pancreatic cancer and may be of importance in the differentiation of chronic pancreatitis from pancreatic cancer. Combination of biomarker testing with KRAS mutation in cfDNA may improve early stage diagnosis; however future studies are needed for validation of this hypothesis and for the development of non-invasive biomarkers of premalignant lesions and pancreatic cancer (Table 2). Further, it has been reported that the presence of mutant KRAS in the circulation is associated with poor prognosis in patients with pancreatic cancer [108, 109]. Multivariate analysis has also showed that KRAS mutations in plasma cfDNA were a stronger prognostic factor for survival ($HR=7.39, p < 0.001$) than elevated CA-19.9 ($HR = 2.49$, $P = 0.087$ [104]. Thus, KRAS mutant cfDNA could be useful as a predictive biomarker for treatment decisions.

One of the major potential applications of cfDNA is disease monitoring. The changes in mutant KRAS levels in the circulation correlates with radiological imaging data and CA-19.9 levels during the course of disease and tumor progression in pancreatic cancer [110]. Methylation of several genes (including NPTX2, SFRP1, and SPARK) has been detected in pancreatic juice samples, and allows distinction of patients with chronic pancreatitis or normal individuals from cases with cancer [111]. Detecting tumor-specific epigenetic alterations in cfDNA could be an attractive option for diagnosis of pancreatic cancer by means of a liquid biopsy approach, since epigenetic markers, including aberrant DNA methylation, can be also found in cfDNA. Yi et al. (2013) demonstrated this possibility by detecting promoter methylation of BNC1 and ADAMTS1 [112]. In the future, it may be worth investigating the feasibility of utilizing combinatorial approaches with multiple blood-based biomarkers, including genomic mutations in cfDNA and epigenetic alterations in cfDNA, as a strategy to improve sensitivity and specificity in the diagnosis of early-stage pancreatic cancer (Table 3).

HEPATOCELLULAR CARCINOMA

Diagnosis of HCC is frequently based on radioimaging data and sometimes tumour tissue may not be obtained. cfDNA analysis may provide genetic information of tumor, where liver biopsy is not done. A number of studies have evaluated cfDNA in HCC (Table 1). Tumor biomarkers showed limited sensitivity & specificity in diagnosis of HCC. A number of studies have reported significantly higher cfDNA levels in HCC compared to controls and cases with chronic liver disease [22, 113, 114]. Yan et al. (2018) detected significantly higher levels of cfDNA in cases of HCC as compared to non HCC cases and these were significantly associated with the degree of liver inflammation, body mass index, and alpha-fetoprotein (AFP) level [115]. Iizuka et al. (2006) indicated that the diagnostic power of cfDNA was superior to that of AFP in hepatitis C virus-related HCC. cfDNA level was also significantly associated with tumor differentiation and tumor size and with 3-year disease-free survival [114, 116]. On multivariate analysis, in cases with chronic HCV infection related HCC, cfDNA levels have been reported to be an independent predictor of shorter overall survival and distant metastases after hepatectomy [117]. Contrarily, Iida et al. (2008), did not find cfDNA association with tumor size, stage, or overall survival [118]. Hypomethylation of LINE1 sequence in cfDNA also acts as an independent predictor of shorter overall survival, associated HBV infection, large tumor size, and advanced CLIP score [24]. Increased cfDNA levels have been detected in HCC and liver cirrhosis when compared to normal controls. These were also found to be associated with tumor size and degree of differentiation. In HCC conflicting data exists in literature on the association of cfDNA levels with TNM stage as well as age, gender, levels of alpha-fetoprotein (AFP), or protein induced by vitamin K absence (PIVKA-II) [114, 116]. Increased cfDNA has been observed in patients with HCV infection, and may be helpful in monitoring the effect of viral infection in chronic liver disease and hepatic carcinoma.

cfDNA integrity is also significantly associated with HCC diagnosis and higher cfDNA integrity correlates well with overall survival, tumor size, TMN stage, vascular and lymphatic invasion, and distant metastases [119]. Fu et al. (2012) analyzed cfDNA levels in HBV related HCC and found that serum cfDNA was significantly higher in HBV related HCC without cirrhosis as compared to 280 HBV infected noncirrhotic controls [120]. Methylation of LINE-1 has also been found to be associated with large tumor size, advanced tumor stage and with the presence of HBsAg.

Table 4. Microsatellite Instability Analysis in cfDNA in Gastro Intestinal malignancies

Cancer	Gene	No. of Patient	Clinical association	Concordance CfDNA vs. Tissue	Reference
Colorectal Cancer	Microsatellite Instability	27	Diagnosis & monitoring neoplasm	59% vs. 96%	Kölble et al. 1999 [69]
Hepato-Cellular Carcinoma	Microsatellite Instability: D8S258 and D8S264	79	Independent predictor for prognosis		Ren et al. 2006 [116]
	rs894151 and rs1243808	209	Prognostic factorfor recurrence and survival		Hu et al. 2011 [180]
GIST	Microsatellite Instability	12	Recurrence	32.6%	Rawnaq et al. 2011 [96]

It is considered as an independent prognostic factor for overall survival. In HCC, promoter methylation of RASSF1A and p16 has been detected in serum/plasma samples. Patients with higher RASSF1A concentrations at diagnosis or one year after tumor resection show a poorer disease-free survival. TP53 mutation is a very common mutation in solid tumors and has been incriminated for drug resistance and poor prognosis (Table 3). Circulating concentration of RASFF1A has been reported to be significantly higher with the disease development in HBV carriers who develop HCC [121,122]. Microsatellite instability (MSI) along with loss of heterozygosity (LOH) also plays an important role in hepatocarcinogenesis. Specifically, MSI & LOH of D8S277, D8S298, and D8S1771 at chromosome 8p have been detected in the plasma DNA of HCC patients (Table 4) [122].

GALLBLADDER CARCINOMA

We have done pioneering work on cfDNA analysis in GBC. In a first time analysis of cfDNA in GBC, we have analyzed cfDNA levels in cases of GBC and compared them to cases with cholecystitis as a disease control as well as normal controls. The mean (±SD) cfDNA levels in control, cholecystitis and cancer patients were 90.67 ng/ml (±52.87), 170.6 ng/ml (±94.13) and 1241.15 ng/ml (±870.27). The cfDNA level in GBC was significantly higher as compared to cholecystitis and healthy control group ($p < 0.0001$). The cfDNA levels were significantly associated with TNM stage, lymph node involvement, distant metastasis, and jaundice (0.001, 0.031, 0.010 and 0.020 respectively). Area under curve in ROC analysis for cancer group versus normal and cholecystitis groups was 1.00 and 0.988 with the sensitivity of 100%, 90.00% and specificity of 100% respectively [23]. Further we analyzed the cfDNA integrity in 60 cases of GBC and 36 controls including diseased controls (cholecystitis) and healthy subjects by utilizing levels of long DNA fragments (ALU247) derived from tumor necrosis, short apoptotic fragments (ALU115) denoting total cfDNA and cfDNA integrity denoting ratio of ALU247 and ALU115. Median levels of ALU115, ALU247 and cfDNA integrity were significantly different in GBC at 1790.88, 673.75, 0.4718 vs. controls at 840.73, 165.03, 0.1989 ng/ml respectively. The sensitivity and specificity of ALU247 in discriminating GBC from controls was highest with a sensitivity, specificity and diagnostic accuracy of 80.0%, 86.1% and 82.2% respectively. CfDNA integrity showed significant diagnostic discrimination of the tumor size I&II vs. III&IV, presence and absence of lymphovascular invasion, stage I&II vs. III&IV and low grade vs. high-grade histology (Table 1). However, in our cases global cfDNA methylation in GBC did not show any significant diagnostic discrimination [123] (Table 3). Further large-scale and prospective studies are needed to confirm the clinical utility of serum cfDNA integrity in the diagnosis of GBC as well as discriminate poor prognosis cases in terms of stage grade and nodal metastasis. Detection of specific mutations and their use as diagnostic and prognostic markers needs to be evaluated.

METHODS FOR QUANTIFICATION OF CFDNA

There is no standardized universal protocol for cfDNA measurement from plasma or serum. Different analytical practices are followed and hence variations occur in literature in the reported cfDNA values for different patient materials and healthy individuals [28, 124]. Methods of blood sampling, processing and sample type (plasma or serum) has been reported to have an effect on the results of the cfDNA concentration. Gormally et al. (2007) reported the impact of heterogeneous sampling procedures on the measurement outcomes in plasma of patients collected from 23 different centers [5]. However, importance has now been placed on creating a universally acceptable cfDNA isolation and quantification protocol, especially in prenatal diagnostics and in the field of clinical practice [125, 126]. The commercially available cfDNA-extraction kits capture and recover only a specific size range of DNA, typically genomic-sized DNA or low-molecular-weight DNA. Such biased results may affect the outcome in case and control studies where the cases and the controls may have differential cfDNA size profiles. Heterogeneity between the studies may be attributed to different isolation and quantification protocols and this difference in the cfDNA yield can be up to 50%. This difference is more remarkable when using silica-gel columns where small DNA fragments of less than 100 bp in size are lost during the isolation process. The use of an isolation kit that captures all DNA fragment is of important for reliability of all downstream experiments [127] especially when small DNA fractions (<100 bp) are considered most informative, since they represent the characteristic qualitative changes in the tumors [51].

Most reliable cfDNA quantification methods rely on Real-time quantitative PCR – a method which typically requires a pre-analytical DNA purification step and is considered as a gold standard. Researchers have also reported significantly increased concentration of cfDNA in plasma and serum by a radioimmunoassay method [128], DNA DipStick-TM Kit [11], Pico Green assay [129], direct nick translation DNA labeling [130] and spectrophotometry [131]. Fluorescence-based commercially available kits that employ a fluorescent DNA-intercalating dye were recently adopted to directly quantify the total cfDNA level in the plasma and serum. However, the detection limit of these kits is very low (typically 20-25 pg/ml) and may have contamination of salt and proteins [5]. There is a need for an assay which eliminates the DNA extraction step and is regardless of its sequence, fragment size or structure. Degradation of almost 30% of cfDNA in the plasma or in extracted samples with respect to preanalytical standardization may occur [132]. Sjöholm et al. (2005) analyzed changes due to storage on stored plasma/serum samples and demonstrated reliable results by using PCR [133]. Lee et al. (2001) found a 20 fold higher cfDNA in serum as compared to fresh plasma, pointing out the difference of leukocyte contamination during the process of clotting [134]. For detection of Epstein-Barr virus (EBV) specific sequences an equivalent amount of cfDNA from plasma and serum sample was observed. [135]. Further, there is no agreement on the suitability of serum or plasma as a more suitable sample type. CfDNA concentration in serum is reported to be higher, up to 24 fold, as compared to plasma [27, 134]. The majority of cfDNA in serum comes from in vitro lysis of leukocytes during the processes of coagulation and/or fibrinolysis [134]. Although some studies have propagated the use of plasma for detection of cfDNA in cancer, the possibility of contamination of plasma with leukocyte DNA after storage is a drawback with the use of plasma.

REFERENCES

[1] Goebel, G., Zitt, M., Zitt, M., & Müller, H. M. (2005). Circulating nucleic acids in plasma or serum (CNAPS) as prognostic and predictive markers in patients with solid neoplasias. *Disease markers*; 21(3):105-120.

[2] Ellinger, J., Wittkamp, V., Albers, P., Perabo, F. G., Mueller, S. C., von Ruecker, A., et al. (2009). Cell-free circulating DNA: diagnostic value in patients with testicular germ cell cancer. *The Journal of urology*; 181(1):363-371.

[3] Cohen, S. J., Punt, C. J., Iannotti, N., Saidman, B. H., Sabbath, K. D., Gabrail, N. Y., et al. (2008). Relationship of circulating tumor cells to tumor response, progression-free survival, and overall survival in patients with metastatic colorectal cancer. *Journal of clinical oncology*; 26(19):3213-3221.

[4] Krebs, M. G., Hou, J. M., Sloane, R., Lancashire, L., Priest, L., Nonaka, D., et al. (2012). Analysis of circulating tumor cells in patients with non-small cell lung cancer using epithelial marker-dependent and-independent approaches. *Journal of Thoracic Oncology*; 7(2):306-315.

[5] Gormally, E., Caboux, E., Vineis, P., & Hainaut, P. (2007). Circulating free DNA in plasma or serum as biomarker of carcinogenesis: practical aspects and biological significance. *Mutation Research/Reviews in Mutation Research*; 635(2):105-117.

[6] Holdenrieder, S., Burges, A., Reich, O., Spelsberg, F. W., & Stieber, P. (2008). DNA integrity in plasma and serum of patients with malignant and benign diseases. *Annals of the New York Academy of Sciences*; 1137(1):162-170.

[7] Su, Y. H., Wang, M., Brenner, D. E., Ng, A., Melkonyan, H., Umansky, S., et al. (2004). Human urine contains small, 150 to 250 nucleotide-sized, soluble DNA derived from the circulation and may be useful in the detection of colorectal cancer. *The Journal of Molecular Diagnostics*; 6(2):101-107.

[8] Botezatu, I., Serdyuk, O. G., Potapova, G., Shelepov, V., Alechina, R., Molyaka, Y., et al. (2000). Genetic analysis of DNA excreted in urine: a new approach for detecting specific genomic DNA sequences from cells dying in an organism. *Clinical chemistry*; 46(8):1078-1084.

[9] Merlo, A., Herman, J. G., Mao, L., Lee, D. J., Gabrielson, E., Burger, P. C., et al. (1995). 5′ CpG island methylation is associated with transcriptional silencing of the tumour suppressor p16/CDKN2/MTS1 in human cancers. *Nature medicine;* 1(7):686-692.

[10] Herman, J. G., Umar, A., Polyak, K., Graff, J. R., Ahuja, N., Issa, J. P. J., et al. (1998). Incidence and functional consequences of hMLH1 promoter hypermethylation in colorectal carcinoma. *Proceedings of the National Academy of Sciences;* 95(12):6870-6875.

[11] Sozzi, G., Conte, D., Mariani, L., Vullo, S. L., Roz, L., Lombardo, C., et al. (2001). Analysis of circulating tumor DNA in plasma at diagnosis and during follow-up of lung cancer patients. *Cancer research*; 61(12):4675-4678.

[12] Chang, H. W., Lee, S. M., Goodman, S. N., Singer, G., Cho, S. K., Sokoll, L. J., et al. (2002). Assessment of plasma DNA levels, allelic imbalance, and CA 125 as diagnostic tests for cancer. *Journal of the National Cancer Institute*; 94(22):1697-1703.

[13] Trejo-Becerril, C., Pérez-Cárdenas, E., Treviño-Cuevas, H., Taja-Chayeb, L., García-López, P., Segura-Pacheco, B., et al. (2003). Circulating nucleosomes and response to chemotherapy: an in vitro, in vivo and clinical study on cervical cancer patients. *International journal of cancer*; 104(6):663-668.

[14] Gal, S., Fidler, C., Lo, Y. M. D., Taylor, M., Han, C., Moore, J., et al. (2004). Quantitation of circulating DNA in the serum of breast cancer patients by real-time PCR. *British journal of cancer*; 90(6):1211-1215.

[15] Umetani, N., Kim, J., Hiramatsu, S., Reber, H. A., Hines, O. J., Bilchik, A. J., et al. (2006). Increased integrity of free circulating DNA in sera of patients with colorectal or periampullary cancer: direct quantitative PCR for ALU repeats. *Clinical chemistry*; 52(6):1062-1069.

[16] Jiang, W. W., Zahurak, M., Goldenberg, D., Milman, Y., Park, H. L., Westra, W. H., et al. (2006). Increased plasma DNA integrity index in head and neck cancer patients. *International journal of cancer*; 119(11):2673-2676.

[17] Sai, S., Ichikawa, D., Tomita, H., Ikoma, D., Tani, N., Ikoma, H., et al. (2007). Quantification of plasma cell-free DNA in patients with gastric cancer. *Anticancer research*; 27(4C):2747-2751.

[18] Ellinger, J., Bastian, P. J., Ellinger, N., Kahl, P., Perabo, F. G., Büttner, R., et al. (2008). Apoptotic DNA fragments in serum of patients with muscle invasive bladder cancer: a prognostic entity. *Cancer letters*; 264(2):274-280.

[19] Altimari, A., Grigioni, A. D. E., Benedettini, E., Gabusi, E., Schiavina, R., Martinell, A., et al. (2008). Diagnostic role of circulating free plasma DNA detection in patients with localized prostate cancer. *American Journal of Clinical Pathology;* 129(5):756-762.

[20] Gornik, I., Wagner, J., Gašparović, V., Lauc, G., & Gornik, O. (2009). Free serum DNA is an early predictor of severity in acute pancreatitis. *Clinical biochemistry;* 42(1):38-43.

[21] Lavon, I., Refael, M., Zelikovitch, B., Shalom, E., & Siegal, T. (2010). Serum DNA can define tumor-specific genetic and epigenetic markers in gliomas of various grades. *Neuro-oncology*; 12(2):173-180.

[22] Chen, K., Zhang, H., Zhang, L. N., Ju, S. Q., Qi, J., Huang, D. F., et al. (2013). Value of circulating cell-free DNA in diagnosis of hepatocelluar carcinoma. *World Journal of Gastroenterology*: WJG; 19(20):3143.

[23] Kumari, S., Tewari, S., Husain, N., Agarwal, A., Pandey, A., Singhal, A., et al. (2017). Quantification of Circulating Free DNA as a Diagnostic Marker in Gall Bladder Cancer. *Pathology & Oncology Research*; 23(1):91-97.

[24] Tangkijvanich, P., Hourpai, N., Rattanatanyong, P., Wisedopas, N., Mahachai, V., & Mutirangura, A. (2007). Serum LINE-1 hypomethylation as a potential prognostic marker for hepatocellular carcinoma. *Clinica chimica acta*; 379(1):127-133.

[25] Tsang, J. C., & Dennis Lo, Y. M. (2007). Circulating nucleic acids in plasma/serum. *Pathology*; 39(2):197-207.

[26] Jahr, S., Hentze, H., Englisch, S., Hardt, D., Fackelmayer, F. O., Hesch, R. D., et al. (2001). DNA fragments in the blood plasma of cancer patients: quantitations and evidence for their origin from apoptotic and necrotic cells. *Cancer research;* 61(4):1659-1665.

[27] Jung, K., Fleischhacker, M., & Rabien, A. (2010). Cell-free DNA in the blood as a solid tumor biomarker-a critical appraisal of the literature. *Clinica Chimica Acta;* 411(21):1611-1624.

[28] Schwarzenbach, H., Hoon, D. S., & Pantel, K. (2011). Cell-free nucleic acids as biomarkers in cancer patients. *Nature Reviews Cancer*; 11(6):426-437.

[29] Takeshita, H., Ichikawa, D., Komatsu, S., Tsujiura, M., Kosuga, T., Deguchi, K., et al. (2010). Prediction of CCND1 amplification using plasma DNA as a prognostic marker in oesophageal squamous cell carcinoma. *British journal of cancer*, 102(9), 1378.

[30] Sorenson, G. D. (2000). Detection of mutated KRAS2 sequences as tumor markers in plasma/serum of patients with gastrointestinal cancer. *Clinical cancer research;* 6(6):2129-2137.

[31] Ichikawa, D., Koike, H., Ikoma, H., Ikoma, D., Tani, N., Otsuji, E., et al. (2004). Detection of aberrant methylation as a tumor marker in serum of patients with gastric cancer. *Anticancer research*; 24(4):2477-2482.

[32] Ikoma, D., Ichikawa, D., Ueda, Y., Tani, N., Tomita, H., Sai, S., et al. (2007). Circulating tumor cells and aberrant methylation as tumor markers in patients with esophageal cancer. *Anticancer research*; 27(1B):535-539.

[33] McCabe, D. C., & Caudill, M. A. (2005). DNA methylation, genomic silencing, and links to nutrition and cancer. *Nutrition reviews*; 63(6):183-195.

[34] Chalitchagorn, K., Shuangshoti, S., Hourpai, N., Kongruttanachok, N., Tangkijvanich, P., Thong-ngam, D., et al. (2004). Distinctive pattern of LINE-1 methylation level in normal tissues and the association with carcinogenesis. *Oncogene;* 23(54):8841-8846.

[35] Hsiung, D. T., Marsit, C. J., Houseman, E. A., Eddy, K., Furniss, C. S., McClean, M. D., et al. (2007). Global DNA methylation level in whole blood as a biomarker

in head and neck squamous cell carcinoma. *Cancer Epidemiology and Prevention Biomarkers*; 16(1):108-114.

[36] Ringhoffer, M., Schmitt, M., Karbach, J., Jager, E., Oesch, F., & Arand, M. (2001). Quantitative assessment of the expression of melanoma-associated antigens by non-competitive reverse transcription polymerase chain reaction. *International journal of oncology;* 19(5):983-990.

[37] Markowitz, S. D., & Bertagnolli, M. M. (2009). Molecular basis of colorectal cancer. *New England Journal of Medicine*; 361(25):2449-2460.

[38] Frattini, M., Gallino, G., Signoroni, S., Balestra, D., Lusa, L., Battaglia, L., et al. (2008). Quantitative and qualitative characterization of plasma DNA identifies primary and recurrent colorectal cancer. *Cancer letters*; 263(2):170-181.

[39] Boni, L., Cassinotti, E., Canziani, M., Dionigi, G., Rovera, F., & Dionigi, R. (2007). Free circulating DNA as possible tumour marker in colorectal cancer. *Surgical oncology;* 16:29-31.

[40] Schwarzenbach, H., Stoehlmacher, J., Pantel, K., & Goekkurt, E. (2008). Detection and Monitoring of Cell-Free DNA in Blood of Patients with Colorectal Cancer. *Annals of the New York Academy of Sciences*; 1137(1):190-196.

[41] Danese, E., Montagnana, M., Minicozzi, A. M., De Matteis, G., Scudo, G., Salvagno, G. L., et al. (2010). Real-time polymerase chain reaction quantification of free DNA in serum of patients with polyps and colorectal cancers. *Clinical chemistry and laboratory medicine*; 48(11):1665-1668.

[42] Mead, R., Duku, M., Bhandari, P., & Cree, I. A. (2011). Circulating tumour markers can define patients with normal colons, benign polyps, and cancers. *British journal of cancer*; 105(2):239.

[43] Hao, T. B., Shi, W., Shen, X. J., Qi, J., Wu, X. H., Wu, Y., et al. (2014). Circulating cell-free DNA in serum as a biomarker for diagnosis and prognostic prediction of colorectal cancer. *British journal of cancer*; 111(8):1482.

[44] El-Gayar, D., El-Abd, N., Hassan, N., & Ali, R. (2016). Increased free circulating DNA integrity index as a serum biomarker in patients with colorectal carcinoma. *Asian Pac J Cancer Prev*; 17(3):939-944.

[45] Misale, S., Yaeger, R., Hobor, S., Scala, E., Janakiraman, M., Liska, D., et al. (2012). Emergence of KRAS mutations and acquired resistance to anti-EGFR therapy in colorectal cancer. *Nature;* 486(7404):532-536.

[46] Diaz Jr, L. A., Williams, R. T., Wu, J., Kinde, I., Hecht, J. R., Berlin, J., et al. (2012). The molecular evolution of acquired resistance to targeted EGFR blockade in colorectal cancers. *Nature*; 486(7404):537-540.

[47] Lecomte, T., Berger, A., Zinzindohoué, F., Micard, S., Landi, B., Blons, H., et al. (2002). Detection of free-circulating tumor-associated DNA in plasma of colorectal cancer patients and its association with prognosis. *International journal of cancer;* 100(5):542-548.

[48] Thierry, A. R., Mouliere, F., El Messaoudi, S., Mollevi, C., Lopez-Crapez, E., Rolet, F., et al. (2014). Clinical validation of the detection of KRAS and BRAF mutations from circulating tumor DNA. *Nature medicine*; 20(4):430.

[49] Taly, V., Pekin, D., Benhaim, L., Kotsopoulos, S. K., Le Corre, D., Li, X., et al. (2013). Multiplex picodroplet digital PCR to detect KRAS mutations in circulating DNA from the plasma of colorectal cancer patients. *Clinical chemistry;* 59(12):1722-1731.

[50] Laurent-Puig, P., Béroud, C., & Soussi, T. (1998). APC gene: database of germline and somatic mutations in human tumors and cell lines. *Nucleic acids research;* 26(1):269-270.

[51] Diehl, F., Li, M., Dressman, D., He, Y., Shen, D., Szabo, S., et al. (2005). Detection and quantification of mutations in the plasma of patients with colorectal tumors. *Proceedings of the National Academy of Sciences of the United States of America*; 102(45):16368-16373.

[52] Baker, S. J., Preisinger, A. C., Jessup, J. M., Paraskeva, C., Markowitz, S., Willson, J. K. V., et al. (1990). p53 gene mutations occur in combination with 17p allelic deletions as late events in colorectal tumorigenesis. *Cancer research*; 50(23):7717-7722.

[53] Ito, T., Kaneko, K., Makino, R., Konishi, K., Kurahashi, T., Ito, H., et al. (2003). Clinical significance in molecular detection of p53 mutation in serum of patients with colorectal carcinoma. *Oncology reports*; 10(6):1937-1942.

[54] Lilleberg, S. L., Durocher, J., Sanders, C., Walters, K., & Culver, K. (2004). High Sensitivity Scanning of Colorectal Tumors and Matched Plasma DNA for Mutations in APC, TP53, K-RAS, and BRAF Genes with a Novel DHPLC Fluorescence Detection Platform. *Annals of the New York Academy of Sciences;* 1022(1):250-256.

[55] Perrone, F., Lampis, A., Bertan, C., Verderio, P., Ciniselli, C. M., Pizzamiglio, S., et al. (2014). Circulating free DNA in a screening program for early colorectal cancer detection. *Tumori*; 100(2): 115-21.

[56] Tetzner, R., Model, F., Weiss, G., Schuster, M., Distler, J., Steiger, K. V., et al. (2009). Circulating methylated SEPT9 DNA in plasma is a biomarker for colorectal cancer. *Clinical chemistry*; 55(7):1337-1346.

[57] Tóth, K., Wasserkort, R., Sipos, F., Kalmár, A., Wichmann, B., Leiszter, K., et al. (2014). Detection of methylated septin 9 in tissue and plasma of colorectal patients with neoplasia and the relationship to the amount of circulating cell-free DNA. *PloS one*; 9(12): e115415.

[58] Lofton-Day, C., Model, F., DeVos, T., Tetzner, R., Distler, J., Schuster, M., et al. (2008). DNA methylation biomarkers for blood-based colorectal cancer screening. *Clinical chemistry*; 54(2):414-423.

[59] Grützmann, R., Molnar, B., Pilarsky, C., Habermann, J. K., Schlag, P. M., Saeger, H. D., et al. (2008). Sensitive detection of colorectal cancer in peripheral blood by septin 9 DNA methylation assay. *PloS one;* 3(11):e3759.

[60] Ahlquist, D. A., Taylor, W. R., Mahoney, D. W., Zou, H., Domanico, M., Thibodeau, S. N., et al. (2012). The stool DNA test is more accurate than the plasma septin 9 test in detecting colorectal neoplasia. *Clinical Gastroenterology and Hepatology;* 10(3):272-277.

[61] Ladabaum, U., Allen, J., Wandell, M., & Ramsey, S. (2013). Colorectal cancer screening with blood-based biomarkers: cost-effectiveness of methylated septin 9 DNA versus current strategies. *Cancer Epidemiology and Prevention Biomarkers*; 22(9):1567-1576.

[62] Morgan, S. R., Whiteley, J., Donald, E., Smith, J., Eisenberg, M. T., Kallam, E., et al. (2012). Comparison of KRAS mutation assessment in tumor DNA and circulating free DNA in plasma and serum samples. *Clinical medicine insights Pathology;* 5: 15.

[63] Mouliere, F., El Messaoudi, S., Gongora, C., Guedj, A. S., Robert, B., Del Rio, M., et al. (2013). Circulating cell-free DNA from colorectal cancer patients may reveal high KRAS or BRAF mutation load. *Translational oncology*; 6(3):319IN6-328IN8.

[64] Lee, B. B., Lee, E. J., Jung, E. H., Chun, H. K., Chang, D. K., Song, S. Y., et al. (2009). Aberrant methylation of APC, MGMT, RASSF2A, and Wif-1 genes in plasma as a biomarker for early detection of colorectal cancer. *Clinical cancer research;* 15(19):6185-6191.

[65] Grady, W. M., Rajput, A., Lutterbaugh, J. D., & Markowitz, S. D. (2001). Detection of aberrantly methylated hMLH1 promoter DNA in the serum of patients with microsatellite unstable colon cancer. *Cancer research*; 61(3):900-902.

[66] Wang, Y. C., Yu, Z. H., Liu, C., Xu, L. Z., Yu, W., Lu, J., ... & Ji, H. Z. (2008). Detection of RASSF1A promoter hypermethylation in serum from gastric and colorectal adenocarcinoma patients. *World journal of gastroenterology: WJG;* 14(19):3074.

[67] Lefebure, B., Charbonnier, F., Di Fiore, F., Tuech, J. J., Le Pessot, F., Michot, F., et al. (2010). Prognostic value of circulating mutant DNA in unresectable metastatic colorectal cancer. *Annals of surgery*; 251(2):275-280.

[68] Lecomte, T., Ceze, N., Dorval, E., & Laurent-Puig, P. (2010). Circulating free tumor DNA and colorectal cancer. *Gastroenterologie clinique et biologique*; 34(12):662-681.

[69] Kölble, K., Ullrich, O. M., Pidde, H., Barthel, B., Diermann, J., Rudolph, B., et al. (1999). Microsatellite alterations in serum DNA of patients with colorectal cancer. *Laboratory investigation; a journal of technical methods and pathology*; 79(9):1145-1150.

[70] Lecomte, T., Berger, A., Zinzindohoué, F., Micard, S., Landi, B., Blons, H., et al. (2002). Detection of free-circulating tumor-associated DNA in plasma of colorectal cancer patients and its association with prognosis. *International journal of cancer*; 100(5):542-548.

[71] Wallner, M., Herbst, A., Behrens, A., Crispin, A., Stieber, P., Göke, B., et al. (2006). Methylation of serum DNA is an independent prognostic marker in colorectal cancer. *Clinical Cancer Research*; 12(24):7347-7352.

[72] Ryan, B. M., Lefort, F., McManus, R., Daly, J., Keeling, P. W. N., Weir, D. G., et al. (2003). A prospective study of circulating mutant KRAS2 in the serum of patients with colorectal neoplasia: strong prognostic indicator in postoperative follow up. *Gut;* 52(1):101-108.

[73] Reinert, T., Schøler, L. V., Thomsen, R., Tobiasen, H., Vang, S., Nordentoft, I., et al. (2016). Analysis of circulating tumour DNA to monitor disease burden following colorectal cancer surgery. *Gut;* 65(4):625-634.

[74] Bettegowda, C., Sausen, M., Leary, R. J., Kinde, I., Wang, Y., Agrawal, N., et al. (2014). Detection of circulating tumor DNA in early-and late-stage human malignancies. *Science translational medicine*; 6(224):224ra24-224ra24.

[75] Morelli, M. P., Overman, M. J., Dasari, A., Kazmi, S. M. A., Mazard, T., Vilar, E., et al. (2015). Characterizing the patterns of clonal selection in circulating tumor DNA from patients with colorectal cancer refractory to anti-EGFR treatment. *Annals of Oncology*; 26(4):731-736.

[76] Spindler, K. L. G., Pallisgaard, N., Vogelius, I., & Jakobsen, A. (2012). Quantitative cell free DNA, KRAS and BRAF mutations in plasma from patients with metastatic colorectal cancer during treatment with cetuximab and irinotecan. *Clinical Cancer Research*; clincanres-0564.

[77] Spindler, K. L. G., Pallisgaard, N., Andersen, R. F., & Jakobsen, A. (2014). Changes in mutational status during third-line treatment for metastatic colorectal cancer-Results of consecutive measurement of cell free DNA, KRAS and BRAF in the plasma. *International journal of cancer*; 135(9):2215-2222.

[78] Tabernero, J., Lenz, H. J., Siena, S., Sobrero, A., Falcone, A., Ychou, M., et al. (2015). Analysis of circulating DNA and protein biomarkers to predict the clinical activity of regorafenib and assess prognosis in patients with metastatic colorectal cancer: a retrospective, exploratory analysis of the CORRECT trial. *The Lancet Oncology*; 16(8):937-948.

[79] Zitt, M., Müller, H. M., Rochel, M., Schwendinger, V., Zitt, M., Goebel, G., et al. (2008). Circulating cell-free DNA in plasma of locally advanced rectal cancer patients undergoing preoperative chemoradiation: a potential diagnostic tool for therapy monitoring. *Disease markers*; 25(3):159-165.

[80] Agostini, M., Pucciarelli, S., Enzo, M. V., Del Bianco, P., Briarava, M., Bedin, C., et al. (2011). Circulating cell-free DNA: a promising marker of pathologic tumor

response in rectal cancer patients receiving preoperative chemoradiotherapy. *Annals of surgical oncology*; 18(9):2461-2468.

[81] Rustgi, A. K., & El-Serag, H. B. (2014). Esophageal carcinoma. New England *Journal of Medicine*; 371(26):2499-2509.

[82] Tomochika, S., Iizuka, N., Watanabe, Y., Tsutsui, M., Takeda, S., Yoshino, S., et al. (2010). Increased serum cell-free DNA levels in relation to inflammation are predictive of distant metastasis of esophageal squamous cell carcinoma. *Experimental and therapeutic medicine*; 1(1):89-92.

[83] Banki, F., Mason, R. J., Oh, D., Hagen, J. A., DeMeester, S. R., Lipham, J. C., et al. (2007). Plasma DNA as a molecular marker for completeness of resection and recurrent disease in patients with esophageal cancer. *Archives of Surgery*; 142(6):533-539.

[84] Takeshita, H., Ichikawa, D., Komatsu, S., Tsujiura, M., Kosuga, T., Deguchi, K., et al. (2010). Prediction of CCND1 amplification using plasma DNA as a prognostic marker in oesophageal squamous cell carcinoma. *British journal of cancer;* 102(9):1378.

[85] Hsieh, C. C., Hsu, H. S., Chang, S. C., & Chen, Y. J. (2016). Circulating cell-free DNA levels could predict oncological outcomes of patients undergoing esophagectomy for esophageal squamous cell carcinoma. *International journal of molecular sciences;* 17(12):2131.

[86] Ling, Z. Q., Zhao, Q., Zhou, S. L., & Mao, W. M. (2012). MSH2 promoter hypermethylation in circulating tumor DNA is a valuable predictor of disease-free survival for patients with esophageal squamous cell carcinoma. *European Journal of Surgical Oncology*; 38(4):326-332.

[87] Kolesnikova, E. V., Tamkovich, S. N., Bryzgunova, O. E., Shelestyuk, P. I., Permyakova, V. I., Vlassov, V. V., et al. (2008). Circulating DNA in the blood of gastric cancer patients. *Annals of the New York Academy of Sciences;* 1137(1):226-231.

[88] Park, J. L., Kim, H. J., Choi, B. Y., Lee, H. C., Jang, H. R., Song, K. S., et al. (2012). Quantitative analysis of cell-free DNA in the plasma of gastric cancer patients. *Oncology letters;* 3(4):921-926.

[89] Kim, K., Shin, D. G., Park, M. K., Baik, S. H., Kim, T. H., Kim, S., et al. (2014). Circulating cell-free DNA as a promising biomarker in patients with gastric cancer: diagnostic validity and significant reduction of cfDNA after surgical resection. *Annals of surgical treatment and research*; 86(3):136-142.

[90] Van Cutsem, E., Sagaert, X., Topal, B., Haustermans, K., & Prenen, H. (2016). Gastric cancer. *The Lancet;* 388(10060):2654-2664.

[91] Ling, Z. Q., Zhao, Q., Zhou, S. L., & Mao, W. M. (2012). MSH2 promoter hypermethylation in circulating tumor DNA is a valuable predictor of disease-free

survival for patients with esophageal squamous cell carcinoma. *European Journal of Surgical Oncology*; 38(4):326-332.

[92] Leung, W. K., To, K. F., Chu, E. S. H., Chan, M. W. Y., Bai, A. H. C., Ng, E. K. W., et al. (2005). Potential diagnostic and prognostic values of detecting promoter hypermethylation in the serum of patients with gastric cancer. *British journal of cancer*; 92(12):2190.

[93] Han, J., Lv, P., Yu, J. L., Wu, Y. C., Zhu, X., Hong, L. L., et al. (2014). Circulating methylated MINT2 promoter DNA is a potential poor prognostic factor in gastric cancer. *Digestive diseases and sciences*; 59(6):1160-1168.

[94] Demetri, G. D., Jeffers, M., Reichardt, P., Kang, Y. K., Blay, J. Y., Rutkowski, P., et al. (2013). Mutational analysis of plasma DNA from patients (pts) in the phase III GRID study of regorafenib (REG) versus placebo (PL) in tyrosine kinase inhibitor (TKI)-refractory GIST: *Correlating genotype with clinical outcomes.*

[95] Maier, J., Lange, T., Kerle, I., Specht, K., Brugel, M., Wickenhauser, C., et al. (2013). Detection of mutant free circulating tumor DNA in the plasma of patients with gastrointestinal stromal tumor (GIST) harboring activating mutations of CKIT or PDGFRA. *Clinical Cancer Research; clincanres*-0765.

[96] Rawnaq, T., Schwarzenbach, H., Schurr, P. G., Freise, K., Brandl, S., Izbicki, J. R., et al. (2011). Monitoring of loss of heterozygosity in serum microsatellite DNA among patients with gastrointestinal stromal tumors indicates tumor recurrence. *Journal of Surgical Research*; 169(1):31-35.

[97] Yoo, C., Ryu, M. H., Na, Y. S., Ryoo, B. Y., Park, S. R., & Kang, Y. K. (2014). Analysis of serum protein biomarkers, circulating tumor DNA, and dovitinib activity in patients with tyrosine kinase inhibitor-refractory gastrointestinal stromal tumors. *Annals of oncology;* 25(11):2272-2277.

[98] Wada, N., Kurokawa, Y., Takahashi, T., Hamakawa, T., Hirota, S., Naka, T., et al. (2016). Detecting secondary C-KIT mutations in the peripheral blood of patients with imatinib-resistant gastrointestinal stromal tumor. *Oncology*; 90(2):112-117.

[99] Ryan, D. P., Hong, T. S., & Bardeesy, N. (2014). Pancreatic adenocarcinoma. New *England Journal of Medicine;* 371(11):1039-1049.

[100] Sorenson, G. D., Pribish, D. M., Valone, F. H., Memoli, V. A., Bzik, D. J., & Yao, S. L. (1994). Soluble normal and mutated DNA sequences from single-copy genes in human blood. *Cancer Epidemiology and Prevention Biomarkers*; 3(1):67-71.

[101] Kinugasa, H., Nouso, K., Miyahara, K., Morimoto, Y., Dohi, C., Tsutsumi, K., et al. (2015). Detection of K-ras gene mutation by liquid biopsy in patients with pancreatic cancer. *Cancer*; 121(13):2271-2280.

[102] Maire, F., Micard, S., Hammel, P., Voitot, H., Lévy, P., Cugnenc, P. H., et al. (2002). Differential diagnosis between chronic pancreatitis and pancreatic cancer: value of the detection of KRAS2 mutations in circulating DNA. *British journal of cancer;* 87(5):551.

[103] Däbritz, J., Preston, R., Hänfler, J., & Oettle, H. (2009). Follow-up study of K-ras mutations in the plasma of patients with pancreatic cancer: correlation with clinical features and carbohydrate antigen 19-9. *Pancreas;* 38(5):534-541.

[104] Chen, H., Tu, H., Meng, Z. Q., Chen, Z., Wang, P., & Liu, L. M. (2010). K-ras mutational status predicts poor prognosis in unresectable pancreatic cancer. *European Journal of Surgical Oncology*; 36(7):657-662.

[105] Bettegowda, C., Sausen, M., Leary, R. J., Kinde, I., Wang, Y., Agrawal, N., et al. (2014). Detection of circulating tumor DNA in early-and late-stage human malignancies. *Science translational medicine*; 6(224):224ra24-224ra24.

[106] Kinugasa, H., Nouso, K., Miyahara, K., Morimoto, Y., Dohi, C., Tsutsumi, K., et al. (2015). Detection of K-ras gene mutation by liquid biopsy in patients with pancreatic cancer. *Cancer;* 121(13):2271-2280.

[107] Gao, Y., Zhu, Y., & Yuan, Z. (2016). Circulating tumor cells and circulating tumor DNA provide new insights into pancreatic cancer. *International journal of medical sciences;* 3(12):902.

[108] Kinde, I., Wu, J., Papadopoulos, N., Kinzler, K. W., & Vogelstein, B. (2011). Detection and quantification of rare mutations with massively parallel sequencing. *Proceedings of the National Academy of Sciences*; 108(23):9530-9535.

[109] Sausen, M., Phallen, J., Adleff, V., Jones, S., Leary, R. J., Barrett, M. T., et al. (2015). Clinical implications of genomic alterations in the tumour and circulation of pancreatic cancer patients. *Nature communications*; 6:7686.

[110] Tjensvoll, K., Lapin, M., Buhl, T., Oltedal, S., Steen-Ottosen Berry, K., Gilje, B., et al. (2016). Clinical relevance of circulating KRAS mutated DNA in plasma from patients with advanced pancreatic cancer. *Molecular oncology*; 10(4):635-643.

[111] Matsubayashi, H., Canto, M., Sato, N., Klein, A., Abe, T., Yamashita, K., et al. (2006). DNA methylation alterations in the pancreatic juice of patients with suspected pancreatic disease. *Cancer research;* 66(2):1208-1217.

[112] Yi, J. M., Guzzetta, A. A., Bailey, V. J., Downing, S. R., Van Neste, L., Chiappinelli, K. B., et al. (2013). Novel methylation biomarker panel for the early detection of pancreatic cancer. *Clinical Cancer Research*; 19(23):6544-6555.

[113] Hosny, G., Farahat, N., Tayel, H., & Hainaut, P. (2008). Ser-249 TP53 and CTNNB1 mutations in circulating free DNA of Egyptian patients with hepatocellular carcinoma versus chronic liver diseases. *Cancer letters;* 264(2):201-208.

[114] Iizuka, N., Sakaida, I., Moribe, T., Fujita, N., Miura, T., Stark, M., et al. (2006). Elevated levels of circulating cell-free DNA in the blood of patients with hepatitis C virus-associated hepatocellular carcinoma. *Anticancer research*; 26(6C):4713-4719.

[115] Yan, L., Chen, Y., Zhou, J., Zhao, H., Zhang, H., & Wang, G. (2018). Diagnostic value of circulating cell-free DNA levels for hepatocellular carcinoma. *International Journal of Infectious Diseases*; 67:92-97.

[116] Ren, N., Qin, L. X., Tu, H., Liu, Y. K., Zhang, B. H., & Tang, Z. Y. (2006). The prognostic value of circulating plasma DNA level and its allelic imbalance on chromosome 8p in patients with hepatocellular carcinoma. *Journal of cancer research and clinical oncology*; 132(6):399-407.

[117] Tokuhisa, Y., Iizuka, N., Sakaida, I., Moribe, T., Fujita, N., Miura, T., et al. (2007). Circulating cell-free DNA as a predictive marker for distant metastasis of hepatitis C virus-related hepatocellular carcinoma. *British journal of cancer;* 97(10):1399.

[118] Iida, M., Iizuka, N., Sakaida, I., Moribe, T., Fujita, N., Miura, T., & Tokuhisa, Y. (2008). Relation between serum levels of cell-free DNA and inflammation status in hepatitis C virus-related hepatocellular carcinoma. *Oncology reports*; 20(4):761-765.

[119] El-Shazly, S. F., Eid, M. A., El-Sourogy, H. A., Attia, G. F., & Ezzat, S. A. (2010). Evaluation of serum DNA integrity as a screening and prognostic tool in patients with hepatitis C virus-related hepatocellular carcinoma. *International Journal of Biological Markers*; 25(2):79.

[120] Fu, X., Wan, S., Hann, H. W., Myers, R. E., Hann, R. S., Au, J., et al. (2012). Relative telomere length: a novel non-invasive biomarker for the risk of non-cirrhotic hepatocellular carcinoma in patients with chronic hepatitis B infection. *European Journal of Cancer*; 48(7):1014-1022.

[121] Wong, I. H., Zhang, J., Lai, P. B., Lau, W. Y., & Lo, Y. D. (2003). Quantitative analysis of tumor-derived methylated p16INK4a sequences in plasma, serum, and blood cells of hepatocellular carcinoma patients. *Clinical Cancer Research*; 9(3):1047-1052.

[122] Chan, K. A., Lai, P. B., Mok, T. S., Chan, H. L., Ding, C., Yeung, S. W., et al. (2008). Quantitative analysis of circulating methylated DNA as a biomarker for hepatocellular carcinoma. *Clinical Chemistry*; 54(9):1528-1536.

[123] Kumari, S., Husain, N., Agarwal, A., Neyaz, A., Gupta, S., Chaturvedi, A., et al. (2018). Diagnostic Value of Circulating Free DNA Integrity and Global Methylation Status in Gall Bladder Carcinoma. *Pathology & Oncology Research*; 1-12.

[124] Gormally, E., Hainaut, P., Caboux, E., Airoldi, L., Autrup, H., Malaveille, C., et al. (2004). Amount of DNA in plasma and cancer risk: a prospective study. *International journal of cancer*; 111(5):746-749.

[125] Butt, A. N., & Swaminathan, R. (2008). Overview of circulating nucleic acids in plasma/serum. *Annals of the New York Academy of Sciences*; 1137(1):236-242.

[126] Wagner, J. (2012). Free DNA–new potential analyte in clinical laboratory diagnostics?. *Biochemia medica*; 22(1):24-38.

[127] Samuels, Y., Wang, Z., Bardelli, A., Silliman, N., Ptak, J., Szabo, S., et al. (2004). High frequency of mutations of the PIK3CA gene in human cancers. *Science;* 304(5670):554-554.

[128] Leon, S. A., Shapiro, B., Sklaroff, D. M., & Yaros, M. J. (1977). Free DNA in the serum of cancer patients and the effect of therapy. *Cancer research*; 37(3):646-650.

[129] Park, J. L., Kim, H. J., Choi, B. Y., Lee, H. C., Jang, H. R., Song, K. S., et al. (2012). Quantitative analysis of cell-free DNA in the plasma of gastric cancer patients. *Oncology letters*; 3(4):921-926.

[130] Fournié, G. J., Courtin, J. P., Laval, F., Chalé, J. J., Pourrat, J. P., Pujazon, M. C., et al. (1995). Plasma DNA as a marker of cancerous cell death. Investigations in patients suffering from lung cancer and in nude mice bearing human tumours. *Cancer letters*; 91(2):221-227.

[131] Stroun, M., Anker, P., Maurice, P., Lyautey, J., Lederrey, C., & Beljanski, M. (1989). Neoplastic characteristics of the DNA found in the plasma of cancer patients. *Oncology;* 46(5):318-322.

[132] Sozzi, G., Roz, L., Conte, D., Mariani, L., Andriani, F., Verderio, P., et al. (2005). Effects of prolonged storage of whole plasma or isolated plasma DNA on the results of circulating DNA quantification assays. *Journal of the National Cancer Institute*; 97(24):1848-1850.

[133] Sjöholm, M. I., Hoffmann, G., Lindgren, S., Dillner, J., & Carlson, J. (2005). Comparison of archival plasma and formalin-fixed paraffin-embedded tissue for genotyping in hepatocellular carcinoma. *Cancer Epidemiology and Prevention Biomarkers;* 14(1):251-255.

[134] Lee, T. H., Montalvo, L., Chrebtow, V., & Busch, M. P. (2001). Quantitation of genomic DNA in plasma and serum samples: higher concentrations of genomic DNA found in serum than in plasma. *Transfusion*; 41(2):276-282.

[135] Jones, K., Nourse, J. P., Keane, C., Crooks, P., Gottlieb, D., Ritchie, D. S., et al. (2012). Tumor-specific but not nonspecific cell-free circulating DNA can be used to monitor disease response in lymphoma. *American journal of hematology;* 87(3):258-265.

[136] Frattini, M., Gallino, G., Signoroni, S., Balestra, D., Battaglia, L., Sozzi, G., et al. (2006). Quantitative analysis of plasma DNA in colorectal cancer patients. *Annals of the New York Academy of Sciences*; 1075(1):185-190.

[137] Czeiger, D., Shaked, G., Eini, H., Vered, I., Belochitski, O., Avriel, A., et al. (2011). Measurement of circulating cell-free DNA levels by a new simple fluorescent test in patients with primary colorectal cancer. *American journal of clinical pathology*; 135(2):264-270.

[138] Cassinotti, E., Boni, L., Segato, S., Rausei, S., Marzorati, A., Rovera, F., et al. (2013). Free circulating DNA as a biomarker of colorectal cancer. *International Journal of Surgery*; 11:S54-S57.

[139] Tomita, H., Ichikawa, D., Ikoma, D., Sai, S., Tani, N., Ikoma, H., et al. (2007). Quantification of circulating plasma DNA fragments as tumor markers in patients with esophageal cancer. *Anticancer research*; 27(4C):2737-2741.

[140] Kolesnikova, E. V., Tamkovich, S. N., Bryzgunova, O. E., Shelestyuk, P. I., Permyakova, V. I., Vlassov, V. V., et al. (2008). Circulating DNA in the blood of gastric cancer patients. *Annals of the New York Academy of Sciences;* 1137(1):226-231.

[141] Hamakawa, T., Kukita, Y., Kurokawa, Y., Miyazaki, Y., Takahashi, T., Yamasaki, M., et al. (2015). Monitoring gastric cancer progression with circulating tumour DNA. *British journal of cancer*; 112(2):352.

[142] Fang, W. L., Lan, Y. T., Huang, K. H., Liu, C. A., Hung, Y. P., Lin, C. H., et al. (2016). Clinical significance of circulating plasma DNA in gastric cancer. *International journal of cancer*; 138(12):2974-2983.

[143] Yang, Y. J., Chen, H., Huang, P., Li, C. H., Dong, Z. H., & Hou, Y. L. (2011). Quantification of plasma hTERT DNA in hepatocellular carcinoma patients by quantitative fluorescent polymerase chain reaction. *Clinical & Investigative Medicine*; 34(4):238-244.

[144] Huang, Z., Hua, D., Hu, Y., Cheng, Z., Zhou, X., Xie, Q., et al. (2012). Quantitation of plasma circulating DNA using quantitative PCR for the detection of hepatocellular carcinoma. *Pathology & Oncology Research*; 18(2):271-276.

[145] Piciocchi, M., Cardin, R., Vitale, A., Vanin, V., Giacomin, A., Pozzan, C., et al. (2013). Circulating free DNA in the progression of liver damage to hepatocellular carcinoma. *Hepatology international*; 7(4):1050-1057.

[146] Huang, A., Zhang, X., Zhou, S. L., Cao, Y., Huang, X. W., Fan, J., et al. (2016). Plasma circulating cell-free DNA integrity as a promising biomarker for diagnosis and surveillance in patients with hepatocellular carcinoma. *Journal of Cancer;* 7(13):1798.

[147] Kopreski, M. S., Benko, F. A., Borys, D. J., Khan, A., McGarrity, T. J., & Gocke, C. D. (2000). Somatic mutation screening: identification of individuals harboring K-ras mutations with the use of plasma DNA. *Journal of the National Cancer Institute;* 92(11):918-923.

[148] Salbe, C., Trevisiol, C., Ferruzzi, E., Mancuso, T., Nascimbeni, R., Di, F. F., et al. (2000). Molecular detection of codon 12 K-RAS mutations in circulating DNA from serum of colorectal cancer patients. *The International journal of biological markers;* 15(4):300-307.

[149] Borchers, R., Heinzlmann, M., Zahn, R., Witter, K., Martin, K., Loeschke, K., & Folwaczny, C. (2002). K-ras mutations in sera of patients with colorectal

neoplasias and long-standing inflammatory bowel disease. *Scandinavian journal of gastroenterology*; 37(6):715-718.

[150] Lindforss, U., Zetterquist, H., Papadogiannakis, N., & Olivecrona, H. (2005). Persistence of K-ras mutations in plasma after colorectal tumor resection. *Anticancer research*; 25(1B):657-661.

[151] Kuo, Y. B., Chen, J. S., Fan, C. W., Li, Y. S., & Chan, E. C. (2014). Comparison of KRAS mutation analysis of primary tumors and matched circulating cell-free DNA in plasmas of patients with colorectal cancer. *Clinica Chimica Acta*; 433:284-289.

[152] Spindler, K. L. G., Pallisgaard, N., Appelt, A. L., Andersen, R. F., Schou, J. V., Nielsen, D., et al. (2015). Clinical utility of KRAS status in circulating plasma DNA compared to archival tumour tissue from patients with metastatic colorectal cancer treated with anti-epidermal growth factor receptor therapy. *European Journal of Cancer*; 51(17): 2678-2685.

[153] Wu, J., Zhao, L. R., Lin, X. Q., Feng, F., Chen, Y. C., Deng, W. Y., et al. (2017). Association of RAS mutations in circulating cell-free DNA in the plasma with clinicopathological features of colorectal cancer. Nan fang yi ke da xue xue bao= *Journal of Southern Medical University*; 37(7):962-966.

[154] Church, T. R., Wandell, M., Lofton-Day, C., Mongin, S. J., Burger, M., Payne, S. R., et al. (2014). Prospective evaluation of methylated SEPT9 in plasma for detection of asymptomatic colorectal cancer. *Gut;* 63(2):317-325.

[155] Ueda, M., Iguchi, T., Masuda, T., Nakahara, Y., Hirata, H., Uchi, R., et al. (2016). Somatic mutations in plasma cell-free DNA are diagnostic markers for esophageal squamous cell carcinoma recurrence. *Oncotarget;* 7(38):62280.

[156] Hamakawa, T., Kukita, Y., Kurokawa, Y., Miyazaki, Y., Takahashi, T., Yamasaki, M., et al. (2015). Monitoring gastric cancer progression with circulating tumour DNA. *British journal of cancer*; 112(2):352.

[157] Shoda, K., Masuda, K., Ichikawa, D., Arita, T., Miyakami, Y., Watanabe, M., et al. (2015). HER2 amplification detected in the circulating DNA of patients with gastric cancer: a retrospective pilot study. *Gastric Cancer*; 18(4):698-710.

[158] Mulcahy, H. E., Lyautey, J., Lederrey, C., qi Chen, X., Anker, P., Alstead, E. M., et al. (1998). A prospective study of K-ras mutations in the plasma of pancreatic cancer patients. *Clinical cancer research*; 4(2):271-275.

[159] Castells, A., Puig, P., Móra, J., Boadas, J., Boix, L., Urgell, E., et al. (1999). K-ras mutations in DNA extracted from the plasma of patients with pancreatic carcinoma: diagnostic utility and prognostic significance. *Journal of Clinical Oncology;* 17(2):578-578.

[160] Takai, E., Totoki, Y., Nakamura, H., Morizane, C., Nara, S., Hama, N., et al. (2015). Clinical utility of circulating tumor DNA for molecular assessment in pancreatic cancer. *Scientific reports*; 5: 18425.

[161] Le Calvez-Kelm, F., Foll, M., Wozniak, M. B., Delhomme, T. M., Durand, G., Chopard, P., et al. (2016). KRAS mutations in blood circulating cell-free DNA: a pancreatic cancer case-control. *Oncotarget*; 7(48):78827.

[162] Nakano, Y., Kitago, M., Matsuda, S., Nakamura, Y., Fujita, Y., Imai, S., (2018). KRAS mutations in cell-free DNA from preoperative and postoperative sera as a pancreatic cancer marker: a retrospective study. *British journal of cancer*.

[163] Hadano, N., Murakami, Y., Uemura, K., Hashimoto, Y., Kondo, N., Nakagawa, N., et al. (2016). Prognostic value of circulating tumour DNA in patients undergoing curative resection for pancreatic cancer. *British journal of cancer*; 115(1):59.

[164] Kirk, G. D., Lesi, O. A., Mendy, M., Szymañska, K., Whittle, H., Goedert, J. J., et al. (2005). 249 ser TP53 mutation in plasma DNA, hepatitis B viral infection, and risk of hepatocellular carcinoma. *Oncogene;* 24(38):5858.

[165] Tóth, K., Sipos, F., Kalmár, A., Patai, Á. V., Wichmann, B., Stoehr, R., et al. (2012). Detection of methylated SEPT9 in plasma is a reliable screening method for both left-and right-sided colon cancers. *PloS one*; 7(9):e46000.

[166] Herbst, A., Rahmig, K., Stieber, P., Philipp, A., Jung, A., Ofner, A., et al. (2011). Methylation of NEUROG1 in serum is a sensitive marker for the detection of early colorectal cancer. *The American journal of gastroenterology*; 106(6):1110.

[167] Tham, C., Chew, M., Soong, R., Lim, J., Ang, M., Tang, C., et al. (2014). Postoperative serum methylation levels of TAC1 and SEPT9 are independent predictors of recurrence and survival of patients with colorectal cancer. *Cancer;* 120(20):3131-3141.

[168] Kawakami, K., Brabender, J., Lord, R. V., Groshen, S., Greenwald, B. D., Krasna, M. J., et al. (2000). Hypermethylated APC DNA in plasma and prognosis of patients with esophageal adenocarcinoma. *Journal of the National Cancer Institute*; 92(22):1805-1811.

[169] Zhai, R., Zhao, Y., Su, L., Cassidy, L., Liu, G., & Christiani, D. C. (2012). Genome-wide DNA methylation profiling of cell-free serum DNA in esophageal adenocarcinoma and Barrett esophagus. *Neoplasia*; 14(1):IN4.

[170] Balgkouranidou I, Karayiannakis A, Matthaios D (2013). Assessment of SOX17 DNA methylation in cell free DNA from patients with operable gastric cancer. Association with prognostic variables and survival. *Clin Chem Lab Med* 2013;51:1505–10.

[171] Shin, D. G., Kim, K. C., Choi, S., Joo, H. J., & Baik, S. H. (2016). A methylation profile of circulating cell free DNA for the early detection of gastric cancer and the effects after surgical resection. *J Clin Exp Oncol*; 5: 1-2.

[172] Liggett, T., Melnikov, A., Yi, Q. L., Replogle, C., Brand, R., Kaul, K., et al. (2010). Differential methylation of cell-free circulating DNA among patients with pancreatic cancer versus chronic pancreatitis. *Cancer*; 116(7):1674-1680.

[173] Melnikov, A. A., Scholtens, D., Talamonti, M. S., Bentrem, D. J., & Levenson, V. V. (2009). Methylation profile of circulating plasma DNA in patients with pancreatic cancer. *Journal of surgical oncology*; 99(2):119-122.

[174] Wong, I. H., Lo, Y. D., Yeo, W., Lau, W. Y., & Johnson, P. J. (2000). Frequent p15 promoter methylation in tumor and peripheral blood from hepatocellular carcinoma patients. *Clinical cancer research*; 6(9):3516-3521.

[175] Chu, H. J., Heo, J., Seo, S. B., Kim, G. H., Kang, D. H., Song, G. A., et al. (2004). Detection of aberrant p16INK4A methylation in sera of patients with liver cirrhosis and hepatocellular carcinoma. *Journal of Korean medical science*; 19(1):83-86.

[176] Yeo, W., Wong, N., Wong, W. L., Lai, P., Zhong, S., & Johnson, P. J. (2005). High frequency of promoter hypermethylation of RASSF1A in tumor and plasma of patients with hepatocellular carcinoma. *Liver International*; 25(2):266-272.

[177] Tan, S. H., Ida, H., Lau, Q. C., Goh, B. C., Chieng, W. S., Loh, M., et al. (2007). Detection of promoter hypermethylation in serum samples of cancer patients by methylation-specific polymerase chain reaction for tumour suppressor genes including RUNX3. *Oncology reports;* 18(5):1225-1230.

[178] Zhang, Y. J., Wu, H. C., Shen, J., Ahsan, H., Tsai, W. Y., Yang, H. I., et al. (2007). Predicting hepatocellular carcinoma by detection of aberrant promoter methylation in serum DNA. *Clinical Cancer Research*; 13(8):2378-2384.

[179] Iyer, P., Zekri, A. R., Hung, C. W., Schiefelbein, E., Ismail, K., Hablas, A., et al. (2010). Concordance of DNA methylation pattern in plasma and tumor DNA of Egyptian hepatocellular carcinoma patients. *Experimental and molecular pathology;* 88(1):107-111.

[180] Hu, J., Wang, Z., Fan, J., Dai, Z., He, Y. F., Qiu, S. J., et al. (2011). Genetic variations in plasma circulating DNA of HBV-related hepatocellular carcinoma patients predict recurrence after liver transplantation. *PLoS One*; 6(10):e26003.

In: Research on New Generation …
Editor: Yildiz Dincer
ISBN: 978-1-53614-367-6

Chapter 5

MICRORNAS IN CANCER MANAGEMENT

Bahadır Batar[1,*], PhD, Emre Ozoran[2], MD and Mehmet Guven[3], PhD

[1]Department of Medical Biology,
Namik Kemal University Medical Faculty, Tekirdag, Turkey
[2]Department of General Surgery, Koc University Hospital,
Istanbul, Turkey
[3]Department of Medical Biology, Cerrahpasa Medical Faculty,
Istanbul University, Istanbul, Turkey

ABSTRACT

MicroRNAs (miRNAs) are small non-coding single-stranded RNA molecules of 19-25 nucleotides in length. miRNAs' main function is the post-transcriptional regulation of gene expression through messenger RNA (mRNA) repression or degradation by binding to 3' untranslated region (3'UTR) of target mRNAs. More than 60% of human protein-coding genes contain miRNA-binding regions within their 3′ UTRs. miRNA-mRNA interactions are mediated by a 6 to 8 nucleotide long seed sequence in the miRNA 5′ end. miRNAs can target different mRNAs. miRNAs have been linked to the etiology, initiation, progression, and prognosis of all kinds of human cancers. Experimental evidence showed that dysregulated miRNAs affect the hallmarks of malignant cells including, self-sufficiency in growth signals, evading apoptosis, sustaining proliferative signals, inducing angiogenesis, activating invasion, and metastases. miRNAs can have oncogenic or tumor suppressive function in the majority of cancers. Tumor suppressor miRNAs target cellular oncogenes and are downregulated in cancer. Oncogenic miRNAs regulate tumor suppressor genes and are overexpressed in cancer. Numerous human miRNA genes are located at fragile sites or in genomic regions that are deleted, amplified or translocated in cancer. miRNA expression is dysregulated in human

* Corresponding Author Email: bbatar@nku.edu.tr.

malignancies through various mechanisms, containing miRNA gene amplification or deletion, aberrant transcriptional control of miRNAs, epigenetic alterations and defects in the miRNA biogenesis machinery. In addition to genomic alterations, dysregulated miRNA expression could be due to alterations in tumor suppressor or oncogenic transcription factors that function as transcriptional activators or repressors to control pri-miRNA transcription, such as c-myc and p53. Genome-wide profiling studies revealed that miRNA expression signatures are related with tumor type, tumor grade and clinical outcomes, so miRNAs could be potential candidates for diagnostic and prognostic biomarkers and development of therapeutic strategies.

Keywords: miRNAs, cancer, lung, breast, prostate, colorectal

1. MicroRNA Biogenesis and Function

MicroRNAs (miRNAs) are small single-stranded endogenous non-coding RNAs, which are 20- 25 nucleotides in length [1]. Almost all eukaryotes have miRNAs in their genome [2]. To date, Release 22 of the miRBase database contains 38589 entries representing hairpin precursor miRNAs, expressing 48885 mature miRNA products, in 271 species [3]. miRNA sequences have been located in intergenic DNA regions and within the introns of protein-coding genes. miRNAs are transcribed as primary-miRNAs (pri-miRNAs) by RNA polymerase II enzyme [4, 5]. The pri-miRNAs are long nucleotide sequences and have single or multiple hairpin structures with a core of 33 to 35 base-pairs [1]. 5′ cap and 3′ poly(A) are often added to the pri-miRNAs, similar to protein-coding mRNAs [6]. The pri-miRNA precursors are processed by Drosha and Dicer RNase III endonucleases [7]. In the nucleus, Drosha cleaves the pri-miRNAs to produce pre-miRNAs, which are exported to the cytoplasm via an exportin-5 (a member of the Ran transport receptor family) [8]. In the cytoplasm, the pre-miRNAs are additionally processed by Dicer to generate a mature, functional, and double-stranded miRNAs [9]. Then, mature miRNAs are incorporated into a multi-protein complex, RNA-induced silencing complex (RISC), which contains the argonaute (AGO) protein. AGO protein plays a key role in silencing of RNA. The main function of mature miRNAs is the post-transcriptional regulation of gene expression through mRNA translational repression or degradation by binding to 3' untranslated region (3' UTR) of target mRNAs [10]. Generally, mammalian miRNAs are exactly complementary in the seed region (2-8 nucleotides in the 5' end), but not in the remaining miRNA sequence, resulting in a deficient miRNA-mRNA match.

In addition to the conventional 3′-UTR mechanism, it is known that miRNAs can be functional in different ways. For example, miR-363 and let-7 suppress during cell proliferation via recruitment of specific micro-RNPs (like AGO2) to AU-rich elements in mRNAs [11]. It has also been demonstrated that miRNAs are able to target to 5′UTR and

3′UTR sequences similarly. miR-10a can bind to the 5′UTR of ribosomal proteins due to starvation and enhance their translation [12]. miRNA dependent mRNA silencing can also occur via binding sites located inside mRNA coding sequences, which regulates embryonic stem cell differentiation [13]. Several studies have suggested that there are non-cytoplasmic miRNA functions in different subcellular compartments. For example, miR-29b carries a different hexanucleotide terminal motif that provides its nuclear transfer and subsequent enrichment in the nucleus [14]. miRNAs act at the promoter level affecting transcription in the nucleus. For example, miR-551b-3p directly upregulates the expression of signal transducer and activator of transcription 3 (*STAT3*) by binding to a complementary *STAT3* promoter sequence, and recruiting RNA polymerase II and the twist family bHLH transcription factor 1 (*TWIST1*) transcription factor for activation of *STAT3* transcription [15]. miRNAs have also been observed in membrane-bound compartments, such as secreted vesicles [16] and mitochondria [17]. Interestingly, muscle-specific miR-1 is able to induce mitochondrial translation of multiple mitochondrial DNA-encoded transcripts, while suppressing its nuclear DNA-encoded targets in the cytoplasm [18].

In mammals, miRNA coding sequences are predicted to account for about 1% of the human genome and more than 60% of protein-coding genes are regulated by miRNAs [19]. Different miRNAs can target a specific mRNA and also a single miRNA can have many mRNA targets.

miRNAs are involved in the communication between cancer cells and the surrounding tumor microenvironment (TME) [20]. Certain databases have experimentally confirmed that there are over 300,000 miRNA-target interactions [21]. This information reveals that miRNAs play important roles in many biological activities such as cell proliferation, differentiation, cell death, cell survival, invasion, and migration [22-24]. Functional miRNA studies have suggested that these non-coding RNAs are related with many diseases, such as cancer, cardiovascular diseases, and neurodegenerative disorders [25].

2. MicroRNAs in Cancer

Carcinogenesis is a multistep process. Genetic changes in normal cells promote the cells through pre-malignant initiation into malignant form [26]. Studies characterizing miRNA over-expression or suppression on mouse models showed the correlations between miRNAs and cancer development [27, 28]. Increasing evidence reveal that dysregulation of miRNA is fundamentally involved in tumor initiation and progression, affecting the expression of critically important genes implicated in cancer processes such as invasion [29], migration and metastasis [30, 31], and cell survival [32]. A study on a genome-wide association by mapping 186 miRNAs found that miRNAs were frequently located at fragile sites, minimal regions of loss of heterozygosity or amplification, or common gap or break regions in human cancers [33]. In addition to structural and genetic alterations, the epigenetic alterations such as DNA promoter hypermethylation or histone hypoacetylation, which leads to silencing of miRNA genes has been determined in some solid tumors and hematological malignancies [34, 35].

Calin and colleagues provided the first evidence of aberrant miRNA expression in human cancer. They showed the loss of miR-15a and miR-16 expression in B-cell chronic lymphocytic leukemia (CLL) due to deletion of the 13q14 chromosomal locus [36]. Many studies have confirmed that miR-15a and miR-16-1 are locked or down-regulated by other mechanisms in about two-thirds of CLL cases [37, 38], and B-cell CLL/lymphoma 2 (BCL2) has been proposed as a downstream effector of miR-15a/16-1 loss in these cases of CLL [39, 40]. Follicular lymphomas frequently express increased levels of *BCL2* due to a chromosomal translocation of the *BCL2* gene into the proximity of immunoglobulin heavy-chain enhancers; this mechanism is not often observed in CLL tumors. Pekarsky and colleagues used luciferase assay experiments to show that miR-15a and miR-16 downregulation is responsible for the enhanced expression of *BCL2* [39]. Further studies have emphasized the role of miR-15a and miR-16 as tumor suppressors in other hematological and solid tumors [41, 42, 43]. In addition, in 2005, Lu et al. found that the expression profile of cellular miRNAs is altered in cancer compared to normal tissue, but also the miRNAs expression pattern correlated with the tumor origin, and even more interesting with the tumor differentiation [44].

The biological roles of miRNAs in cancer can be divided into two groups essentially: tumor suppressor and oncogenic miRNAs (oncomiRs). Tumor suppressor miRNAs generally target cellular oncogenes and are downregulated in cancer; an example of this is the let-7 family in lung cancer [45], targeting e.g., Kirsten rat sarcoma viral oncogene homolog (*KRAS*) [46]. It has been shown that the tumor suppressor miR-34a is positively controlled by tumor protein 53 (*TP53*) [47], repressed by v-myc avian myelocytomatosis viral oncogene homolog (*MYC*) [48], and silenced by aberrant CpG methylation [49]. Methylated miR-9 family genes (miR-9-1, miR-9-2 and miR-9-3) have been identified in several metastatic cancer cell lines [50]. Methylation of miR-9 family genes simultaneously occurs in gastric cancer [51]. Methylation of miR-9-1 is related with lymph node metastasis in colorectal cancer [52], and methylation of miR-9-1 and miR-9-3 is positively correlated with metastatic recurrence of renal cell carcinoma [53]. Oncogenic miRNAs such as the miR-17/92 cluster, also known as oncomir-1 [54], regulate tumor suppressor genes and have increased expression in cancer. miR-21 is one of the commonly studied oncomiRs and has been reported at high expression levels in glioblastoma [55], pancreatic cancer [56], breast cancer [57] and colon cancer [58]. It has an antiapoptotic effect by targeting the tumor suppressors such as phosphatase programmed cell death 4 (*PDCD4*) [59].

Several studies detected the circulating miRNAs in cancer patients and analyzed their potential association with the primary tumors [60]. The circulating miRNAs could be detected in body fluids (blood, urine, tears, saliva, seminal fluid, cerebrospinal fluid, and extracellular fluid) in a stable form [61]. It was reported that tumor-suppressor miRNA expression levels are increased in circulation and are involved in immune responses [62]. Extracellular miRNAs released from normal or tumor cells may function as mediators of paracrine or endocrine signaling pathways among different kinds of tumor cells [63, 64].

It has been well established that miRNAs play an important role in intracellular processes, also recently, some findings support an extracellular role of miRNAs that are generated by microenvironment cells. As hormones, a donor cell releases miRNAs as 'free' molecules or vesicles secreted by active mechanisms [65]. Then, the cells of the other parts of the body take up these miRNAs and regulation of the protein expression profile occurs in these recipient cells. It has been demonstrated that miR-181c released from cancer-derived extracellular vesicles triggers brain metastasis by stimulating the breakdown of the blood brain barrier via the downregulation of its target gene 3-phosphoinositide dependent protein kinase 1 (*PDPK1*) [66]. It has been reported that circulating miRNAs play a role in transforming fibroblasts into cancer-associated fibroblasts (CAFs) [67]. Reduced expression of miR- 31 and miR-214 and increased expression of miR-155 have been found to be related with reprograming quiescent fibroblasts to CAFs in ovarian cancers. It has been revealed that miR-214 directly targets the *CCL5* (C-C motif ligand 5) chemokine important for CAF function [68]. Furthermore, miR-31 was reported to be the most downregulated miRNA in CAFs isolated from endometrial cancer when in comparison to normal endometrial fibroblasts. miR-31 directly targets the homeobox gene SATB homeobox 2 (*SATB2*), which is significantly elevated in CAFs and is responsible for chromatin remodeling and gene expression regulation [69]. miR-155 or miR-146a-containing exosomes modifies the endogenous immune response to endotoxins in vivo [70]. Fundamentally, these studies reveal that secreted miRNAs represent a novel regulatory mechanism by which donor cells can affect the gene expression of recipient cells, and thus impact physiological and pathological processes.

More than half of miRNAs genes are located in cancer-related genomic regions, indicating that these miRNAs may play a more important key role in the etiology, tumorigenesis, development and prognosis of human cancers by regulating the expression of tumor-associated genes [71].

Though there are more than 200 types of cancer, the four most common cancers occurring worldwide are lung, female breast, colorectal (also called bowel cancer) and prostate cancer. These account for more than half (53%) of all new cases.

2.1. Lung Cancer

Lung cancer (LC) is a serious health burden constituting the majority of cancers in both genders. It is divided into two major histological groups: small cell lung cancer (SCLC, 15% of all lung cancers) and non-SCLC (NSCLC, 85% of all lung cancers). Adenocarcinoma, squamous cell carcinoma (SqCC), and large cell carcinoma are the three main histological subtypes of NSCLC. Atypical miRNA expression was reported in lung cancer by multiple studies, backing up the association of miRNAs in the carcinogenesis of lung cancer [72, 73].

Plasma or serum miRNA signatures probable clinical value in early detection of LC had been reported by various studies. The results of these studies could be of paramount importance in the initial screening of LC in vast populations [74-76]. Yang et al. performed a meta-analysis study including 134 studies from 65 articles encompassing 6,919 patients with LC and 7,064 healthy controls [77]. According to their findings, miR-21-5p, miR-223-3p, miR-155-5p, miR-210-5p, miR-205-5p, miR-25-5p and miR-19b-3p showed upregulation while miR-126-3p, miR-148b-3p, let-7a and let-7d showed downregulation; which could be utilized as probable biomarkers. In addition, miR-21-5p, miR-223-3p, miR-155-5p and miR-126-3p panel was thought to be used as potential biomarkers for LC. An extensive meta-analysis encompassing 20 previously published miRNA expression studies in LC, with 598 cancer patients and 528 healthy controls, was performed by Vosa et al. [78]. They reported seven upregulated miRNA meta-signatures (miR-21, miR-210, miR-182, miR-31, miR-200b, miR-205 and miR-183) and eight downregulated miRNAs (miR-126-3p, miR-30a, miR-30d, miR-451a, miR-126-5p, miR-143, miR-145 and miR-486-5p). Four-miRNA signatures (miR-200b, miR-301, miR-141, and miR-193b) having the potential of differentiation of NSCLC patients from the healthy population were identified by Nadal et al. [79].

The expression of Let-7 is decreased in adenocarcinoma in situ than adjacent normal lung tissue. Diminished expression of Let-7, which is among the initially reported tumor suppressor microRNAs in LC, is related with poor prognosis [45]. This finding shows the decreased expression of let-7 at the initial period of lung carcinogenesis [80]. *MYC*, *RAS*, and *HMGA2* are among the oncogenes enrolled in the proliferation of the cell, whose expression is inhibited by Let-7 [81]. Interestingly, let-7 directly downregulates DICER1's expression is directly downregulated by let-7, showing the potential control of let-7 over the global production of miRNAs [82]. Lct-7 inhibits the expression of cell division protein kinase 6 (*CDK6*) controlling the cell cycle [81]. Let-7b-5p, let-7e-5p, miR-23a-3p, and miR-486-5p based miRNA profile was formed by Jin et al. [83]. This profile diagnosed 43 NSCLC out of 60 cases with a sensitivity of 80.25%, a specificity of 92.31% and an AUC (under the roc curve) value of 0.899. MiR-143/145cluster was related to p53 and KRAS [84]. MiR-451a's upregulation was related with an enhanced response to cisplatin in LC patients [85]. Although miR- 486-5p had not been studied in

LC in detail, it was reported to target CD40, a pro-survival cell surface receptor, in ductal adenocarcinoma cells [86].

miR-21 is among the established oncogenic miRNAs. It inhibits the negative regulators of RAS/MEK/ERK signaling pathway, leading to inhibition of apoptosis thus promotes carcinogenesis. MAP Kinase (MAPK) pathway serves a major function in LC. Flavanone was shown to inhibit MAPK, preventing the occurrence of metastasis of LC cells [87]. Wang et al. study reported 2 miRNAs that were upregulated (miR-21-5p and miR-233-3p) and 9 miRNAs that were downregulated (miR-126-3p, miR-133a-3p, miR-140-5p, miR-143-5p, miR-145-5p, miR-30a-5p, miR-30d-3p, miR-328-3pn, and miR-451); constituting eleven important miRNAs in NSCLC [88]. In addition, miR-21, miR-20a, miR-145, miR-223, and miR-221 were reported by Geng et al. to be utilized as promising biomarkers for the early screening of NSCLC [88].

miR-221/222 influences lung carcinogenesis by downregulating *TIMP3* and *PTEN* tumor suppressor genes [90]. Upregulation of miR-221/222 enhances cell migration while inhibiting apoptosis by targeting TIMP3 and PTEN.

Bianchi et al. reported 34-microRNA signature which has the potential to determine patients at the early stages of the disease [74]. The microRNA signature was later lightened up to a 13-microRNA signature (miR-92a-3p, miR-30b-5p, miR-191-5p, miR-484, miR-328-3p, miR-30c-5p, miR-374a-5p, let-7d-5p, miR-331-3p, miR-29a-3p, miR-148a-3p, miR-223-3p, and miR-140-5p). The level of plasma miR-92a-2 was statistically significantly higher in SCLC patients when compared with healthy controls according to the study by Yu et al. [91]. Ranade et al. also showed that higher tumor miR-92a-2 levels were associated with chemoresistance and with decreased survival in patients with SCLC; in their study involving 34 SCLC tumor specimens [92].

Hu et al. aimed to show the value of promising miRNAs as biomarkers which could differentiate NSCLC from normal cases, with the utilization of miRNA array datasets [93]. Five microarray datasets were chosen and used in order to locate the promising miRNAs which could differentiate between NSCLC and control. According to sensitivity and specify top 14 miRNAs were defined. Five were upregulated (miR-9, miR-140-5p, miR-708, miR-296-3p, miR-892b) while nine were downregulated (miR-584, miR-218, miR-30b, miR-522, miR486-5p, miR-34c-3p, miR-34b, miR-516b, miR-592) in NSCLC. In another study, aberrant expression of miR-9 aids in the invasion of tumor cells by the direct downregulation of the expression of CBX7 protein [94]. MiR-140-5p significantly reduces macrophage differentiation-associated protein levels in NSCLC cells are lowered by miR-140-5p, thus Erk1/2 signaling regulation results in the inhibition of cell proliferation [95]. Polley et al. in their study analyzing the response to drugs and miRNA expression; also reported SCLC lines that responded to etoposide showed increased miR-200c-3p and decreased low miR-140-5p and miR-9-5p expression [96].

119 NSCLC specimens were examined by Wang et al. [97]. They discovered that miR-216a was significantly downregulated in NSCLC tissue in comparison with the

neighboring lung tissue. The downregulation was related with the tumor stage and the development of metastasis in NSCLC patients. miR-216a was reported to inhibit cell growth by its effects on cell proliferation and apoptosis both in vitro and in vivo conditions as reported by various experimental studies.

Seven miRNA panel's (miR-34a/b/c, miR-21, miR-29b, miR-155, and let-7a) expression status in 31 SCLC tumors; 14 SCLC cell lines and 26 NSCLC cell lines were studied by Lee et al. [98]. The expression of the aforementioned panel was not associated with overall and progression-free survival. Also, it was not a predictor for the response to treatment. In 2017, Fadejeva also showed that the expression of miR-34a was down-regulated in NSCLC tissues [99]. p53 is involved in the expression of miR-34a, which is enrolled in the control of cell cycle, apoptosis and differentiation [100]. miR-34a was found to be upregulated in NSCLC tumor tissues and NSCLC cell lines treated with CDDP; showing that CDDP could act in the induction of the expression of miR-34a.

In the study by Nadal et al., examining microRNA profiling of lung adenocarcinoma subtypes portrayed different microRNA expression profiles were present with different morphological subtypes of lung adenocarcinoma [101]. The upregulation of miR-27a, miR-132 and miR-212 was strongly related to solid adenocarcinoma [101]. On the other hand, miR-27a-5p, miR-34b-3p, miR-485-3p, and miR-574-3p were found to be significantly markedly downregulated in cancer tissues, by Mizuno et al. in their study analyzing autopsy specimens of patients with SCLC [102]. The ectopic expression of miR-27a-5p and miR-34b-3p inhibited the aggressiveness of cancer cells. Thus it could be hypothesized that they serve a tumor suppressor function in SCLC cells. Further analysis showed that topoisomerase 2 alpha (*TOP2A*), maternal embryonic leucine zipper kinase (*MELK*) and centromere protein F (*CENPF*) genes were regulated by miR-27a-5p- and miR-34b-3p. Since these oncogenes are expressed in an increased fashion in the autopsy specimens of SCLC patients; these miRNAs present clinical importance.

Although smoking tobacco products is accepted as the major risk factor in the development of LC; not all of the smokers are subject to it. In a study, the expression of 484 miRNAs was studied in the lungs of rats following 28 days of exposure to environmental cigarette smoke. 126 out of 484 miRNAs were downregulated [103]. Since microRNA levels and cellular differentiation are closely related, most miRNAs are downregulated in tumor tissue. The decrease in the expression of microRNA in cancer cells is proportionally related with the levels of cellular differentiation. The level of reduction is larger in differentiated tumors [104]. Let-7, miR-10, miR-26, miR-30, miR-34, miR-99, miR-122, miR-123, miR-124, miR-125, miR-140, miR-146, miR-191, mi-192, miR-219, miR-222 and miR-223 are among the well-known downregulated microRNA families. Proliferation, angiogenesis, apoptosis, stress response and gene expression are among the various functions of the cell that are related to the mentioned microRNAs. When expression of miR-126 is decreased in LC, PIK3R2, its target, is reduced. This change affects the PTEN/PI3K/AKT signaling pathway resulting in the suppression of cell growth, migration, and invasion [105]. A variant of adenocarcinoma, enteric adenocarcinomas possesses microRNAs that are the same with decisive pancreatic ductal adenocarcinoma-associated microRNAs associated with tumor aggressiveness (miR-31, miR-126, miR-506, miR-508-3p, and miR-514. The similarity could account for the aggressive behavior of lung enteric adenocarcinoma, thereby aiding possible future tailored-therapeutic approaches [106].

In another study, epigenetic downregulation of miR-487b expression occurs as the condensate of cigarette smoke reaches epithelial cells of the airway. The rise in the expression of miR-487b oncogenic targets (*SUZ12, BMI1, WNT5A, MYC* and *KRAS*) and the aforementioned mechanism give way to a rise in the proliferation and invasion of LC cells. It could be concluded that the inhibition of miR-487b enhances tumorigenesis, proliferation and invasion; while its expression decreases the growth and metastatic potential of LC [107]. Carcinogens found in tobacco smoke contribute to epigenetic silencing. Epithelial-mesenchymal transition (EMT) which had strong links with LC, had been shown to be induced by epigenetic mechanisms like the downregulation of miR-200 and miR-205 [108]. The change in the expression of miR-218, a microRNA found in the epithelium in the human airway that is among the most often affected by smoking, was enough to cause a proportionate shift in the expression of assumed miR-218 mRNA targets in primary bronchial epithelial cells and H1299 cells. However, the change in the expression of miR-218 could affect the expression *MAFG* (MAF BZIP Transcription Factor G) gene targets; as MAFG's binding sites are overrepresented in the epithelial cells of smokers [104].

In the study performed by Mancuso et al., miR-200c, miR-205, and miR- 192 were found to be either directly or indirectly related to tumorigenesis in SCLC [109]. miR-200c's overexpression inhibits *PTEN45* and activates phosphoinositide 3-kinase/Akt/mammalian target of rapamycin pathway which in turn enhances tumor growth and survival in SCLC as seen in quite a few cancers [110]. miR-192 affects DNA repair and damage. miR-192 could cause downregulation in *ERCC3* and *ERCC4*, the proteins enrolled the nucleotide excision repair pathway; leading to failure during nucleotide excision and genetic instability in cancer [111].

A polarized epithelial cell's transformation into mesenchymal cell phenotype via various biochemical changes is called EMT. MiR-200 has an important part in the induction of EMT. The decline in the expression of miR-200 upregulates *CDH1* (also known as E-cadherin), *VIM* (also known as vimentin), *ZEB1*, and *ZEB2*, leading to EMT during the progression of LC [112]. Tejero et al. showed that increased levels of miR-200c were related to decreased overall survival in a cohort of NSCLC patients with adenocarcinoma [113]. In addition, miR-218 via the modulation Slug/ZEB2 signals plays a major function in the regulation of EMT-related traits and the metastasis of LC [114]. Increasing evidence has shown other miRNAs, including miR-124, miR-135a, miR-148a and miR-193a-3p/5p were reported as strong suppressors of EMT and they are usually downregulated in LC [115].

Lei et al. reported that miR-205 showed over-expression in NSCLC cell lines and tissues [116]. miR-205 promotes cellular growth, metastasis and chemoresistance to cisplatin by targeting *PTEN*. The increased expression of miR-205 was portrayed to be unique to SqCC by Labanony et al. In addition the expression of miR-205 was a discriminatory factor between SqCCs and non-SqCC NSCLCs [117]. miRNA profiling had been utilized by various studies to discriminate SqCC from non-SqCC [118]. Bishop et al. showed that the differentiation between SqCCs from non-SqCC NSCLCs could be made by the expression of miR-205 in small biopsies or aspirates [119]. The expression of miR-205, miR-196b, and miR-375 was shown by Hamamoto et al. as a method to differentiate SqCCs from non-SqCC NSCLCs [118].

LC cells could generate miR-494 giving way to tumor angiogenesis. As a result of activity, HIF-1α-induced upregulation of miR-494 enhances the development of the tumor in a hypoxic setting [120]. On the contrary, miR-494 also downregulates the cellular proliferation in LC. The constitutive expression of miR-494 in A549 LC cells was shown to result in the suppression of cell proliferation and induction of senescence.

Various studies have proposed miR-155 as a major oncogenic miRNA that could function as a potential therapeutic target in NSCLC [121, 122]. miR-155 targets TP53 which controls the expression of miR-155. miR-155/TP53 feedback loop is shown to be related with resistance to chemotherapy [121]. In the study of Xue et al., miR-21 and miR-155 were shown to enhance the progression of NSCLCs. This progression was to a certain degree achieved by downregulation of tumor suppressor genes like *SOCS1*, *SOCS6*, and *PTEN* [122]. miR-155 was shown to regulate cellular apoptosis and DNA damage via an Apaf-1-mediated pathway in A549 LC line [123].

The activation of p53, the main facilitator of CDDP-induced cell death, is inhibited by miR-630. This action prohibits the initial response to DNA damage in LC cells [124]. miRNA-630 plays a role in disrupting the pro-apoptotic pathways controlled by p53. In another study, miR-92b was found have upregulation in NSCLC tissue and cell lines according to several studies. Overexpression of miR-92b promoted A549/CDDP cell growth; inhibited apoptosis. This targeted *PTEN* (tumor suppressor gene) leading to the induction of CDDP-resistance [125]. On the other hand, miR-203 was reported to downregulate LIN28B an upregulate let-7 biogenesis promoting apoptosis and inhibiting cellular proliferation of LC, by Zhou et al. A major regulatory network in LC between miR-203, LIN28B, and let-7 was unveiled with their study [126]. Also, miR-34, a major actor in the tumor suppression of TP53, is transcribed by TP53 in response to oncogenic stress and DNA damage [127]. In LC, the reduction in the expression of miR-34 leads to induction of expression of *BCL2*, *MET*, *PDGFRA* and *PDGFRB*, miR-34 target genes; eventually resulting TNF-related apoptosis-inducing ligand (TRAIL)-induced cell death. miR-195 is a suppressor of cancer growth. Its expression had been shown to be related to decreased mortality in various cancer types, including NSCLC [128]. The expression of miR-195 is also reduced in NSCLC when compared with normal tissues. It had been found to be related with worse overall survival in these patients. BAP1 (BRCA1 associated protein-1) is a deubiquitinating enzyme found in the nucleus, functioning as a tumor suppressor in LC. miR-31 targets this enzyme. The expression of BAP1 is decreased with the rise in miR-31's expression. This results in the inhibition of apoptosis and proliferation of cells [129]. Studies on miR-31 are controversial. miR-31 had been reported as an oncogenic miRNA by some studies while others define it as a tumor suppressor in LC [130, 131].

2.2. Breast Cancer

BC (BC) is the most common cancer among females around the world that is the most prevalent cause of death among females suffering from cancer. A large body of studies has been done to investigate the expression level of miRNAs which their results indicate prognostic and predictive values of cancer-related circulating miRNAs or tissue-specific miRNAs in BC patients. There is a growing body of evidence showing the aberrant activation of some known circulating miRNAs as well as tissue specific-miRNAs in patients affected by BC [132].

In a large cohort study, encompassing 1280 serum samples from BC patients and healthy controls; analyzing the level of expression of serum miRNA, a combination of five miRNA (miR-1246, miR-1307-3p, miR-4634, miR-6861-5p and miR-6875-5p) was associated with the capacity for the detection of BC at early stages [133].

MiR-21 had been reported to be upregulated in the serum of BC patients [134]. miR-21 had been reported to aim genes involved in cellular growth and migratory behavior of cells, such as tumor suppressor tropomyosin 1 (*TPM1*), *PTEN*, and TIMP metallopeptidase inhibitor 3 (*TIMP3*) [59]. In the meta-analysis by Li et al., encompassing six studies, evaluating the diagnostic ability of miR-21 in BC patients; miR-21 was reported to possess the capacity to be a specific and sensitive biomarker for early diagnosis of BC [135]. Asaga et al. displayed that miR-21 had prognostic value for the delineation of BC patients from healthy individuals and patients with metastases from those without metastasis [136]. Another research report that miR-21, miR-16, and miR-451 upregulated in the plasma of BC patients; suggesting the combination of miR-451 and miR-145 as biomarkers in plasma, to possess a capacity to have a potential and specific test for screening BC [137].

The function of miR-155 as an oncomiR through targeting cytokine signaling 1 (SOCS1) and STAT3 pathways in breast carcinoma were reported by Jiang et al. [148]. Zhu et al. reported that the importance of miR-155 in the serum of women with and without BC with progesterone receptor (PR) positivity [139]. miR-155 expression levels were higher in women with PR positive tumors. Sun et al. reported that serum miR-155 levels were higher in BC patients when compared with healthy women [140]. The levels of miR-155 decreased after surgery and chemotherapy. This finding portrays its possible value as a predictive biomarker for the evaluation of response to therapy [140].

Aguilara et al. conducted miRNA profiling on the serum of BC patients, studying the sensitivity and specificity of miRNAs by the construction of Receiver Operating Characteristic (ROC) curves [141]. miR-155, miR-21, miR-10b, miR-125b, miR-145, miR-191, miR-382 had different expression levels in the serum of cancer patients compared to normal controls. miR-145, miR-155 and miR-382 were suggested to be novel noninvasive biomarkers for BC detection after the completion of ROC curve analysis [141]. The level of expression of miR155, miR-10b, and miR195 in the serum of BC patients was markedly upregulated in comparison with the healthy controls in the study by Hagrass et al., highlighting their ability to become potential novel biomarkers of BC [142].

In another study, Heneghan et al. reported that miR-195 had high levels of expression in BC patients. In addition, they showed that the circulating levels of miR-195 and let-7a decreased in patients after surgery; the latter was also related to the presence of lymph node metastasis and estrogen receptor positivity (ER) [143]. In other studies, Zhao et al. found that miR-589 was significantly upregulated and let-7c was significantly downregulated in BC patients. They reported 26 different miRNAs in BC patients [144]. miR-122 in the serum was related with recurrence and metastatic recurrence in patients [145]. Chen et al. reported that miR-10b and miR-373 in the circulation could be used to differentiate patients with lymph node metastasis in patients without lymph node metastasis [146]. Cuk et al., in their study on miRNA expression profiling in plasma of early stage BC patients and healthy controls; reported that miR-148b, miR-376c, miR-409-3p, and miR-801 were significantly over-expressed in BC patients [147]. In addition, Van Schooneveld et al. found that several miRNAs, especially the levels of circulating miR-411, miR-299, miR-215 were markedly lower in untreated metastatic BC patients [63]. In a study by Zheng et al. circulating miR-106b was reported to be a potential prognostic biomarker, which was associated with tumor size, and presence of lymph-node metastasis; in addition to shorter disease-free survival times and overall survival times [148].

There is also a growing body of data showing the evaluating the expression pattern of miRNAs in the tissue of BC patients. Iorio et al. identified 29 miRNAs with significant alterations in expression in BC tissue compared to normal tissue. Among these miRNAs; miR-21 and miR-155 were upregulated while miR-10b, miR-125b and mR-145 were downregulated in the BC tissue. miR-203 and miR-213 were shown to be associated with tumor stage or with ER and PR status [57]. In the study by Camps et al, in 2008, miR-21 was reported to be associated with disease-free and overall survival; while miR-210 was reported to be an independent predictor of overall survival and disease-free survival

[149]. In another study, Iorio et al. reported that let-7, miR-9-3, miR-10b, miR-21, and miR-145 had prognostic value, as they were associated with tumor grade, degree of invasion, and progression to lymph node metastasis [150]. Wang et al. also found that the expression levels of miR-21, miR-106a, and miR-155 were upregulated while miR-126, miR-199a, and miR-335 were downregulated in BC tumor tissues in comparison with healthy people [151]. In addition, the expression levels of previously known miRNAs (miR-21, miR-126, miR-155, miR-199a, and miR-335) were associated with sex hormone receptor expression status and histopathologic grade of the BC [151].

Although miR-17-5p shows overexpression in cancers, it functions as a tumor suppressor in BC through inhibition of AIB1 translation, disrupting cancer cell proliferation and progression [152]. Quesne et al. reported the relationship of miR-205 and let-7b with prognosis in BC [153]. In another study, Foekens et al. analyzed 249 miRNAs in BC tissue, reporting that miR-7, miR-210, miR-128a, and miR-516–3p were associated with lymph node-negative/estrogen receptor-positive (ER/LNN) BC [154]. miR-210 was the only marker related with early recurrence in ER/LNN breast tumor. In addition, miRNA-210 had a statistically significant association with worse outcome in the 69 triple-negative BC patients [154]. Xu et al. reported that miR-154 was downregulated in BC tissues and it had functions in targeting E2F transcription factor 5 protein [155]. Also, miR-497 was reported to be downregulated in BC tissues, portraying an important function to become an independent marker for prediction of BC prognosis [156].

2.3. Prostate Cancer

Prostate cancer (PC) is the most common male malignancy excluding skin cancers. Despite its prevalence, up to date no biomarker has been able to display its aggressiveness precisely. Various miRNA profiles were developed in PC [157, 158]. Differential expression of miRNAs in PC has the potential ability to differentiate benign from malignant disease. Comparative miRNA signatures possess the ability to detect silent and aggressive disease, thus establishing miRNAs as prognostic and diagnostic markers in PC [159]. In the development of PC, various cell signaling pathways and cellular processes were held responsible. The proliferation of cell, the progression of cell cycle, cellular death, EMT, angiogenesis, metabolism of tumor cell and metastasis are among the cellular pathways involved in these processes [160, 161].

The first attempts to find a miRNA profile of human PC was performed on PC cell lines [157, 162]. Initial studies on miRNA profiles in PC reported that miR-20, miR-21, miR-99a, miR-141, miR-182, miR-198, let-7a were overexpressed and miR-145 and miR-155 were downregulated in PC cell lines [162]. In another study, miR-16, miR-21, miR-34a, miR-141, miR-143, miR-145, miR-155, miR-125b, miR-221, miR-375, miR-425 and let-7a were analyzed in PC patients [163]. miR-21, miR-32, miR-221, miR-222, miR-181, miR-18a, and miR-429 are the most commonly overexpressed miRNAs in PC [164-169]. These miRNAs, will be analyzed in detail, have crucial functions in PI3K/AKT pathway, EMT, cell proliferation, apoptosis and androgen receptor (AR) expression [165].

The regulation of miRNAs by the AR signaling pathway as well as post-transcriptional regulation in PC is important in the development of castration-resistant PC (CRPC) and its therapeutic options [170]. Massillo et al. studied the pathogenesis of PC, reporting the role of miRNAS in at least two different pathways [171]. miR-205 was involved in metastasis, apoptosis and AR signaling; miR-125b was related with p53 and AR; miR-21 had roles in AR signaling, bone metastasis while miR-221/222 was involved in AR signaling, cell cycle. miR-135a, miR-32, miR-125b, miR-27a, miR-141 and miR-21 were reported as the most important miRNAs induced by AR; miR-221/222 cluster and miR-99a are the major miRNAs directly downregulated by AR [171]. miR-99a had been shown to be underexpressed in the PC tissue, metastatic PC and CRPC samples compared to normal prostate tissues [172, 173]. Androgen by its direct bonding to miR-99a/let7c/125b-2 cluster gene inhibits its expression in AR-positive PC cells.

miR-21 targets the reversion inducing cysteine-rich protein with kazal motifs, also known as *RECK*, gene constituting important functions in tumorigenesis and invasiveness. In human PC DU145 cells, miR-21 directly represses RECK; a putative tumor suppressor gene that is thought to regulate the expression of matrix metallopeptidase 9 (MMP9); enhancing cancer invasion and metastasis [174]. In addition, miR-21 aims and downregulates tumor suppressor genes like PTEN and programmed cell death (PDCD4) via p53 network [165]. TGFβ is among the most important pathways disrupted in PC. Mishra et al. showed that the combined effects of miR-21 and AR signaling were found to have a crucial role in the repression of the expression of TGFβ receptor II (*TGFBR2*) in PC cells via a positive feedback loop, repressing growth [175].

Vlaemick-Guillem et al. showed that miR-21, miR-141 and miR-221 were commonly observed as the most frequently deregulated miRNAs in PC [176]. miR-141 in serum was found to be increased in men with high risk PC, metastatic PC and castrate resistant PC. In addition, these miRNAs had been reported to portray changed levels in tissue samples [163, 177]. The levels of miR-21 are also altered in urine [177]. The merger of miR-19a and miR-19b, other miRNAs found in increased values in urine; with miR-21 has better diagnostic power than prostate specific antigen (PSA) [177]. The value of miR-141, the pioneering miRNA to be considered as an important biomarker for the diagnosis of PC was confirmed by several studies. Increased miR-141 levels were associated with poor prognosis and the presence of metastasis in PC patients [178].

miR-33a was reported to have decreased expression in tumor samples by the study of Karatas et al. [179]. Lowered miR-33a activity was associated with the proliferative, invasive, and anchorage-independent growth potential of PC cells in vitro while higher miR33a levels had the opposite effect; defining miR-33a as a tumor suppressor in PC.

In another study, the overexpression of miR-141 increased the proliferation of LNCaP cells; its inhibition by anti-miR-141 suppressed cell growth [180]. In the study by Huang et al., the expression of miR-141-3p was shown to be decreased in the tissues of PC with metastasis to the bone in comparison with tissues of PC without bone metastasis [181]. Decreased miR-141-3p expression is associated with increased serum PSA levels, Gleason grade and presence of bone metastasis in PC patients. Moreover, upregulation of miR-141-3p suppressed EMT, invasion and migration of PC cells in vitro. In addition, it was shown that miR-141-3p inhibits the activation of NF-κB signaling through direct aiming of tumor necrosis factor receptor-associated factor 5 (TRAF5) and 6 (TRAF6). The latter further suppresses invasion, migration and bone metastasis of PC cells [181].

Xu et al. reported the downregulation of miR-129 in PC which is associated with the development and progression of PC [182]. miR-129 was considered to play a tumor suppressor role in PC as, its capacity in the inhibition of cellular growth, migration, induction of apoptosis was displayed. *ETS1 (*electron transport chain 1) was considered to be the target of mir-129 and it can inhibit the ETS1 protein expression through binding to the corresponding 3′-UTR site on *ETS1* gene.

miR-32 targets tumor suppressor genes enrolled proliferation, survival, and migration of the PC cells during progression of PC [183]. Controlled by androgens via stimulation of dihydrotestosterone (DHT) stimulation, it shows increased expression in CRPC specimens B-cell translocation gene 2 (*BTG2*) and phosphoinositide-3-kinase interacting protein 1 (*PIK3IP1*) are both targeted by miR-32, constituting its classic oncogenic characteristic effect. *BTG2* and *PIK3IP1* control inhibition and anti-proliferation of PI3K. PI3K is an established controller of proliferation, migration and survival of the cell. The overexpression of miR-32 enhances the growth of LNCaP cells growth, reduces apoptosis; in accordance with BTG2-downregulated expression [184].

miR-221, miR-222, and miR-429 are among the various miRNAs affecting the expression of the cell cycle regulator p27Kip1 [166, 167, 185]. miR-221 and miR-222 demonstrate their oncogenic activity via targeting various tumor suppressor genes. They are reported to be overexpressed in PC. miR-221/miR-222's ability to upregulate S-phase kinase-associated protein 2 (Skp2), cyclin A and cyclin D1 through targeting p27Kip1 suppression were reported in many studies. This effects the progression of the cell cycle, increases clonogenicity and enhances tumorigenicity [166]. The abnormal expression and paradoxical roles of miR-221-5p in human PC tissues have been reported by gene microarray data [186]. It was Shao et al.'s study that suggested the major role miR-221-5p has in the progression of PC [187]. They based their hypothesis on the fact that the upregulation and silencing of the expression of miR-221-5p are related to cell proliferation, migration and tumorigenesis. Their in vitro and in vivo experiments lead to the finding that miR-221-5p increases cell proliferation and metastasis by post-transcriptional regulation of *SOCS1* (Suppressor of Cytokine Signalling 1) which is a key inhibitory molecule of MAPK/ ERK signaling, and is tumor suppressor genes [187].

Greco et al. showed that miR-221 has the ability to be utilized as a biomarker in the early diagnosis and novel therapy target in benign prostate hyperplasia [188]. Sun et al. reported that androgens downregulate the miR-221/222 cluster and this cluster was markedly upregulated in CRPC cells. By the downregulation of *HECTD2* (HECT Domain E3 Ubiquitin Protein Ligase 2) and *RAB1A* (member RAS oncogene family) the targets of miR-221/222, overexpression of miR-221/222 in LNCaP cells enhances the development of the CRPC phenotype [189]. The expression of p27Kip1, cell cycle regulator, is upregulated in human PC cell lines. p27Kip1 has an inverse correlation with miR-429. The downregulation of miR-429 expression inhibited cellular proliferation and arrested cancer cells in the G1 phase of the cell cycle, probably through direct targeting of p27Kip1 [167].

The loss of chromosome (chr) 8p21 is the most common change in the oncogenome of the prostate. It has been related with the loss of NKX3.1 homeobox gene. The deletions of Chr8p21 rise markedly with the increase in tumor grade. They are related to poor prognosis in PC. These findings could be related with the critical involvement of this region in tumor progression. That specific locus is related with a miRNA gene cluster-miR-3622a/b-, which has a major role as a suppressor in PC. Bucay et al. reported that frequent loss of miR-3622a at chr8p21 region by genetic and epigenetic mechanisms results in the induction of EMT induction via upregulation of *ZEB1* and *SNAI2* leading to enhancement of PC progression and metastasis [190].

The downregulation of miR-383, which is situated in the chr8p22 region within intron 3 of the sarcoglycan zeta (*SGCZ*) gene, is reported to be related with male infertility. It has been shown to promote testicular embryonal carcinoma cell proliferation through targeting interferon regulatory factor-1 (*IRF1*) [191]. Restoration of miR-383 expression could constitute a therapeutic option in PC based on the in vivo data by Bucay et al. portraying a tumor-suppressive role of miR-383 in primary and metastatic PC mouse models. In that study, the regulation of CD44 was the main area affected by miR-383 on PC cell lines. miR-320, another miRNA component of this region, has also been shown to downregulate in PC and has suppressive role over the cancer stem cell-like characteristics of PC cells through downregulation of the Wnt/beta-catenin signaling pathway [192].

Also, the deletions at chr5q region (particularly chr5q14-q23) are commonly seen in PC according to genomic studies, underlining the importance of this region in the carcinogenesis of PC. miR-3607 is located in this region. Expression analysis of miR-3607, which is located in the aforementioned region, in a clinical cohort of PC specimens studied by Saini et al. showed that miR-3607 is markedly attenuated [193]. Decreased miR-3607 expression was related correlated with tumor progression and poor survival outcome in PC. In the same study, miR-3607 expression showed a significant association with serum PSA levels in PC patients. The downregulation of miR-3607 expression in PC clinical samples was suggestive of the tumor-suppressive activity of this miRNA. miR-

3607 was hypothesized to regulate Src family kinases- Lyn and Src. Src kinase is aberrantly activated in PC [193]. Src kinases controls the proliferation, apoptosis, invasion and migration of PC cells [194].

miR-17-92 cluster, several of which are upregulated in PC, is a candidate diagnostic and prognostic biomarker for PC. Its use with serum PSA could increase the accuracy for diagnosis of PC. Feng et al. reported that miR-17-92 cluster was upregulated in PC tissue samples when compared with BPH control tissue [195]. This finding portrays the contribution of miRNAs to oncogenesis in PC. Farran et al. reported a new circulating miRNA signature consisting of a combination of interacting miRNAs (miR-17/miR-192) and an independent miRNA (miR-181a), which has the ability of differentiating aggressive and non-aggressive PC patients [196]. Various studies have reported that miR-17 is increasingly founding PC cells, exosomes, tissue and the serum of high-risk patients [156, 195, 197]. miR-17, also known as oncomir-1, exhibits its oncogenic effect through inhibition of a complicated network of tumor suppressors; genes involved in anti-proliferative and anti-apoptotic mechanisms and cell cycle including *PTEN*, *HIF1A* (Hypoxia-inducible factor 1-alpha) and *CDKN1A* (cyclin-dependent kinase inhibitor 1A) [198-200]. The loss of PTEN is among one of the hallmarks of PC [201]. p53 induces miR-192 directly [202]. When induced miR-192 works as a tumor suppressor in the control of the angiogenic switch and the cell cycle [203, 204]. BCL-2 and MYLK (myosin light chain kinase) that are overexpressed in aggressive PC are the target of miR-192 [204]. miR-18a, an oncogene in the miR-17–92 cluster, is upregulated in PC [169]. The overexpression of serine/threonine protein kinase 4 (*STK4*) a direct target of miR-18a, downregulates the expression of *STK4* leads to inhibition of apoptosis and promotion of tumor survival [169].

In cancer cells downregulation of tumor suppressor miRNA can lead to proliferation [205], EMT, invasiveness and metastasis [186]. The downregulation of miR-145, miR-200, miR-29b, miR-205, and miR-940 in PC may end up in increases in the expression of EMT. In PC, the miR-143/miR-145 cluster is downregulated as in many solid tumors. It functions as a tumor suppressor responsible for the reduction in cellular migration and invasion. The expression of E-cadherin is decreased, EMT phenotype and cellular proliferation are promoted by this cluster [206]. The expression of miR-145, which functions as a tumor suppressor, is regulated by p53 and RREB1 (RAS-responsive element binding protein 1), FoxO (Forkhead box O), and C/EBP- β (CCAAT/enhancer-binding protein beta), other transcriptional factors [207]. Avgeris et al. supported the tumor-suppressor role of miR-145, underlining its clinical use in PC. By silencing MYC (MYC-c), PPP3CA, EGFR, NUDT1, TNSF10, SWAP70, DEFA, CBFB, CLINT1, and RTKN it promotes apoptosis [207]. Szczyrba et al. utilized a deep sequencing technique discovering 33 dysregulated miRNAs in PC vs. normal tissues [208]. miR-148a, miR-200c, miR-375 (all upregulated) and miR-143, miR-145 and miR-223 (all downregulated) were the most markedly changed miRNAs portraying diagnostic abilities

[208]. miR-143-3p, miR-196a-5p, miR-34a-5p, miR-501-3p and miR-92a-1-5p were reported by Rodriguez et al. as exosomes markedly downregulated in PC patients [209]. The methylation of the promoter region or p53 mutation of miR-145 in PC cell lines had been held responsible for the decreased expression of mirR-145 [209]. The pro-apoptotic gene, TNFSF10 was found to be markedly upregulated by overexpression of miR-145 according to the results of target gene research [210]. Although Even though miR-145 levels in PC tissues are usually lower than non-tumor tissues; there are variations in the expression of miR-145 expression with a 10- to 30-fold range in tumor tissues [186, 210]. These findings suggest the presence of a heterogeneous pattern which could be utilized to differentiate PC patients on the basis of tumor aggressiveness. Chen et al. in their study on PC and benign prostate tissue samples, reported lowered miR-145 expression in PC samples, especially in patients with tumor progression, hypothesizing the role miR-145 could play in the prognosis of PC [211]. This finding was not validated in the study by Kang et al., suggesting that these types of miRNA may not possess prognostic values in PC patients [212].

Hu et al. found that miR-212 was downregulated in PC tissues and several PC cell lines [213]. Resumption expression of miR-212 attenuated cell proliferation, invasion and increased the apoptosis of PC. miR-212 targeted mitogen-activated protein kinase 1 (MAPK1), a well-known oncogene. The expression of MAPK1 was upregulated in PC tissues which were inversely proportional to the expression of miR-212 [213].

In the study by Chen et al., miRNA-199a was found to be downregulated in PC, in addition this was characteristic of paclitaxel (PTX) resistance in PC [214]. Another function of miRNA-199a was the expression of Yamaguchi sarcoma viral homolog 1 (YES1), an important molecule related with chemoresistance to multiple cancers, by targeting the 3′-UTR [215].

miR-200 family controls the genes enrolled EMT [216]. miR-200 family inhibits cancer invasion though direct inhibition of the translation factors zinc-finger E-box binding homeobox 1 and 2 (*ZEB1* and *ZEB2*) [217]. Various studies showed that that miR-200c was overexpressed in PC tissues [217, 218]. miR-200b was downregulated in both prostate tumors and cell lines [219]. In silico analysis was done using The Cancer Genome Atlas (TCGA) database in order to discover circulating mRNA and miRNA as potential markers of PC [219]. 2,267 genes and 49 miRNAs were differentially expressed among normal and tumoral samples. Souza et al. reported 11 genes and 8 miRNAs based on their prediction analyses of target genes and integrative analysis of mRNA and miRNA expression [219]. *OR51E2* (Olfactory receptor 51E2) and *SIM2* (Single-minded homolog 2) genes and miR-200c and miR-200b miRNAs displayed a marked relationship with PC. SIM2 and OR51E2, circulating mRNAs which are overexpressed in the plasma of patients with PC, have a major role in tumor biology as well as the development and progression of PC [220]. A member of the family of transcription factors with basic helix-loop-helix/per-Arnt-Sim (bHLH/PAS) domains, the *SIM2* gene, has been playing a major role in the pathogenesis of PC [220, 221].

A tumor suppressor, miR-29b is downregulated in human PC cells [222]. Myeloid cell leukemia 1 (Mcl-1) and matrix metalloproteinase-2 (MMP-2) protein expression were repressed by the exogenous expression of miR-29b in PC cells. It postulated pleiotropic effects through metastatic cascade encompassing invasion, motility, cellular survival, and proliferation which resulted in the metastatic dissemination of PC cells [223]. Overexpression of miR-29b in metastatic PC cells affects the EMT process. miR-29b overexpression induces the expression of E-cadherin and increases the expression of N-cadherin, Twist, and Snail, mesenchymal markers, prompting the reversal of EMT and decreased the level of invasion.

miR-205-5p was reported to be markedly downregulated in naive PC and CRPC specimens by Yamada et al. [224]. 37 genes were putative targets of miR-205-5p regulation as reported by Genome-wide gene expression analyses and in silico miRNA database searches. *HMGB3, SPARC, MKI67, CENPF, CDK1, RHOU*, and *POLR2D* were seven specific genes showing elevated expression that are involved in PC pathogenesis based on TCGA database analyses. In naive PC specimens, *CENPF* showed overexpression which functioned as an oncogenic gene in PC cells [225]. Cancer cell migration and invasion in PC cells were markedly inhibited by the ectopic expression of miR-205-5p [225]. Thus, it could be concluded that antitumor miR-205-5p directly targets oncogenic *HMGB3* in naive PC and CRPC cells. Moreover, via inhibition of ZEB (zinc-finger E-box binding homeobox) 1 and *ZEB2* and increasing E-cadherin levels, miR-205 prevents EMT, enhancing metastasis [226]. The loss of anti-apoptotic protein Bcl-2, a direct target of miR-205, has a major function in androgen-independent expression related with the higher level of Bcl-2 and perhaps in docetaxel resistance in PC [227]. Schaefer et al. discovered 15 dysregulated miRNAs in PC samples (10 downregulated and 5 upregulated) [228]. miR-205, miR-149 and miR-181b (all downregulated in PC) and miR-96, miR-182, miR-375 (all upregulated in PC) were the miRNAs with the best diagnostic potential [228]. miR-205 and miR-214 were reported to possess the potential of becoming non-invasive molecular biomarkers in PC [229].

Downregulated in PC cells; miR-940 suppresses the migration and invasion properties of cells, thus attenuating their anchorage-independent growth and increasing E-cadherin expression by implicating in EMT. Loss of miR-940 in PC cells resulted in a rise in the expression of migration and enhancer 1 (*MIEN1*), the latter made the progression of PC possible [230]. Nonetheless, the upregulation of miR-940 in PC cells is the main controller of *MIEN1*. *MIEN1's* function in the migration and invasion of cancer cells is via increasing filopodia formation and cytoskeletal rearrangement. This results in the upregulation of Akt-dependent Nuclear Factor-kappaB (NF-κB) target genes expression [230].

miR-488 aims AR expression directly, eliciting its tumor-suppressing activity. Expression of AR protein was downregulated in androgen-responsive and androgen-refractory PC cells by miR-488. The overexpression of miR-488 in PC cells increased apoptosis and inhibited cellular growth via direct downregulation of AR expression, forming its tumor suppressor effect [231].

miR-135a is considered to have tumor suppressor function since as it is upregulated in androgen-responsive PC cells. Its in vivo overexpression inhibits invasion abilities of PC cells via downregulation of Rho-associated protein kinase 1 *(ROCK1)* and *ROCK2* gene expression, through direct targeting of these genes. Thus, androgen regulated miR-135a working over the downregulation of ROCK1 and ROCK2 proteins inhibits PC cell migration and invasion, portraying its tumor suppressor role in androgen-sensitive PC [232].

Abnormal activation of the MYC oncogene could be pathologically activated by various mechanisms, which is currently defined as a major event in many cellular pathways; resulting in the development of various types of neoplasia, including PC [233]. miRNAs were suggested to regulate c-Myc expression [234]. Microarray analysis identified miR-27a-5p, miR-570, and miR-1292 as overexpressed in MYC upregulated PC samples. This finding is suggestive of possible targets of c-Myc regulation. Barros Silva et al. found that miR-27a-5p was positively regulated by MYC; the silencing by aberrant promoter methylation occurs early in the carcinogenesis of the prostate at the same time with the loss of MYC regulatory activity [235]. During the progression of PC, the promoter of miR-27a-5p is hypomethylated, allowing for MYC to resume its regulatory activity. miR-27a is an androgen-regulated oncomiR in PC, targeting the tumor suppressor and AR corepressor prohibitin, a known tumor-suppressor gene, and co-repressor of the AR. miR-27a regulates the downregulation of prohibitin protein product, leading to increased expression of AR target genes and increase in the cellular growth of PC cell [236]. The overexpression of miR-27a decreased the proliferation of PC, migration, induced PC cell cycle arrest and apoptosis [237].

In a series of primary PC and visceral/bone metastatic lesions low expression of miR-15 and miR-16, initially discovered tumor suppressor miRNA, are related with the presence of bone metastasis and poor clinical outcome. The downregulation of miR-15a and miR-16-1 lead to inhibition of apoptosis via BCL2; regulation of the cell cycle via cyclin D1 protein; promotion of bone metastasis via Wnt signaling in PC [222]. Additionally, the upregulated miR-21 expression is in synchronization with miR-15 and miR-16 loss; enhancing the progression of cancer in murine PC models. In accordance with this finding, deregulation of miR-15/miR-16 and miR-21 was observed in a subset of PC patients with progressive disease; and this finding was associated with poor prognosis and decreased disease-free survival [222, 238]. Another study reported that miR-16 was upregulated in the plasma of metastatic PC patients, however, it was downregulated in primary or metastatic PC tissues [239].

Various miRNAs have been associated with the progression of PC and its metastasis [240, 241]. Loss of miR-101 through epigenetic reprogramming is involved in overexpression of enhancer of zeste polycomb repressive complex 2 subunit (*EZH2*), associated with the aggressive phenotype [241]. miR101-1 and miR-101-2's loss from either single or both genomic loci had been reported in an important number primary PC and up to two-thirds of metastatic PC.

Decreased miR-34b, a potential biomarker candidate for progressive PC, levels are associated with clinicopathological progression of PC, including increased Gleason score, perineural invasion and lymph node metastasis. PC Patients with tumors expressing low miR-34b levels are at significantly increased the risk of biochemical and lower overall survival probability [242]. The downregulation of miR-224 was associated with advanced clinical stage and presence of metastasis by several studies, making it another probable biomarker candidate in progressive PC. Decreased expression of miR-224 was related to decreased progression-free survival in PC patients [243].

The let-7 miRNA family seem to act as tumor suppressors. Let-7 a, b, c, d, f, g were downregulated in PC while let-7 b, d, i were upregulated [244]. Regulation of interleukin (IL)-6, Myc, Lin28 and AR are among the various pathways utilized by Let-7c to repress prostate tumor growth. Downregulation of let-7 results in the activation of protein synthesis of insulin-like growth factor mRNA binding protein 1 (IMP-1). IMP-1 increases levels of multidrug resistance-1 (MDR1) protein in cancer cells. Let-7a targets E2F transcription factor 2 (E2F2), cyclin D2 (CCND2), and insulin-like growth factor 1 receptor (IGF1R) in PC cells. miRNA let-7a is significantly downregulated in PC tissues and cell lines. CC chemokines receptor 7 (CCR7) had been reported to have functions in the invasion and metastasis of several cancers. Current studies had shown that the expression of CC chemokine receptor type 7 (CCR7) increased in PC, while microRNA (miRNA) Let-7a expression decreased. Tank et al. hypothesized that the level of miRNA Let-7a in PC cells is low, the expression of CCR7 in the target gene is inhibited directly, the enhanced CCR7 binds to the ligand CCLl9/CCL21 and it is activated EMT by the MAPK signal cascade [245]. This hypothesis could account for the theoretical basis of miRNA Let-7a regulation of CCR7 in the pathogenesis and potential targeted therapy of invasion and metastasis of PC.

The use of circulation miRNAs as biomarkers for the diagnosis of PC had been shown by various studies [163, 246, 247]. miR-21's level of expression was increased in androgen-independent PC cells more than in androgen-dependent PC cells [32]. This finding suggests that the expression level of miR-21 could be utilized in the stage-specific classification of PC. miRNA expression profiling could be done in body fluids like urine and serum, in order to find miRNAs that could be utilized as novel biomarkers [248]. In the early stages of PC, miR-21 significantly increases, however it does not reach the same levels in advanced PC. miR-21, miR-141, and miR-221 are among the circulating miRNAs that are most frequently deregulated in PC [246, 249].

In accordance with all types of malignancy, the ability to predict the prognosis of a patient is a crucial part in the clinical management of PC. To this end, various studies tried to find a distinct pattern of miRNA expression to classify patients according to their risk of developing aggressive PC. miR-29c, miR-34a, and miR-141 showed downregulation in higher (grade 5) compared to lower (grade 3) Gleason grades [250]. Patients with an increased risk of biochemical recurrence had lower miR-331-3p and miR-152 levels [251]. Sun et al. showed that miR-221 and miR-222 had increased expression in CRPC tissues while miR-221 was reported to stimulate overexpression of genes involved in metastasis [252]. Ambs et al. reported that miR-32, miR-182, miR-31, and miR-26a were the most significantly overexpressed tumor miRNAs; according to their study on genomic profiling of miRNAs in 60 tumor tissue samples and 16 controls. Their finding suggested direct involvement in tumor progression [253]. Other studies showed the oncogenic function of role of miR-181a and miR-196a, other miRNAs such, through the control of signaling in PC [254, 255].

Bertoli et al., performed a meta-analysis and discovered 29 miRNAs that have diagnostic values in the tissues of PC and other biofluids (let-7a/b/c/i, miR-15b, miR-17, miR-20a, miR-21, miR-24, miR-25, miR-26a/b, miR-31, miR-32, miR-34b, miR-93, miR-106a, miR-141, miR-143, miR-145, miR-148a, miR-155, miR-182, miR-187, miR-200b, miR-218, miR-221, miR-223 and miR-375) [256]. Greco et al. analyzed 64 miRNAs from 37 articles and listed them based on their p values ($p \leq 0.05$) in a meta-analysis [188]. miR-221 had been found to possess a biomarker potential for the early diagnosis and a novel therapeutic target potential of BPH. Song et al. analyzed 104 articles published in 2007-2017, in their meta-analysis [257]. Their pooled or stratified analysis reported 10 up-regulated miRNAs (miR-18a, miR-34a, miR-106b, miR-141, miR-182, miR-183, miR-200a/b, miR-301a, and miR-375) and 14 down-regulated miRNAs (miR-1, miR-23b/27b, miR-30c, miR-99b, miR-139-5p, miR-152, miR-187, miR-204, miR-205, miR-224, miR-452, miR-505, and let-7c), which had good diagnostic and predictive potential to differentiate PC from BPH/normal controls. Decreased expression of let-7c together with increased expression of miR-32 may be utilized to delineate metastatic PC from local/primary PC. Pashaei et al. reported a common signature of a total of 37 DE microRNAs including 15 overexpressed and 22 underexpressed microRNA genes in a meta-analysis including six miRNA expression datasets of recurrent PC [268]. miR-449A and miR-579 were among the overexpressed DE that was associated with increased combined P-values across all studies.

2.4. Colorectal Cancer

Colorectal cancer (CRC) is responsible for 8% of cancer-associated mortality. It is the second most common cancer in women and the third most common cancer in men

[259]. Several studies have assessed the biological roles and relevant mechanisms of diverse miRNAs in CRC tumorigenesis. More than 100 altered miRNAs have been reported studies on cell lines, tumor and normal tissues.

Over-growth, loss of apoptotic regulation, invasion, angiogenesis and maintenance of cancer stem cells (CSCs) are among the processes related to the progression of CRC [260]. The transformation of a benign adenoma or polyp into malignant carcinoma occur depending on changes in important signaling pathways. These signaling pathways include Wnt/ß-catenin signaling pathway, epidermal growth factor receptor (EGFR) pathway, transforming growth factor beta (TGF-ß) and the TP53 network [261]. The miR-135 family and miR-34 family directly targets the UTR of genes involved in Wnt/ß-catenin signaling pathway and have a major contribution to the progression of CRC cancer, mir-150, miR-145, miR-29b and miR-146a are also enrolled in several parts of this process [262].

The sensitivity of the miR-135, miR-17-92 cluster was 74% and the specificity was 79% in expression analysis performed in stool samples as reported by Koga et al. [263]. This finding puts this method forward as a possible screening test for CRC screening. The ability of high levels of miR-135b attenuating the effects of doxorubicin in the induction of apoptosis in CRC cell lines [264]. Kanaan et al. studied another panel of miRNAs containing miR-135b, miR-31, miR-1, and miR-133a; concluding that this method had high detection accuracy with 100% sensitivity and 85% specificity [265]. When the promoters of miR-34a, miR-34b, miR-34c, miR-9, miR-129 and miR-137 were hypermethylated the expression of miRNAs in CRC tissues decreased. This could portray a contribution to transcriptional downregulation of miRNAs [52]. MiR-150 regulates the interaction between Wnt/ß-catenin and CREB CAMP (Responsive Element Binding Protein 1) signaling pathways. The overexpression of miR-150 leads to EMT which could enhance invasion and migration of cells in CRC. miR-150 had been reported as a prognostic biomarker for response to chemotherapy response by Feng et al. [266]. They showed its anti-cancer effects which were exerted through downregulation of c-myb.

miR-17, miR-92a-1, miR-18a, miR-19a, miR-20a, and miR-19b-1 from miR-17-92 cluster have been shown to be dysregulated in several cancers [267] The miR-17-92 cluster had been reported to be overexpressed in various cancers. It is involved in the promotion of unrestrained cell growth giving its name, oncomiR. miR-17-92 had been reported to show overexpression in the tissues and serum of CRC patients [268, 269]. A validation study utilizing 180 independent sets portrayed the ability established by miR-92 levels in plasma to differentiate between colorectal and gastric cancers, inflammatory bowel disease or normal subjects [270]. miR-29a and miR-92a levels in CRC patients were markedly upregulated in a study analyzing the miRNA expression in plasma of 157 patients (with advanced adenomas or carcinomas) and gastric cancer [271]. The most predictive miRNA was miR-92a, with AUC values ranging from 0.786 to 0.890, from 65.5% to 89% sensitivity, and from 70% to 82.5% specificity. Thus, the evidence stands behind the oncogenic function of the miR-17-92 cluster in CRC [271]. miR-17-5p has the highest expression in epithelial colon cells. Its level of expression rises in the transitional zone from normal to adenoma to adenocarcinoma (N-A-AC), thus hinting at the possibility of the function of miR-17-5p in the sequential evolution of early colon cancer [271]. miR-17-5p had been reported to show overexpression in CRC tissue samples by various studies [272, 273]. During the early embryo phase of the colon epithelium miR-17-5p expression levels are high, although it is sustained only in the proliferative crypt progenitor compartment. Downregulation of E2F1 by miR-17-5p has major functions for proliferation both during embryonic colon development and colon carcinogenesis [274].

The deficiency in the control mechanisms of important miRNAs like miR-34 family, miR-125b and miR-215 in TP53 function deficiency could lead to cellular growth and proliferation in CRC. The downregulation of miR-125a and miR-125b have been shown to be associated with chemoresistance to paclitaxel. The aforementioned miRNAs may enhance survival of CSCs [275]. miR-215, miR-224, miR-140, and miR-20a had been associated to increase drug resistance. New studies are being performed to uncover whether these miRs could be utilized to figuring out the efficacy of chemotherapy or as direct treatment targets.

MiR-143, miR-145, mir-7, miR-18a, miR-126, miR-21, miR-32, miR-92a and miR-181a are microRNAs affecting the activation of EGFR signaling pathway by various ways. Michael et al. reported the relationship between miRNAs and CRC, reporting the decreased levels of miR-143 and miR-145 in CRC tissue [276]. Lowered miR-143 levels are related to progression of colon cancer, poor prognosis, the presence of metastasis and resistance to chemotherapy [277, 278]. MiR-145 was suggested to function as a tumor suppressor [279]. Ahmed et al. collected stool samples from CRC patients and healthy controls comparing the level of miRNAs [280]. They reported that miR-143, miR-145, miR-320, miR-126, miR-484-5p, miR-16, and miR-125b, were downregulated in the stool samples of CRC patients. 35 miRNAs had been reported to portray different regulation in CRC cells when compared with normal cells. Among these miRNAs, miR-143, miR-145, miR-1, miR-126, miR-30a, miR-191 and miR-192 were downregulated; while miR-106, miR-31, miR-21, miR-25, miR-20a, miR-93, miR-183 and miR-203 were upregulated in CRC [280].

Yokobori et al. reported that miR-7 functions as a tumor suppressor through modulation of the EGFR signaling pathway in CRC cases [281]. They discovered that the addition of miR-7 to cetuximab could benefit the patients resistant to targeted anti-EGFR therapy [288]. In patients with decreased expression of miR-181a, the clinical outcome was worse in patients treated with EGFR antibodies [282].

miR-21, which is among the researched miRNAs, possesses a high degree of stability making it a major biomarker candidate of CRC [283]. miR-21 had been hypothesized to function as a potential oncogene. Various studies had shown the upregulation of miR-21 in CRC. Increased levels of serum miR-21 and miR-92a could become potential biomarkers for the early detection of CRC and advanced adenoma [284]. Also, miR-21 had been hypothesized to be an independent biomarker of the prognosis of CRC prognosis as it has the ability to delineate adenoma and CRC from healthy controls [285]. Increased levels of miR-21 expression could suggest poor survival via an association with Stage III/IV CRC according to a meta-analysis. The pooled hazard ratio after elimination of heterogeneity was 2.32 (95% CI 1.82–2.97, P = 0.0001), indicating the increased stability in the prediction of worse survival [286]. Postoperative CRC serum miR-21 levels were found to be markedly decreased than preoperative levels [287, 288]. Increased levels of miR-21 in primary CRC tissues and matched serum samples were also related with increased tumor size and presence of distant metastasis. Rises in the expression of miR-21 were related with resistance to 5-fluorouracil (5-FU) chemotherapy as portrayed by cohort studies [58]. The resistance was induced by the inhibition of the DNA repair protein MutS homolog2 (MSH2) in this in vitro study. Serum miRNA panels could be used in CRC patients. In one research, six serum miRNAs (miR-21, let-7g, miR-31, miR-92a, miR-181b, and miR-203) showed 93% sensitivity and 91% specificity in comparison with carcinoembryonic antigen (CEA) and CA19-9 in CRC cases [289]. Xu et al. performed a meta-analysis encompassing seven studies with 676 CRC patients and 417 controls; reporting that the value of 16.89 indicated that circulating miR-21 could be beneficial to be used as a biomarker for the diagnosis of CRC [286]. miR-21's diagnostic and prognostic value in CRC patients was analyzed in a meta-analysis [290]. The diagnostic sensitivity was 64%, specificity was 85%, and AUC was 0.85; portraying the possible function of circulating miR-21 has as a predictor for detecting CRC, while tissue miR-21 was a useful marker for predicting CRC survival [290].

The miR-29 family has a major function in the biology of CRC. miR-29a, which is a member of the miR-29 family (miR-29a, miR-29b, miR-29c) was reported to be upregulated in CRC patients by various studies [270, 291, 292]. Huang et al. reported that that plasma levels of miR-92a, in addition to miR-29a, could differentiate CRC and advanced adenomas from other colorectal diseases [270]. In the light of these findings, miR-92a could be a potential biomarker in the early diagnosis of CRC. In addition to blood samples, stool samples could be used for the detection of CRC-specific miRNA. In a study, the risk of recurrence was related to increased levels of miR-29a in patients with stage II CRC, however, this was not observed in patients with stage I CRC [293]. On the other hand, miR-29b was downregulated in CRC patients in two studies [294, 295]. miR-29b had been reported to act as a tumor suppressor in various cancers and diseases [296]. It has a major function in the control of the progression of CRC, and could be used as a suitable prognostic marker and therapeutic target [293]. In the meta-analysis by Zhi et al. the pooled specificity and positive likelihood ratio of serum miR-29a were 0.89 and 4.66, respectively [297]. In the light of these results, patients with CRC have an approximately 5-fold higher chance of being miR-29a-positive compared to those without CRC.

Zhang et al. showed that miR-181a suppressed *PTEN*, which affects proliferation and cell viability in CRC cells [298]. The oncogenic effect of miR-181b in CRC was proved by the identification of *PDCD4* as its target gene. The control of PDCD4 by miR-181b occurs at the posttranscriptional level [299]. The role of miR-181a in influencing the invasive potential of CRC cells was proposed in a study evaluating the relationship between miR-181a and the *WIF-1* gene [300]. The expression of miR-181 family members in CRC was markedly related to survival.

miR-200a, miR-200b, miR-200c, miR-141 and miR-429 are the members of the miR-200 family. The miR-200 family had been thoroughly studied to uncover its function in EMT in tumor metastasis. It has the potential to become an important marker for both prognostic and therapeutic management of CRC. Decreased levels of expression of miR-200a, miR-200c, and miR-429 in the primary tumor tissue in CRC could be related to poor survival. Toiyama et al. showed that although the expression of miR-200c in the serum samples increased with tumor stage, the expression in the primary tumor tissue decreased [301]. Their hypothesis that the alteration could be explained by the rise in the expression of miR-200c by the metastatic tumor. Rises in the blood levels of miR-141, miR-200a and miR-200c could be related with worse outcomes. The use of plasma miR-141 as a supplement to CEA in the detection of a patient with CRC and distant metastasis was reported [302].

Shang et al. reported that miR-23a could control 5-FU chemosensitivity by targeting APAF-1/caspase-9 apoptotic pathways in colon cancer cell lines in vitro [303]. Ogata-Kawata et al. reported that miR-23a extracted from serum samples portrayed an unexpectedly high diagnostic accuracy of CRC with 92% sensitivity and 100% specificity [288].

miR106a, an oncogenic miRNA controlling the expression of cancer-related genes (RUNX3, Twist1 and pRB) has increased expression is during tumor suppression [304-306]. Its expression is reduced in CRC tissues. The level of miR106a expression seem to be related with patient survival [307]. A meta-analysis study encompassing 14 diagnostic or prognostic studies reported that miR106a had a pooled sensitivity of 0.53, specificity of 0.85, and AUC of 0.72, thus showing that measuring miR106a level is a promising noninvasive method for the diagnosis of CRC [308].

miR-221 showed overexpression in the blood samples of cancer patients, indicating the possibility that miR-221 could possess a major function in tumorigenesis. Thus, it could be beneficial to utilize miR-221 as a biomarker in the diagnosis of CRC. Yau et al., using a miR expression array, reported that the expression of miR-221 and miR-18a in CRC patients increased steadily with tumor initiation and progression [309].

In a meta-analysis study encompassing 34 manuscripts which studied plasma- or serum-based miRNAs in the diagnosis of CRC, reported that a total of 31 single miRNAs were dysregulated in CRC patients [310]. Eight of these miRNAs (six were upregulated: miR-17-3p, miR-18a, miR-21, miR-29a, miR-92, and miR-106a and two were downregulated: miR-29b and miR-145.) were identified as dysregulated and validated by more than one study [310]. Yan et al. also performed a meta-analysis study encompassing103 studies from 36 articles with a total of 3124 CRC patients and 2579 healthy individuals, reported a panel of circulating miRNAs with 45 members (let-7g, miR-106a, miR-106b, miR-1246, miR-129-3p, miR-133a, miR-139-3p, miR-143, miR-145, miR-155, miR-15b, miR-17-3p, miR-181b, miR-18a, miR-193a-3p, miR-19a, miR-19a-3p, miR-19b, miR-200c, miR-203, miR-20a, miR-21, miR-221, miR-223-3p, miR-23a, miR-29a, miR-31, miR-331, miR-338-5p, miR-375, miR-378, miR-409-3p, miR-422a, miR-431, miR-532-5p, miR-601, miR-634, miR-7, miR-760, miR-767-3p, miR-877*, miR-92, miR-92a, miR-92a-3p, miR-93) with an overall diagnostic accuracy with sensitivity of 77.1% and specificity of 81.6% [311].

3. MicroRNAs as Biomarkers of Diagnosis and Prognosis

miRNAs are concerned with many aspects of cancer and they control a large number of targets rather than a few specific genes due to their mechanism of action. It means that patterns of microRNAs may give insights on unknown complex processes in several target genes, helping researchers to find novel useful biomarkers.

The biomarker is defined as "a characteristic that is objectively measured and evaluated as an indicator of normal biological processes, pathogenic processes, or pharmacologic responses to a therapeutic intervention" [312]. miRNAs have most of the characteristics of the biomarker, assessing analytical criteria and clinical utility. miRNAs show specificity to the pathology of interest and sensitivity to physiological or pathological changes [313]. Classifiers have started to evaluate the miRNA expression patterns in the first decade of 2000. Lu et al. performed a bead-based miRNA profiling method to assess miRNA expression levels in normal and tumor tissues. They determined that expression patterns of miRNAs can not only distinguish tumor origin, also the degree of differentiation and classify poorly undifferentiated tumor tissues [44]. Other evidence is about miRNA signatures that could differentiate between lung tumor tissues and non-tumor tissues. Also, it was observed different expression between adenocarcinoma (AD) and squamous cell carcinoma (SCC) tissues and distinct prognosis [314]. A comprehensive analysis of 22 different types of tumor tissue, suggested a signature of 48 miRNAs able to reach a classification accuracy >90% [315]. miRNA expression profiles may be used for classification of patients to detect the right strategy in the clinical practice.

miRNA expression signatures are usually preferred to deliver a specific diagnosis. Nine microRNAs signatures were able to differentiate between breast cancer tissues and normal cancer tissues collected by TCGA, with a high precision value [316]. Another study results showed that five miRNAs (miR-424, miR326, miR-511, miR-125b-2 and miR-451) provide high diagnostic accuracy of hepatocellular carcinoma from 377 patients with hepatocellular carcinoma [317].

Each cancer type is composed of several subtypes derived from different cellular origins and each of them has to be treated specifically. Thus, it is important to differentiate these cancer types and numerous studies pointed at this goal. A study on muscle-invasive bladder cancer found a signature of 63 miRNAs could be differentiated between basal and luminal tumors [318]. In accordance with these results, Blenkiron et al. performed a model-based discriminant analysis for basal-like and luminal A breast tumors and determined that those groups showed differences a set of miRNAs [319]. In addition to these results, 10 miRNAs were differentiated between lung neuroendocrine (NE) tumors histotypes [320].

A study in breast cancer compared to circulating miRNA profiles in healthy women with early-stage breast cancer suggested that downregulation of miR-181a and miR-1304 is related with the disease [146]. miR-195 and let-7a expression levels were higher in the plasma of breast cancer patients than in healthy individuals [146]. Investigators analyzing the miRNA plasma levels in 74 patients diagnosed with distinct types of lung cancer determined three miRNAs (miR-155, miR-197, and miR-182) that were at an increased concentration in patients with cancer than in control individuals [321]. However, the expression levels of these miRNAs were not associated with any specific type of lung cancer or any particular stage, so the miRNAs are exactly diagnostic biomarkers. The investigators also observed that the serum concentrations of miR-155, miR-197 and miR-182 decreased owing to chemotherapy in most clinically responsive patients [321].

Stable miRNAs can be detectable in urine. A study revealed that miR-126, miR-182 and miR-199a expression levels were significantly increased in the urine of stage 1 bladder cancer patients, as well as in those with stage 2 and stage 3 disease [322]. Also, urine miRNA pattern can identify miR-96 and miR-183 as biomarkers for urothelial carcinoma.

miRNAs have a great prognostic signature due to their stability in clinical samples and their potent expression profiles [44]. Many signatures have been suggested, and their use is under investigation for clinical trials.

It has been performed a meta-analysis of 43 studies in 20 cancer types and shown different microRNA signatures, which can be related with many reasons, such as sample preparation, assay methodology, and characteristics of patients [323]. miR-21 (overexpressed in cancer) and let-7 (reduced in cancer) were the most common miRNAs associated with patient outcome within all miRNAs [324].

Expression signatures of miRNAs mostly reflect the embryonic or developmental origin of the tumor type [44]. This provides tumor classification based on miRNAs. A study on 22 different tumor types revealed miRNA expression classified tumors in terms of origins with accuracy higher than 90% [316]. Leukemia classification by World Health Organization (WHO) is done according to progenitor cell lineage [325], and miRNA expression pattern across leukemia subtypes also reflects the cell of origin. Association between miRNA expression profiles and acute myeloid leukemia (AML) revealed specific miRNA expression patterns in each subtype of disease [326]. Furthermore, miRNA patterns were different between cellular subsets when stem and progenitor cells were isolated from prostate tumors. This shows that microRNA expression profiles are determinative of the cellular populations in a tumor [327].

Recently, different patterns of miRNA expression have also been defined in luminal (epithelial origin), basal-like (myoepithelial origin), and human epidermal growth factor receptor 2 (HER2) breast cancers [328].

Several cancer subtypes have been identified using the TCGA data, and miRNAs have been used to create subtypes [329]. It has been shown that miRNAs play a role in cancer progression [330] and miRNAs might be useful for the estimation of metastatic outcomes for patient management. Specific miRNAs support endothelial recruitment to metastases in breast cancer and might act as effective biomarkers for predicting this event [331].

Numerous studies have associated miRNAs with known biomarkers for therapy decisions, but not used in the clinical decision yet. For example, in CML, cell levels with the BCR-ABL rearrangement, which is the characterization of the disease, reduce over time with treatment of imatinib. It has been found that miR-451 levels inversely correlate with BCR-ABL levels of diagnosis and on treatment [332]. Single nucleotide polymorphisms (SNPs) in miRNA genes may also be predictors of response. Base excision repair genes play an important role in treatment resistance, and 3' UTRs variations in the miRNA genes have been suggested to reflect colorectal cancer prognosis and treatment response [333].

Using of circulating miRNAs as biomarkers in distinct cancer types is a rapidly developing area [343]. Tumor cells can release miRNAs, stabilized by their association into micro-vesicles [60]. It has been shown that miRNAs provide their stability in other bodily fluids, such as urine and saliva [334]. However, most studies have worked with serum microRNAs as biomarkers. A study in NSCLC patients (n=391) determined 35 highly expressed miRNAs. Of these, 17 miRNAs were related to patient survival [335].

At first, the instance of detection of cancer miRNAs in the serum was reported in diffuse large B-cell lymphoma patients [336]. The serum levels of three miRNAs, miR-155, miR-210, and miR-2, were well-known to be increased in these tumors, and furthermore, the researchers revealed increased expression of miR-21 in the serum of patients with improved relapse-free survival [336].

Examples of dysregulated circulating miRNA expression levels in several cancers proposed to have either diagnostic and/or prognostic value.

3.1. let-7 Family

let-7 family contains 13 different miRNAs in humans: *let-7a-1*, *7a-2*, *7a-3*, *7b*, *7c*, *7d*, *7e*, *7f-1*, *7f-2*, *7g*, *7i*, *miR-98*, and *miR-202* [337]. It has been shown that expressions of *let-7* family members are downregulated in a wide variety of cancers, including melanoma, pancreatic cancer, prostate cancer, and sarcoma. On the other hand, some miRNAs, are upregulated in lymphoma, mesothelioma, and breast cancer [338]. *let-7* has been suggested to be a direct regulator of several important oncogenes, such as the three *RAS* genes [46], high mobility group AT-hook 2 *(HMGA2)* [339], *STAT3* [340], ubiquitin like with PHD and ring finger domains 2 *(UHRF2)* [341], and *MYC* [342]. Also, *let-7* family targets cell cycle and cell proliferation genes [343, 82]. One of the direct target of let-7 family is also RNase III nuclease, processes pre-miRs, so let-7 family members might regulate their own processing [83].

Distinct expression of *let-7* family miRNAs have been identified and thus let-7 family members have diagnostic value in serum/plasma of many cancer types, such as lung cancer (*let-7a*, *let-7c*, *let-7f*) [344], prostate cancer (*let-7a*) [167], gastric cancer (*let-7a*, *let-7c*, *let-7i*, *let-7f*) [345], ovarian cancer (*let-7b*, *let-7f*, *let-7i*) [346], hepatocellular carcinoma (HCC) (*let-7b*, *let-7f*) [347], breast cancer (*let-7c*, *let-7b*, *let-7g*) [348], acute myeloid leukemia (AML) (*let-7b*, *let-7d*) [349], thyroid carcinoma (*let-7e*) [350], and CRC (*let-7a*) [295]. Furthermore, they play a role as prognostic biomarkers in many cancer types including myelodysplasia (*let-7a*) [351], lung cancer (*let-7b*, *let-7f*, *let-7i*) [355], hepatocellular carcinoma (*let-7b*, *let-7f*) [347], multiple myeloma (*let-7e*) [352], prostate cancer (*let-7d*) [353], ovarian cancer (*let-7f*) [354], and breast cancer (*miR-202*) [355].

3.2. miR-10b

miR-10b acts as a metastatic in several types of cancers including breast cancer [356], glioma [357], and esophageal cancer [358]. Syndecan 1 *(SDC1)* [369], homeobox D10 *(HOXD10)* [360], Kruppel-like factor 4 *(KLF4)* [361], MHC class I polypeptide-related sequence B *(MICB)* [362], and cadherin 1 *(CDH1)* [363] genes are targeted by this miRNA. Furthermore, it regulates E2F transcription factor 1 *(E2F1)*-mediated transcription [364] and cell cycle regulators such as BUB1 mitotic checkpoint serine/threonine kinase *(BUB1)*, polo-like kinase 1 *(PLK1)*, and cyclin A2 *(CCNA2)* [365].

It has been shown that expression levels of circulating *miR-10b* are upregulated in patients with ovarian cancer [366], lung cancer [367], esophageal [368], and glioblastoma [369] compared to healthy controls. Consistent with this, *miR-10b* is highly expressed in plasma from metastatic breast cancer patients [370].

3.3. miR-16

It has been suggested that circulating *miR-16 levels* are significantly differentially expressed in patients compared to healthy controls in many types of cancers, such as oral cancer [371], breast cancer [372], gastric cancer [373, 374], osteosarcoma [375], liver carcinoma [376], and esophageal carcinoma [377]. Also, *miR-16* is related to prognosis and tumor size in gastric cancer [375], hepatocellular carcinoma [376], and esophageal squamous cell carcinoma (ESCC) [377].

3.4. miR-17-92 Cluster

It has been observed that miR-17-92 cluster have prognostic value in colorectal cancer [378], sporadic melanoma [379], breast cancer [380], bladder cancer [381], multiple myeloma [382], lung cancer [383], and prostate cancer [384]. Also, circulating *miR-17~92* miRNA expression levels have been associated with the response to chemotherapy (*i.e.,* predictive biomarkers) in both breast cancer [385] and multiple myeloma [382].

3.5. miR-29 Family

Distinct expression of *miR-29* family members, miR-29a, -29b, and -29c, in the plasma and/or serum has been shown in several cancer types. All *miR-29* family members' expression levels were upregulated in the serum of osteosarcoma patients. Especially, there was an association between *miR-29a* and *miR-29b* and poor prognosis [386]. *miR-29a* is upregulated in colorectal cancer and thus it has been considered as a potential non-invasive biomarker for early detection of colorectal cancer [299]. Similarly, *miR-29b* serum levels act as potential biomarkers for diagnosis and prognosis of colorectal cancer [387]. However, *miR-29c* acts as a predictor of postoperative early relapse [388]. *miR-29c* was downregulated in the serum of nasopharyngeal carcinoma patients compared to controls [389].

4. MIRNAS-BASED CANCER THERAPIES

miRNAs are used as therapeutic targets to have oncogene or tumor suppressor gene functions [71]. miRNA-based therapy is a new therapeutic approach for cancers [390]. There are currently two strategies based on miRNA-targeted cancer treatment: a) inhibition of proliferation or induction of apoptosis in tumor cells by using exogenous miRNAs, which are tumor suppressor miRNAs. For example, the chemically produced miRNA mimics act as endogenous mature double-stranded miRNA, which aims to restore and/or enhance endogenous miRNA function [391, 392]. Generation of viral vectors, such as adenoviral, lentiviral and retroviral vectors, which express specific miRNAs, could increase the function of miRNAs in tumor cells; or b) inhibition of the function of miRNAs, which are oncogenic miRNAs, by antisense oligonucleotides (ASOs) strategy.

4.1. miRNA Inhibition Therapy

It has been shown an increased expression of oncogenic miRNAs in human cancers and inhibition of oncogenic miRNAs provides normal expression and function of its target tumor suppressive genes. miRNA inhibitors are complementary single-stranded oligonucleotides and cause sequestration of a specific endogenous miRNA in an unrecognized conformation [393]. Thus, the RISC cannot process mature miRNA and miRNA is excluded from the RISC. These contain antisense anti-miR oligonucleotides (AMOs), locked nucleic acids (LNAs) anti-miRNAs, antagomirs, miRNA sponges, miRNA masks and small molecule inhibitors of miRNAs [393]. AMOs are single-stranded, modified anti-sense oligonucleotides (ASOs), which are 17-22 nucleotide in length and selected complementary to a target miRNA [394]. AMOs inhibit the miRNAs by annealing to the mature miRNA and block the interaction of that miRNA with its target mRNAs. As a result, it is produced ASO-miRNA duplex through Watson-Crick binding and is performed RNAse-H-mediated cleavage of the target miRNA gene [393]. For the potential clinical trial, it has been revealed that in vivo miR-122 is silenced by AMOs in mouse liver [395]. However, unmodified AMOs cannot inhibit the function of miRNA in vitro.

The antagomirs have chemical modifications and are cholesterol-conjugated single-stranded 23-nt RNA molecules complementary to the selected miRNAs. The modifications provide the higher stability of the RNA and protect it from degradation [393]. A study in mice suggested that antagomir-122 stimulated a marked, specific, and persistent (up to 23 days) loss of endogenous miR-122 gene expression in liver, lung, kidney, heart, intestine, fat, skin, bone marrow, muscle, ovaries and adrenals [396]. Similar results were for antagomir-16, targeting the ubiquitously expressed miR-16 [396]. Also, silencing of miRNAs by antagomirs leads to other positive physiologic effects, such as the reduction in cholesterol levels of plasma after antagomir-122 administration. One advantage compared to siRNA technology is that antagomirs did not induce an immune response [393]. Another example of modified AMOs are LNAs anti-miRs.

LNA-modified oligonucleotides have increased thermal stability and high-affinity hybridization with their targeted RNA molecules. Also, they show high metabolic stability for in vivo delivery [393]. It has been demonstrated that silencing of miR-21 in vitro by LNA-modified antisense oligonucleotides results with significant reduction in cell viability and elevated intracellular levels of caspases in glioblastomas [397].

In a study, blocking of miR-9, which is overexpressed in breast cancer cells and directly targets *CDH1*, using a 'miRNA sponge' prevented metastasis formation [398]. 'miRNA masks' are gene-specific anti-miRNAs, which can selectively inhibit the interaction between the target miRNA and specific mRNA [399].

Several miRNAs have promising therapeutic properties in clinical trials:

4.1.1. miR-34

The miR-34 family contains miR-34a, miR-34b, and miR-34c. It has been reported that all of these miRNAs have tumor-suppressive properties, such as the inhibition of migration, invasion, proliferation, cell survival, EMT, and drug resistance [400, 401, 402]. miR-34 therapy has been shown in many different preclinical studies in numerous types of cancers [403, 404, 405]. miR-34a replacement can prevent tumor growth in colon [406] and lung cancer models [254]. Most publications have revealed the successful in vivo use of miR-34 family-based therapy alone or in combination with chemotherapy, radiotherapy, or other targeted therapies [407]. Recently, a study suggested that miR-34a replacement with radiotherapy-induced a tumor immune response by targeting programmed death-ligand 1 (*PDL1*) [401]. This finding is interesting in the success of immune therapies, which aims to inhibit *PDL1* in the clinic [408]. Some great preclinical studies have led to the development of miR-34a replacement therapy for clinical use. A miR-34a mimic within liposomes (*MRX34*) was the first miRNA-based cancer therapeutic to begin a phase I clinical study in April 2013 (409). A data on the effect of *MRX34* in advanced solid tumor patients, which were treated between April 2013 and September 2014 [410]. However, the clinical trial was terminated due to immune response-related severe adverse effects in five patients [410].

4.1.2. miR-15/16

It has been suggested that miRNA-15/16 family members are downregulated in various cancers. miR-15/16 regulates many cancer-related pathways such as apoptosis, cell growth, cell cycle progression, angiogenesis, and EMT [410, 42, 411, 412]. Some findings have shown the effectiveness of miR- 15/16 replacement therapy in preclinical models including multiple myeloma [41], chronic lymphocytic leukemia [413], prostate [414], lung [415], gastric [416], colon [417], and ovarian cancer [418]. The last study also revealed that a combination of miR-15a/16 liposomes and cisplatin showed an increased therapeutic efficiency in decreasing tumor weight compared to approach alone [418]. This result indicates that it could also be used combined miR-15/16 replacement and other drugs therapies. Furthermore, in vitro studies demonstrated that a synergistic effect of miR-15a/16 mimics and miR-34a mimics inhibit cell cycle progression in lung adenocarcinoma cells [419]. TargomiRs have been developed according to the consensus sequence of miR-15/16 family members [420]. TargomiRs-based treatment of recurrent malignant pleural mesothelioma (MPM) and NSCLC was initiated in the phase I clinical trial in December 2014 [421].

4.1.3. miR-155

Overexpression of oncogenic miRNA miR-155 is inhibited by an anti-miR in a clinical trial. miR-155 plays a critical role in malignant hematopoiesis, and is overexpressed in numerous hematopoietic cancer in addition to some solid cancers [422]. Several targets of miR-155 regulate many cellular events such as cell growth, invasion, migration, and inhibition of apoptosis [422, 140, 423]. Also, it has been shown that increased expression of miR-155 is related to genomic instability and resistance to therapeutic approaches. Recently, an in vivo study has revealed that the replacement of anti-miR-155 provided the sensitization of miR-155-overexpressing tumors to cisplatin chemotherapy in an orthotropic lung cancer model [122]. Lymphomas and macroglobulinemia were successfully treated with anti-miR-155 in mouse models [424, 425, 426]. However, miR-155 is involved in the activation of immune responses. Also, it is indicated that the blocking of miR-155 in tumor-associated immune cells might promote tumor immune evasion and progress rather than inhibit tumor growth. This view is supported by the fact that the cells of the tumor microenvironment, which have deficient miR-155 promotes tumor growth [427, 428]. In another study, miR-155 mimic was replaced to dendritic cells in the tumor microenvironment of an ovarian cancer xenograft, which resulted in the activation of the anti-tumor immune response, inhibition of tumor growth, and enhanced survival of the mice [429]. These results suggested that miR-155 may serve as a tumor suppressor or an oncogene due to its expression level [430]. A phase I clinical trial with an LNA-modified anti-miR-155, was launched in November 2015 (431). The preliminary data of this trial shows that intratumoral LNA-modified anti-miR-155 injection leads to improvement of the cutaneous lesions [432].

REFERENCES

[1] Bartel DP. MicroRNAs: genomics, biogenesis, mechanism, and function. *Cell* 116: 281–297, 2004.

[2] Wilczynska A, Bushell M. The complexity of miRNA-mediated repression. *Cell Death Differ* 22: 22e33, 2015.

[3] www.mirbase.org.

[4] Lee Y, Kim M, Han JJ, Yeom KH, Lee S, et al. MicroRNA genes are transcribed by RNA polymerase II. *EMBO J* 23: 4051–4060, 2004.

[5] Zeng Y, Cai X, Cullen BR. Use of RNA polymerase II to transcribe artificial microRNAs. *Methods Enzymo* 392: 371–380, 2005.

[6] Cai XZ, Hagedorn CH, Cullen BR. Human microRNAs are processed from capped, polyadenylated transcripts that can also function as mRNAs. *RNA* 10: 1957–1966, 2004.

[7] Kuehbacher A, Urbich C, Dimmeler S. Targeting microRNA expression to regulate angiogenesis. *Trends Pharmacol Sci* 29: 12–15, 2008.

[8] Yi R, Qin Y, Macara IG, Cullen BR. Exportin-5 mediates the nuclear export of pre-microRNAs and short hairpin RNAs. *Genes Dev* 17: 3011–3016, 2003.

[9] Carthew RW, Sontheimer EJ. Origins and mechanisms of miRNAs and siRNAs. *Cell* 136: 642–655, 2009.

[10] Ambros V. The functions of animal microRNAs. *Nature* 431: 350–355, 2004.

[11] Vasudevan S, Tong Y, Steitz JA. Switching from repression to activation: microRNAs can up-regulate translation. *Science* 318: 1931–1934, 2007.

[12] Orom UA, Nielsen FC, Lund AH. MicroRNA-10a binds the 5′UTR of ribosomal protein mRNAs and enhances their translation. *Mol Cell* 30: 460–471, 2008.

[13] Tay Y, Zhang J, Thomson AM, Lim B, Rigoutsos I. MicroRNAs to Nanog, Oct4 and Sox2 coding regions modulate embryonic stem cell differentiation. *Nature* 455: 1124–1128, 2008.

[14] Hwang HW, Wentzel EA, Mendell JT. A hexanucleotide element directs microRNA nuclear import. *Science* 315: 97–100, 2007.

[15] Chaluvally-Raghavan P, Jeong KJ, Pradeep S, Silva AM, Yu S, et al. Direct upregulation of STAT3 by MicroRNA-551b- 3p deregulates growth and metastasis of ovarian cancer. *Cell Rep* 15: 1493–1504, 2016.

[16] Zhang Y, Liu D, Chen X, Li J, Li L, et al. Secreted monocytic miR-150 enhances targeted endothelial cell migration. *Mol Cell* 39: 133–144, 2010.

[17] Das S, Ferlito M, Kent OA, Fox-Talbot K, Wang R, et al. Nuclear miRNA regulates the mitochondrial genome in the heart. *Circ Res* 110: 1596–1603, 2012.

[18] Zhang X, Zuo X, Yang B, Li Z, Xue Y, et al. MicroRNA directly enhances mitochondrial translation during muscle differentiation. *Cell* 158: 607–619, 2014.

[19] Friedman RC, Farh KK, Burge CB, Bartel DP. Most mammalian mRNAs are conserved targets of microRNAs. *Genome Res* 19: 92–105, 2009.

[20] Vannini I, Fanini F, Fabbri M. Emerging roles of microRNAs in cancer. *Curr Opin Genet Dev* 48: 128-133, 2018.

[21] Chou CH, Chang NW, Shrestha S, Hsu SD, Lin YL, et al. miRTarBase 2016: updates to the experimentally validated miRNA-target interactions database. *Nucleic Acids Res* 44(D1): D239–47, 2016.

[22] Giovannetti E, Erozenci A, Smit J, Danesi R, Peters GJ. Molecular mechanisms underlying the role of microRNAs (miRNAs) in anticancer drug resistance and implications for clinical practice. *Crit Rev Oncol Hematol* 81: 103–122, 2012.

[23] Kumar MS, Armenteros-Monterroso E, East P, Chakravorty P, Matthews N, et al. HMGA2 functions as a competing endogenous RNA to promote lung cancer progression. *Nature* 505: 212–217, 2014.

[24] Winter J, Diederichs S. MicroRNA biogenesis and cancer. *Methods Mol Biol* 676: 3–22, 2011.

[25] Esteller, M. Non-coding RNAs in human disease. *Nat Rev Genet* 12: 861–874, 2011.

[26] Ji W, Sun B, Su C. Targeting MicroRNAs in Cancer Gene Therapy. *Genes* (Basel) 8: E21, 2017.

[27] Iorio MV, Croce CM. MicroRNA dysregulation in cancer: Diagnostics, monitoring and therapeutics. A comprehensive review. *EMBO Mol Med* 4: 143–159, 2012.

[28] Kasinski AL, Slack FJ. Epigenetics and genetics. MicroRNAs en route to the clinic: Progress in validating and targeting microRNAs for cancer therapy. *Nat Rev Cancer* 11: 849–864, 2011.

[29] Wu H, Zhu S, Mo Y. Suppression of cell growth and invasion by miR-205 in breast cancer. *Cell Res* 19: 439–448, 2009.

[30] Gregory PA, Bert AG, Paterson EL, Barry SC, Tsykin A, et al. The miR-200 family and miR-205 regulate epithelial to mesenchymal transition by targeting ZEB1 and SIP1. *Nat Cell Biol* 10: 593–601, 2008.

[31] Korpal M, Kang Y. The emerging role of miR-200 family of microRNAs in epithelial-mesenchymal transition and cancer metastasis. *RNA Biol* 5: 115–119, 2008.

[32] Li T, Li D, Sha J, Sun P, Huang Y. MicroRNA-21 directly targets MARCKS and promotes apoptosis resistance and invasion in prostate cancer cells. *Biochem Biophys Res Commun* 383: 280–285, 2009.

[33] Calin GA, Sevignani C, Dumitru CD, Hyslop T, Noch, E, et al. Human microRNA genes are frequently located at fragile sites and genomic regions involved in cancers. *Proc Natl Acad Sci USA* 101: 2999–3004, 2004.

[34] Lujambio A, Ropero S, Ballestar E, Fraga MF, Cerrato C, et al. Genetic unmasking of an epigenetically silenced microRNA in human cancer cells. *Cancer Res* 67: 1424–1429, 2007.

[35] Hackanson B, Bennett KL, Brena RM, Jiang J, Claus R, et al. Epigenetic modification of CCAAT/enhancer binding protein alpha expression in acute myeloid leukemia. *Cancer Res* 68: 3142–3151, 2008.

[36] Calin GA, Dumitru CD, Shimizu M, Bichi R, Zupo S, et al. Frequent deletions and down-regulation of micro-RNA genes miR15 and miR16 at13q14 in chronic lymphocytic leukemia. *Proc Natl Acad Sci USA* 99: 15524–15529, 2002.

[37] Sampath D, Liu C, Vasan K, Sulda M, Puduvalli VK, et al. Histone deacetylases mediate the silencing of miR-15a, miR-16, and miR-29b in chronic lymphocytic leukemia. *Blood* 119: 1162–1172, 2012.

[38] Allegra D, Bilan V, Garding A, Döhner H, Stilgenbauer S, et al. Defective DROSHA processing contributes to downregulation of MiR-15/-16 in chronic lymphocytic leukemia. *Leukemia* 28: 98–107, 2014.

[39] Pekarsky Y, Croce CM. Role of miR-15/16 in CLL. *Cell Death Differentiation* 22: 6–11, 2015.

[40] Cimmino A, Calin GA, FabbriM, IorioMV, FerracinM, et al. miR-15 and miR-16 induce apoptosis by targeting BCL2. *Proc Natl Acad Sci U S A* 102: 13944–13949, 2005.

[41] Calin GA, Cimmino A, Fabbri M, Ferracin M, Wojcik SE, et al. MiR-15a and miR-16-1 cluster functions in human leukemia. *Proc Natl Acad Sci U S A* 105: 5166–5171, 2008.

[42] Bonci D, Coppola V, Musumeci M, Addario A, Giuffrida R, et al. The miR-15a–miR-16-1 cluster controls prostate cancer by targeting multiple oncogenic activities. *Nat Med* 14: 1271–1277, 2008.

[43] Kang W, Tong JH, Lung RW, Dong Y, Zhao J, et al. Targeting of YAP1 by microRNA-15a and microRNA-16-1 exerts tumor suppressor function in gastric adenocarcinoma. *Mol Cancer* 14: 52, 2015.

[44] Lu J. Getz G, Miska EA, Alvarez-Saavedra E, Lamb J, et al. MicroRNA expression profiles classify human cancers. *Nature* 435: 834-838, 2005.

[45] Takamizawa J, Konishi H, Yanagisawa K, Tomida S, Osada H, et al., Reduced expression of the let-7 microRNAs in human lung cancers in association with shortened postoperative survival. *Cancer Res* 64: 3753e3756, 2004.

[46] Johnson SM, Grosshans H, Shingara J, Byrom M, Jarvis R, et al., RAS is regulated by the let-7 microRNA family. *Cell* 120: 635e647, 2005.

[47] Chang TC, Wentzel EA, Kent O., Ramachandran K, Mullendore M, et al. Transactivation of miR-34a by p53 broadly influences gene expression and promotes apoptosis. *Mol Cell* 26: 745–752, 2007.

[48] Chang TC, Yu D, Lee YS, Wentzel EA, Arking DE, et al. Widespread microRNA repression by Myc contributes to tumorigenesis. *Nat Genet* 40: 43–50, 2008.

[49] Lodygin D, Tarasov V, Epanchintsev A, Berking C, Knyazeva T, et al. Inactivation of miR-34a by aberrant CpG methylation in multiple types of cancer. *Cell Cycle* 7: 2591–2600, 2008.

[50] Lujambio A, Calin GA, Villanueva A, Ropero S, Sanchez-Cespedes M, et al. A microRNA DNA methylation signature for human cancer metastasis. *Proc Natl Acad Sci USA* 105: 13556–13561, 2008.

[51] Tsai KW, Liao YL, Wu CW, Hu LY, Li SC, et al. Aberrant hypermethylation of miR-9 genes in gastric cancer. *Epigenetics* 6: 1189–1197, 2011.

[52] Bandres E, Agirre X, Bitarte N, Ramirez N, Zarate R, et al. Epigenetic regulation of microRNA expression in colorectal cancer. *Int J Cancer* 125: 2737–2743, 2009.

[53] Hildebrandt MA, Gu J, Lin J, Ye Y, Tan W, et al. Hsa-miR-9 methylation status is associated with cancer development and metastatic recurrence in patients with clear cell renal cell carcinoma. *Oncogene* 29: 5724–5728, 2010.

[54] Mendell JT. miRiad roles for the miR-17-92 cluster in development and disease. *Cell* 133: 217e222, 2008.

[55] Ciafrè SA, Galardi S, Mangiola A, Ferracin M, Liu CG, et al. Extensive modulation of a set of microRNAs in primary glioblastoma. *Biochem. Biophys. Res Commun* 334: 1351–1358, 2005.

[56] Bloomston M, Frankel WL, Petrocca F, Volinia S, Alder H, et al. MicroRNA expression patterns to differentiate pancreatic adenocarcinoma from normal pancreas and chronic pancreatitis. *JAMA* 297: 1901–1908, 2007.

[57] Iorio MV, Ferracin M, Liu CG, Veronese A, Spizzo R, et al. MicroRNA gene expression deregulation in human breast cancer. *Cancer Res* 65: 7065–7070, 2005.

[58] Schetter AJ, Leung SY, Sohn JJ, Zanetti KA, Bowman ED, et al. MicroRNA expression profiles associated with prognosis and therapeutic outcome in colon adenocarcinoma. *JAMA* 299: 425–436, 2008.

[59] Frankel LB, Christoffersen NR, Jacobsen A, Lindow M, Krogh A, et al. Programmed cell death 4 (PDCD4) is an important functional target of the microRNA miR-21 in breast cancer cells. *J. Biol Chem* 283: 1026–1033, 2008.

[60] Mitchell PS, Parkin RK, Kroh EM, Fritz BR, Wyman SK, et al. Circulating microRNAs as stable blood-based markers for cancer detection. *Proc Natl Acad Sci USA* 105: 10513–10518, 2008.

[61] Ghai V, Wang K. Recent progress toward the use of circulating microRNAs as clinical biomarkers. *Arch Toxicol* 90: 2959–2978, 2016.

[62] Chen G, Wang J, Cui Q. Could circulating miRNAs contribute to cancer therapy? *Trends Mol Med* 19: 71–73, 2013.

[63] Van Schooneveld E, Wouters MC, Van der Auwera I, Peeters DJ, Wildiers H, et al. Expression profiling of cancerous and normal breast tissues identifies microRNAs that are differentially expressed in serum from patients with (metastatic) breast cancer and healthy volunteers. *Breast Cancer Res* 14: R34, 2012.

[64] Moltzahn F, Olshen AB, Baehner L, Peek A, Fong L, et al. Microfluidic-based multiplex qRT-PCR identifies diagnostic and prognostic microRNA signatures in the sera of prostate cancer patients. *Cancer Res* 71: 550–560, 2011.

[65] Shah MY, Calin GA. The mix of two worlds: non-coding RNAs and hormones. *Nucleic Acid Ther* 23: 2–8, 2013.

[66] Tominaga N, Kosaka N, Ono M, Katsuda T, Yoshioka Y, et al. Brain metastatic cancer cells release microRNA- 181c-containing extracellular vesicles capable of destructing blood-brain barrier. *Nat Commun* 6: 6716, 2015.

[67] Carstens JL, Lovisa S, Kalluri R. Microenvironment-dependent cues trigger miRNA-regulated feedback loop to facilitate the EMT/MET switch. *J Clin Invest* 124: 1458–1460, 2014.

[68] Mitra AK, Zillhardt M, Hua Y, Tiwari P, Murmann AE, et al. MicroRNAs reprogram normal fibroblasts into cancer-associated fibroblasts in ovarian cancer. *Cancer Discov* 2: 1100–1108, 2012.

[69] Aprelikova O, Yu X, Palla J, Wei BR, John S. et al. The role of miR-31 and its target gene SATB2 in cancer-associated fibroblasts. *Cell Cycle* 9: 4387–4398, 2010.

[70] Alexander M, Hu R, Runtsch MC, Kagele DA, Mosbruger TL, et al. Exosome-delivered microRNAs modulate the inflammatory response to endotoxin. *Nat Commun* 6: 7321, 2015.

[71] Zhang B, Pan X, Cobb GP, Anderson TA. microRNAs as oncogenes and tumor suppressors. *Dev Biol* 302: 1–12, 2007.

[72] Inamura K, Ishikawa Y. MicroRNA In Lung Cancer: Novel Biomarkers and Potential Tools for Treatment. *J Clin Med* 5: E36, 2016.

[73] MacDonagh L, Gray SG, Finn SP, Cuffe S, O'Byrne KJ, Barr MP. The emerging role of microRNAs in resistance to lung cancer treatments. *Cancer Treat Rev* 41: 160–169, 2015.

[74] Bianchi F, Nicassio F, Marzi M, Belloni E, Dall'olio V. A serum circulating miRNA diagnostic test to identify asymptomatic high-risk individuals with early stage lung cancer. *EMBO Mol Med* 3:495–503, 2011.

[75] Montani F, Marzi MJ, Dezi F, Dama E, Carletti RM. miR-Test: A blood test for lung cancer early detection. *J Natl Cancer Inst* 107: djv063, 2015.

[76] Hou J, Meng F, Chan LW, Cho WC, Wong SC. Circulating Plasma MicroRNAs As Diagnostic Markers for NSCLC. *Front Genet* 7: 193, 2016.

[77] Yang Y, Hu Z, Zhou Y, Zhao G, Lei Y. The clinical use of circulating microRNAs as non-invasive diagnostic biomarkers for lung cancers. *Oncotarget* 8: 90197–90214, 2017.

[78] Võsa U, Vooder T, Kolde R, Vilo J, Metspalu A, Annilo T. Meta-analysis of microRNA expression in lung cancer. *Int J Cancer* 132: 2884-2893, 2013.

[79] Nadal E, Truini A, Nakata A, Lin J, Reddy RM. A Novel Serum 4-microRNA Signature for Lung Cancer Detection. *Sci Rep* 5: 12464, 2015.

[80] Inamura K, Togashi Y, Nomura K, Ninomiya H, Hiramatsu M. let-7 microRNA expression is reduced in bronchioloalveolar carcinoma, a non-invasive carcinoma, and is not correlated with prognosis. *Lung Cancer* 58: 392–396, 2007.

[81] Johnson CD, Esquela-Kerscher A, Stefani G, Byrom M, Kelnar K. The let-7 microRNA represses cell proliferation pathways in human cells. *Cancer Res* 67: 7713–7722, 2007.

[82] Tokumaru S, Suzuki M, Yamada H, Nagino M, Takahashi T. let-7 regulates Dicer expression and constitutes a negative feedback loop. *Carcinogenesis* 29: 2073–2077, 2008.

[83] Jin X, Chen Y, Chen H. Evaluation of Tumor-Derived Exosomal miRNA as Potential Diagnostic Biomarkers for Early-Stage Non-Small Cell Lung Cancer Using Next-Generation Sequencing. *Clin Cancer Res* 23: 5311-5319, 2017.

[84] Takaoka Y, Shimizu Y, Hasegawa H. Forced expression of miR-143 represses ERK5/c-Myc and p68/p72 signaling in concert with miR-145 in gut tumors of Apc(Min) mice. *PLoS One* 7: e42137, 2012.

[85] Bian HB, Pan X, Yang JS. Upregulation of microRNA-451 increases cisplatin sensitivity of non-small cell lung cancer cell line (A549). *J Exp Clin Cancer Res* 30: 20, 2011.

[86] Mees ST, Mardin WA, Sielker S. Involvement of CD40 targeting miR-224 and miR-486 on the progression of pancreatic ductal adenocarcinomas. *Ann Surg Oncol* 16: 2339–2350, 2009.

[87] Zhang JG, Wang JJ, Zhao F, Liu Q, Jiang K, Yang GH. MicroRNA-21 (miR-21) represses tumor suppressor PTEN and promotes growth and invasion in non-small cell lung cancer (NSCLC). *Clin Chim Acta* 411: 846–852, 2010.

[88] Wang K, Chen M, Wu W. Analysis of microRNA (miRNA) expression profiles reveals 11 key biomarkers associated with non-small cell lung cancer. *World J Surg* Oncol 15: 175, 2017.

[89] Geng Q, Fan T, Zhang B, Wang W, Xu Y. Five microRNAs in plasma as novel biomarkers for screening of early-stage non-small cell lung cancer. *Respir Res* 15: 149, 2014.

[90] Garofalo M, Di Leva G, Romano G, Nuovo G, Suh SS. miR-221&222 regulate TRAIL resistance and enhance tumorigenicity through PTEN and TIMP3 downregulation. *Cancer Cell* 16: 498–509, 2009.

[91] Yu Y, Zuo J, Tan Q, Zar Thin K, Li P. Plasma miR-92a-2 as a biomarker for small cell lung cancer. *Cancer Biomark* 18: 319-327, 2017.

[92] Ranade AR, Cherba D, Sridhar S, Richardson P, Webb C. MicroRNA 92a-2*: a biomarker predictive for chemoresistance and prognostic for survival in patients with small cell lung cancer. *J Thorac Oncol* 5: 1273-2178, 2010.

[93] Hu L, Ai J, Long H, Liu W, Wang X, et al. Integrative microRNA and gene profiling data analysis reveals novel biomarkers and mechanisms for lung cancer. *Oncotarget* 7: 8441-8454, 2016.

[94] Xie D, Shang C, Zhang H, Guo Y, Tong X. Up-regulation of miR-9 target CBX7 to regulate invasion ability of bladder transitional cell carcinoma. *Med Sci Monit* 21: 225–230, 2015.

[95] Li W, He F. Monocyte to macrophage differentiation-associated (MMD) targeted by miR-140-5p regulates tumor growth in non-small cell lung cancer. *Biochem Biophys Res Commun* 450: 844–850, 2014.

[96] Polley E, Kunkel M, Evans D, Silvers T, Delosh R, et al. Small Cell Lung Cancer Screen of Oncology Drugs, Investigational Agents, and Gene and microRNA Expression. *J Natl Cancer Inst* 108: djw122, 2016.

[97] Wang RT, Xu M, Xu CX, Song ZG, Jin H. Decreased expression of miR216a contributes to non-small-cell lung cancer progression. *Clin Cancer Res* 20: 4705–4716, 2014.

[98] Lee JH, Voortman J, Dingemans AM, Voeller DM, et al. MicroRNA Expression and Clinical Outcome of Small Cell Lung Cancer. *PLoS One* 6: e21300, 2011.

[99] Fadejeva I, Olschewski H, Hrzenjak A. MicroRNAs as regulators of cisplatin-resistance in non-small cell lung carcinomas. *Oncotarget* 8: 115754-115773, 2017.

[100] Sun F, Wan M, Xu X, Gao B, Zhou Y. Crosstalk between miR-34a and Notch Signaling Promotes Differentiation in Apical Papilla Stem Cells (SCAPs) *J Dent Res* 93:589–595, 2014.

[101] Nadal E, Zhong J, Lin J, Reddy RM, Ramnath N. A MicroRNA cluster at 14q32 drives aggressive lung adenocarcinoma. *Clin Cancer Res* 20:3107–3117, 2014.

[102] Mizuno K, Mataki H, Arai T, Okato A, Kamikawaji K, et al. The microRNA expression signature of small cell lung cancer: tumor suppressors of miR-27a-5p and miR-34b-3p and their targeted oncogenes. *J Hum Genet* 62: 671-678, 2017.

[103] Izzotti A, Calin GA, Arrigo P, Steele VE, Croce CM, De Flora S. Downregulation of microRNA expression in the lungs of rats exposed to cigarette smoke. *Faseb J* 23: 806–812, 2009.

[104] Schembri F, Sridhar S, Perdomo C, Gustafson AM, Zhang X, et al. MicroRNAs as modulators of smoking-induced gene expression changes in human airway epithelium. *Proc Natl Acad Sci USA* 106: 2319–2324, 2009.

[105] Song L, Li D, Gu Y, Wen ZM, Jie J, et al. MicroRNA-126 targeting PIK3R2 inhibits NSCLC A549 cell proliferation, migration, and invasion by regulation of PTEN/PI3K/AKT pathway. *Clin Lung Cancer* 17: e65–e75, 2016.

[106] Garajova I, Funel N, Fiorentino M, Agostini V, Ferracin M, et al. MicroRNA profiling of primary pulmonary enteric adenocarcinoma in members from the same family reveals some similarities to pancreatic adenocarcinoma-a step towards personalized therapy. *Clin Genet* 7: 129, 2015.

[107] Xi S, Xu H, Shan J, Tao Y, Hong JA, et al. Cigarette smoke mediates epigenetic repression of miR-487b during pulmonary carcinogenesis. *J Clin Invest* 123: 1241–1261, 2013.

[108] Tellez CS, Juri DE, Do K, Bernauer AM, Thomas CL, et al. EMT and stem cell-like properties associated with miR-205 and miR-200 epigenetic silencing are early

manifestations during carcinogen-induced transformation of human lung epithelial cells. *Cancer Res* 71: 3087–3097, 2011.

[109] Mancuso G, Bovio E, Rena O, Rrapaj E, Mercalli F, et al. Prognostic impact of a 3-MicroRNA signature in cytological samples of small cell lung cancer. *Cancer Cytopathol* 124: 621-629, 2016.

[110] Li D, Zhang Y, Xie Y, Xiang J, Zhu Y, Yang J. Enhanced tumor suppression by adenoviral PTEN gene therapy combined with cisplatin chemotherapy in small-cell lung cancer. *Cancer Gene Ther* 20: 251-259, 2013.

[111] Xie QH, He XX, Chang Y. MiR-192 inhibits nucleotide excision repair by targeting ERCC3 and ERCC4 in HepG2.2.15cells. *Biochem Biophys Res Commun* 410: 440-445, 2011.

[112] Takeyama Y, Sato M, Horio M, Hase T, Yoshida K, et al. Knockdown of ZEB1, a master epithelial-to-mesenchymal transition (EMT) gene, suppresses anchorage-independent cell growth of lung cancer cells. *Cancer Lett* 296: 216–224, 2010.

[113] Tejero R, Navarro A, Campayo M, Viñolas N, Marrades RM, et al. miR-141 and miR-200c as markers of overall survival in early stage non–small cell lung cancer adenocarcinoma. *PLoS One* 9: e101899, 2014.

[114] Shi ZM, Wang L, Shen H, Jiang CF, Ge X, et al. Downregulation of miR-218 contributes to epithelial-mesenchymal transition and tumor metastasis in lung cancer by targeting Slug/ZEB2 signaling. *Oncogene* 36: 2577–2588, 2017.

[115] Chen Y, Lu L, Feng B, Han S, Cui S, et al. Non-coding RNAs as emerging regulators of epithelial to mesenchymal transition in non-small cell lung cancer. *Oncotarget* 8:36787–36799, 2017.

[116] Lei L, Huang Y, Gong W. miR-205 promotes the growth, metastasis and chemoresistance of NSCLC cells by targeting PTEN. *Oncol Rep* 30: 2897-902, 2013.

[117] Lebanony D, Benjamin H, Gilad S, Ezagouri M, Dov A, et al. Diagnostic assay based on hsa-miR-205 expression distinguishes squamous from nonsquamous non-small-cell lung carcinoma. *J Clin Oncol* 27: 2030–2037, 2009.

[118] Hamamoto J, Soejima K, Yoda S, Naoki K, Nakayama S, et al. Identification of microRNAs differentially expressed between lung squamous cell carcinoma and lung adenocarcinoma. *Mol Med Rep* 8: 456–462, 2013.

[119] Bishop JA, Benjamin H, Cholakh H, Chajut A, Clark DP, Westra WH. Accurate classification of non-small cell lung carcinoma using a novel microRNA-based approach. *Clin*

[120] Mao G, Liu Y, Fang X, Liu Y, Fang L, et al. Tumor-derived microRNA-494 promotes angiogenesis in non-small cell lung cancer. *Angiogenesis* 18: 373–382, 2015.

[121] Van Roosbroeck K, Fanini F, Setoyama T, Ivan C, Rodriguez-Aguayo C, et al. Combining anti-miR-155 with chemotherapy for the treatment of lung cancers. *Clin Cancer Res* 23: 2891-2904, 2017.

[122] Xue X, Liu Y, Wang Y, Meng M, Wang K, et al. MiR-21 and MiR-155 promote non-small cell lung cancer progression by downregulating SOCS1, SOCS6, and PTEN. *Oncotarget* 7: 84508–84519, 2016.

[123] Zang YS, Zhong YF, Fang Z, Li B, An J. MiR-155 inhibits the sensitivity of lung cancer cells to cisplatin via negative regulation of Apaf-1 expression. *Cancer Gene Ther* 19: 773–778, 2012.

[124] Galluzzi L, Morselli E, Vitale I, Kepp O, Senovilla L, et al. miR-181a and miR-630 regulate cisplatin-induced cancer cell death. *Cancer Res* 70: 1793–1803, 2010.

[125] Li Y, Li L, Guan Y, Liu X, Meng Q, Guo Q. MiR-92b regulates the cell growth, cisplatin chemosensitivity of A549 non small cell lung cancer cell line and target PTEN. *Biochem Biophys Res Commun* 440: 604–610, 2013.

[126] Zhou Y, Liang H, Liao Z, Wang Y, Hu X, et al. miR-203 enhances let-7 biogenesis by targeting LIN28B to suppress tumor growth in lung cancer. *Sci Rep* 7:42680, 2017.

[127] He L, He X, Lim LP, de Stanchina E, Xuan Z, et al. A microRNA component of the p53 tumour suppressor network. *Nature* 447: 1130–1134, 2007.

[128] Inamura K. Major Tumor Suppressor and Oncogenic Non-Coding RNAs: Clinical Relevance in Lung Cancer. *Cells* 6: pii: E12, 2017.

[129] Yu M, Liang H, Fu Z, Wang X, Liao Z, et al. BAP1 suppresses lung cancer progression and is inhibited by miR-31. *Oncotarget* 7: 13742–13753, 2016.

[130] Xu H, Ma J, Zheng J, Wu J, Qu C, et al. MiR-31 Functions as a Tumor Suppressor in Lung Adenocarcinoma Mainly by Targeting HuR. *Clin Lab* 62: 711–718, 2016.

[131] Okudela K, Suzuki T, Umeda S, Tateishi Y, Mitsui H, et al. A comprehensive search for microRNAs with expression profiles modulated by oncogenic KRAS: Potential involvement of miR-31 in lung carcinogenesis. *Oncol Rep* 32:1374–1384, 2014.

[132] Deng P, Wang L, Cui L, Feng K, Liu F, Du X, Tong W, Nie X, Ji W, Weining S. Global Identification of MicroRNAs and Their Targets in Barley under Salinity Stress. *PLoS One* 10(9): e0137990, 2015.

[133] Shimomura A, Shiino S, Kawauchi J, Takizawa S, Sakamoto H, et al. Novel combination of serum microRNA for detecting breast cancer in the early stage. *Cancer Sci* 107(3): 326–334, 2016.

[134] Heneghan HM, Miller N, Lowery AJ, et al. Circulating micrornas as novel minimally invasive biomarkers for breast cancer. *Ann Surg* 251:499–505, 2010.

[135] Li S, Yang X, Yang J, Zhen J and Zhang D. Serum microRNA-21 as a potential diagnostic biomarker for breast cancer: a systematic review and meta-analysis. *Clin Exp Med* 16: 29-35, 2016.

[136] Asaga S, Kuo C, Nguyen T, Terpenning M, Giuliano AE, et al. Direct serum assay for microRNA-21 concentrations in early and advanced breast cancer. *Clin Chem* 57: 84-91, 2011.

[137] Ng EK, Li R, Shin VY, Jin HC, Leung CP, et al. Circulating microRNAs as specific biomarkers for breast cancer detection. *PLoS ONE* 8: 53141, 2013.

[138] Jiang S, Zhang HW, Lu MH, He XH, Li Y, et al. MicroRNA-155 functions as an OncomiR in breast cancer by targeting the suppressor of cytokine signaling 1 gene. *Cancer Research* 70: 3119–3127, 2010.

[139] Zhu W, Qin W, Atasoy U, Sauter ER. Circulating microRNAs in breast cancer and healthy subjects. *BMC Research Notes* 19: 89, 2009.

[140] Sun Y, Wang M, Lin G, Sun S, Li X, et al. Serum microRNA-155 as a potential biomarker to track disease in breast cancer. *PLoS ONE* 7: e47003, 2012.

[141] Mar Aguilar F, Mendoza Ramírez JA, Malagón Santiago I, Espino Silva P K, Santuario Facio SK, et al. Serum circulating microRNA profiling for identification of potential breast cancer biomarkers. *Disease Markers* 34: 163–169, 2013.

[142] Hagrass HA, Sharaf S, Pasha HF, Tantawy E A, Mohamed RH, et al. Circulating microRNAs-a new horizon in molecular diagnosis of breast cancer. *Genes & Cancer* 6: 281-287, 2015.

[143] Heneghan HM, Miller N, Kelly R, Newell J, Kerin M J. Systemic miRNA-195 differentiates breast cancer from other malignancies and is a potential biomarker for detecting noninvasive and early stage disease. *The Oncologist* 15: 673–682, 2010.

[144] Zhao H, Shen J, Medico L, Wang D, Ambrosone CB, Liu S. A pilot study of circulating miRNAs as potential biomarkers of early stage breast cancer. *PLoS ONE* 5: e13735, 2010.

[145] Wu X, Somlo G, Yu Y, Palomares MR, Li AX, et al. De novo sequencing of circulating miRNAs identifies novel markers predicting clinical outcome of locally advanced breast cancer. *J Transl Med* 10: 42, 2012.

[146] Chen W, Cai F, Zhang B, Barekati Z, Zhong XY. The level of circulating miRNA-10b and miRNA-373 in detecting lymph node metastasis of breast cancer: Potential biomarkers. *Tumor Biology* 34: 455–462, 2013.

[147] Cuk K, Zucknick M, Madhavan D, Schott S, Golatta M, et al. Plasma microRNA panel for minimally invasive detection of breast cancer. *PLoS One* 8: e76729, 2013.

[148] Zheng R, Pan L, Gao J, Ye X, Chen L, et al. Prognostic value of miR-106b expression in breast cancer patients. *Surg Res* 195: 158-65, 2015.

[149] Camps C, Buffa FM, Colella S, Moore J, Sotiriou C, et al. Hsa-miR-210 Is induced by hypoxia and is an independent prognostic factor in breast cancer. *Clinical Cancer Research* 14: 1340–1348, 2008.

[150] Iorio MV, Casalini P, Tagliabue E, Ménard S, Croce C M. MicroRNA profiling as a tool to understand prognosis, therapy response and resistance in breast cancer. *Eur J Cancer* 44: 2753–2759, 2008.

[151] Wang F, Zheng Z, Guo J, Ding X. Correlation and quantitation of microRNA aberrant expression in tissues and sera from patients with breast tumor. *Gynecol Oncol* 119: 586–593, 2010.

[152] Volinia S, Calin GA, Liu CG, Ambs S, Cimmino, et al. A microRNA expression signature of human solid tumors defines cancer gene targets. *Proc Natl Acad Sci USA* 103, 2257–2261, 2006.

[153] Quesne JL, Jones J, Warren J, Dawson SJ, Ali HR, et al. Biological and prognostic associations of miR-205 and let-7b in breast cancer revealed by in situ hybridization analysis of micro-RNA expression in arrays of archival tumour tissue. *J Pathology* 227: 306–314, 2012.

[154] Foekens JA, Sieuwerts AM, Smid M, Look MP, de Weerd V, et al. Four miRNAs associated with aggressiveness of lymph node-negative, estrogen receptor-positive human breast cancer. *Proc Natl Acad Sci USA* 105: 13021–13026, 2008.

[155] Xu H, Fei D, Zong S, Fan Z. MicroRNA-154 inhibits growth and invasion of breast cancer cells through targeting E2F5. *Am J Transl Res* 8: 2620-630, 2016.

[156] Wang S, Li H, Wang J, Wang D. Retracted article: Expression of microRNA-497 and its prognostic significance in human breast cancer. *Diagn Pathol* 8: 172, 2013.

[157] Porkka KP, Pfeiffer MJ, Waltering KK, Vessella RL, Tammela TL, et al. MicroRNA expression profiling in prostate cancer. *Cancer Res* 67, 6130-6135, 2007.

[158] Ren Q, Liang J, Wei J, Basturk O, Wang J, et al. Epithelial and stromal expression of miRNAs during prostate cancer progression. *Am J Transl Res* 6: 329-339, 2014.

[159] Kanwal R, Plaga AR, Liu X, Shukla GC, Gupta S. MicroRNAs in prostate cancer: Functional role as biomarkers. *Cancer Lett 407*: 9-20, 2017.

[160] Hanahan D, Weinberg RA. Hallmarks of cancer: the next generation. *Cell* 144: 646-674, 2011.

[161] Da Silva HB, Amaral EP, Nolasco EL, De Victo NC, Atique R, et al. Dissecting major signaling pathways throughout the development of prostate cancer, *Prostate Cancer* 2013, 920612, 2013.

[162] Roa W, Brunet B, Guo L, Amanie J, Fairchild A, et al. Identification of a new microRNA expression profile as a potential cancer screening tool. *Clin Investig Med Med Clin Exp* 33: E124–E132, 2010.

[163] Kelly BD, Miller N, Sweeney KJ, Durkan GC, Rogers E, et al. A circulating microRNA signature as a biomarker for prostate cancer in a high risk group. *J Clin Med* 4: 1369–1379, 2015.

[164] Kojima S, Goto Y, Naya Y. The roles of microRNAs in the progression of castration-resistant prostate cancer. *J Hum Genet* 62: 25-31, 2017.

[165] Yang Y, Guo YX, Shao ZQ. miR-21 targets and inhibits tumor suppressor gene PTEN to promote prostate cancer cell proliferation and invasion: an experimental study. *Asian Pac J Trop Med* 10: 87-91, 2017.

[166] Zheng C, Yinghao S, Li J. MiR-221 expression affects invasion potential of human prostate carcinoma cell lines by targeting DVL2. *Med Oncol* 29: 815-822, 2012.

[167] Ouyang Y, Gao P, Zhu B, Chen X, Lin F, et al. Downregulation of microRNA-429 inhibits cell proliferation by targeting p27Kip1 in human prostate cancer cells. *Mol Med Rep* 11: 1435-1441, 2015.

[168] Tong SJ, Liu J, Wang X, Qu LX. microRNA-181 promotes prostate cancer cell proliferation by regulating DAX-1 expression. *Exp Ther Med* 8: 1296-1300, 2014.

[169] Hsu TI, Hsu CH, Lee KH, Lin JT, Chen CS, et al. MicroRNA-18a is elevated in prostate cancer and promotes tumorigenesis through suppressing STK4 in vitro and in vivo. *Oncogenesis* 3: e99, 2014.

[170] Srivastava A, Suy S, Collins SP, Kumar D. Circulating MicroRNA as biomarkers: an update in prostate Cancer. *Mol Cell Pharmacol* 3: 115-124, 2011.

[171] Massillo C, Dalton GN, Farré PL, De Luca P, De Siervi A. Implications of microRNA dysregulation in the development of prostate cancer. *Reproduction* 154: R81-R97, 2017.

[172] Lin S, Gregory RI. MicroRNA biogenesis pathways in cancer. *Nat Rev Cancer* 15 321–333, 2015.

[173] Rane JK, Erb HH, Nappo G, Mann VM, Simms MS, et al. Inhibition of the glucocorticoid receptor results in an enhanced miR-99a/100-mediated radiation response in stem-like cells from human prostate cancers. *Oncotarget* 7: 51965–51980, 2016.

[174] Ribas J, Lupold SE. The transcriptional regulation of miR-21, its multiple transcripts, and their implication in prostate Cancer. *Cell Cycle* 9: 923-929, 2010.

[175] Mishra S, Deng JJ, Gowda PS, Rao MK, Lin CL, et al. Androgen Receptor and MicroRNA-21 axis down-regulates transforming growth factor beta receptor II (TGFBR2) expression in Prostate Cancer. *Oncogene* 33: 4097-4106, 2014.

[176] Vlaeminck-Guillem V. When Prostate Cancer Circulates in the Bloodstream. *Diagnostics* (Basel) 5: 428-474, 2015.

[177] Stuopelytè K, Daniùnaitè K, Jankevicius F, Jarmalaitè S. Detection of miRNAs in urine of prostate cancer patients. *Medicina* 52:116–124, 2016.

[178] Lee DJ, Recabal P, Sjoberg DD, Thong A, Lee JK, et al. Comparative effectiveness of targeted prostate biopsy using magnetic resonance fusion software and visual targeting: A prospective study. *J Urol* 196: 697–702, 2016.

[179] Karatas OF, Wang J, Shao L, Ozen M, Zhang Y, et al. miR-33a is a tumor suppressor microRNA that is decreased in prostate Cancer. *Oncotarget* 8: 60243-60256, 2017.

[180] Waltering KK, Porkka KP, Jalava SE, Urbanucci A, Kohonen PJ, et al. Androgen regulation of micro-RNAs in prostate cancer. *Prostate* 71: 604–614, 2011.

[181] Huang S, Wa Q, Pan J, Peng X, Ren D, et al. Downregulation of miR-141-3p promotes bone metastasis via activating NF-κB signaling in prostate cancer. *J Exp Clin Cancer Res* 36: 173, 2017.

[182] Xu S, Yi XM, Zhang ZY, Ge JP, Zhou WQ. miR-129 predicts prognosis and inhibits cell growth in human prostate carcinoma *Mol Med Rep* 14: 5025-5032, 2016.

[183] Lo UG, Yang D, Hsieh JT, The role of microRNAs in prostate cancer progression, *Transl Androl Urol* 2: 228-241, 2013.

[184] Jalava SE, Urbanucci A, Latonen L, Waltering KK, Sahu B, et al. Androgen-regulated miR-32 targets BTG2 and is overexpressed in castration-resistant prostate cancer. *Oncogene* 31: 4460-4471, 2012.

[185] Sun T, Yang M, Chen S, Balk S, Pomerantz M, et al. The altered expression of MiR-221/-222 and MiR-23b/-27b is associated with the development of human castration resistant prostate cancer. *Prostate* 72: 1093-1103, 2012.

[186] Ozen M, Creighton CJ, Ozdemir M, Ittmann M. Widespread deregulation of microRNA expression in human prostate cancer, *Oncogene* 27: 1788-1193, 2008.

[187] Shao N, Ma G, Zhang J, Zhu W. miR-221-5p enhances cell proliferation and metastasis through post-transcriptional regulation of SOCS1 in human prostate cancer. *BMC Urol* 18:14, 2018.

[188] Greco F, Inferrera A, La Rocca R, Navarra M, Casciaro M, et al. The Potential Role of MicroRNAs as Biomarkers in Benign Prostatic Hyperplasia: A Systematic Review and Meta-analysis. *Eur Urol Focus* pii: S2405-4569(18)30009-9, 2018.

[189] Sun T, Wang Q, Balk S, Brown M, Lee GS, et al. The role of microRNA-221 and microRNA-222 in androgen-independent prostate cancer cell lines. *Cancer Res* 69: 3356-3363, 2009.

[190] Bucay N, Bhagirath D, Sekhon K, Yang T, Fukuhara S, et al. A novel microRNA regulator of prostate cancer epithelial-mesenchymal transition. *Cell Death Differ* 24:1263-1274, 2017.

[191] Bucay N, Sekhon K, Yang T, Majid S, Shahryari V, et al. MicroRNA-383 located in frequently deleted chromosomal locus 8p22 regulates CD44 in prostate cancer. *Oncogene* 36: 2667–2679, 2017.

[192] Hsieh IS, Chang KC, Tsai YT, Ke JY, Lu PJ, et al. MicroRNA-320 suppresses the stem cell-like characteristics of prostate cancer cells by downregulating the Wnt/beta-catenin signaling pathway. *Carcinogenesis* 34: 530-538, 2103.

[193] Saini S, Majid S, Shahryari V, Tabatabai ZL, Arora S, et al. Regulation of SRC kinases by microRNA-3607 located in a frequently deleted locus in prostate cancer. *Mol Cancer Ther* 13: 1952-1963, 2014.

[194] Fizazi K. The role of Src in prostate cancer. *Ann Oncol* 18: 1765–1773, 2007.

[195] Feng S, Qian X, Li H, Zhang X. Combinations of elevated tissue miRNA-17-92 cluster expression and serum prostate-specific antigen as potential diagnostic biomarkers for prostate cancer. *Oncol Lett* 14: 6943-6949, 2017.

[196] Farran B, Dyson G, Craig D, Dombkowski A, Beebe-Dimmer JL, et al. A study of circulating microRNAs identifies a new potential biomarker panel to distinguish aggressive prostate cancer. *Carcinogenesis* 39: 556-561, 2018.

[197] Sapre, N. Hong MKH, Macintyre G, Lewis H, et al. Curated microRNAs in urine and blood fail to validate as predictive biomarkers for high-risk prostate cancer. *PLoS One* 9: e91729, 2014.

[198] Luan Y, Chen M, Zhou L. MiR-17 targets PTEN and facilitates glial scar formation after spinal cord injuries via the PI3K/Akt/mTOR pathway. *Brain Res Bull* 128: 68–75, 2017.

[199] Taguchi A, Yanagisawa K, Tanaka M, Cao K, Matsuyama Y, et al. Identification of hypoxia-inducible factor-1 alpha as a novel target for miR-17-92 microRNA cluster. *Cancer Res* 68: 5540–5545, 2008.

[200] Li S, Wang C, Yu X, Wu H, Hu J, Wang S, Ye Z. miR-3619-5p inhibits prostate cancer cell growth by activating CDKN1A expression. *Oncol Rep* 37: 241–248, 2017.

[201] Wise HM, Hermida MA, Leslie NR. Prostate cancer, PI3K, PTEN and prognosis. *Clin Sci* (Lond.) 131: 197–210, 2017.

[202] Hünten S, Kaller M, Drepper F, Oeljeklaus S, Bonfert T, et al. p53-regulated networks of protein, mRNA, miRNA, and lncRNA expression revealed by integrated pulsed stable isotope labeling with amino acids in cell culture (pSILAC) and next generation sequencing (NGS) analyses. *Mol Cell Proteomics* 14: 2609–2629, 2015.

[203] Wu SY, Rupaimoole R, Shen F, Pradeep S, Pecot CV, et al. A miR-192-EGR1-HOXB9 regulatory network controls the angiogenic switch in cancer. *Nat Commun* 7: 11169, 2016.

[204] Fujita A. Gomes LR, Sato JR, Yamaguchi R, et al. Multivariate gene expression analysis reveals functional connectivity changes between normal/tumoral prostates. *BMC Syst Biol* 2: 106, 2008.

[205] Lin D, Cui F, Bu Q, Yan C. The expression and clinical significance of GTP-binding RAS-like 3 (ARHI) and microRNA 221 and 222 in prostate cancer. *J Int Med Res* 39:1870-1875, 2011.

[206] Kojima S, Enokida H, Yoshino H, Itesako T, Chiyomaru T, et al. The tumor-suppressive microRNA-143/145 cluster inhibits cell migration and invasion by targeting GOLM1 in prostate cancer. *J Hum Genet* 59: 78-87, 2014.

[207] Avgeris M, Stravodimos K, Fragoulis EG, Scorilas A. The loss of the tumour-suppressor miR-145 results in the shorter disease-free survival of prostate cancer patients. *Br J Cancer* 108: 2573–2581, 2013.

[208] Szczyrba J, Loprich E, Wach S, Jung V, Unteregger G, et al. The microRNA profile of prostate carcinoma obtained by deep sequencing. *Mol Cancer Res* 8: 529–538, 2010.

[209] Rodríguez M, Bajo-Santos C, Hessvik NP, Lorenz S, Fromm B, et al. Identification of non-invasive miRNAs biomarkers for prostate cancer by deep sequencing analysis of urinary exosomes. *Mol Cancer* 16: 156, 2017.

[210] Zaman MS, Chen Y, Deng G, Shahryari V, Suh SO, et al. The functional significance of microRNA-145 in prostate cancer. *Br J Cancer* 103: 256–264, 2010.

[211] Chen X, Gong J, Zeng H, Chen N, Huang R, et al.. MicroRNA145 targets BNIP3 and suppresses prostate cancer progression. *Cancer Res* 70: 2728–2738, 2010.

[212] Kang SG, Ha YR, Kim SJ, Kang SH, Park HS, et al. Do microRNA 96, 145 and 221 expressions really aid in the prognosis of prostate carcinoma? *Asian J Androl* 14:752–757, 2012.

[213] Hu B, Jin X, Wang J. MicroRNA-212 targets mitogen-activated protein kinase 1 to inhibit proliferation and invasion of prostate cancer cells. *Oncol Res* 2018 [ahead of print].

[214] Chen L, Cao H, Feng Y. MiR-199a suppresses prostate cancer paclitaxel resistance by targeting YES1. *World J Urol* 36: 357-365, 2018.

[215] Liu L, Yang J, Zhu X, Li D, Lv Z, et al. Long noncoding RNA H19 competitively binds miR-17-5p to regulate YES1 expression in thyroid cancer. *FEBS J* 283:2326–2339, 2016.

[216] Puhr M, Hoefer J, Schäfer G, Erb HHH, Oh SJ, et al. Epithelial-to-mesenchymal transition leads to docetaxel resistance in prostate cancer and is mediated by reduced expression of miR-200c and miR-205. *Am J Pathol* 181: 2188–2201, 2012.

[217] Wach S, Nolte E, Szczyrba J, Stöhr R, Hartmann A, et al. MicroRNA profiles of prostate carcinoma detected by multiplatform microRNA screening. *Int J Cancer* 130: 611–621, 2012.

[218] Cheng HH, Mitchell PS, Kroh EM, Dowell AE, Chéry L, et al. Circulating microRNA Profiling Identifies a Subset of Metastatic Prostate Cancer Patients with Evidence of Cancer-Associated Hypoxia. *PLoS One* 8: e69239, 2013.

[219] Souza MF, Kuasne H, Barros-Filho MC, Cilião HL, Marchi FA, et al. Circulating mRNAs and miRNAs as candidate markers for the diagnosis and prognosis of prostate cancer. *PLoS One* 12: e0184094, 2017.

[220] Lu B, Asara JM, Sanda MG, Arredouani MS. The Role of the Transcription Factor SIM2 in Prostate Cancer. *PLoS One* 6: e28837, 2011.

[221] Halvorsen OJ, Rostad K, Øyan AM, Puntervoll H, Bø TH, et al. Increased expression of SIM2-s protein is a novel marker of aggressive prostate cancer. *Clin Cancer Res* 13: 892–897, 2007.

[222] Bonci D, Coppola V, Patrizii M, Addario A, Cannistraci, FA, et al. A microRNA code for prostate cancer metastasis. *Oncogene* 35: 1180-1192, 2016.

[223] Mott JL, Kobayashi S, Bronk SF, Goresmir GJ. Mir-29 regulates Mcl-1 protein expression and apoptosis. *Oncogene* 26: 6133-6140, 2007.

[224] Yamada Y, Nishikawa R, Kato M, Okato A, Arai T, et al. Regulation of HMGB3 by antitumor miR-205-5p inhibits cancer cell aggressiveness and is involved in prostate cancer pathogenesis. *J Hum Genet* 63: 195-205, 2018.

[225] Nishikawa R, Goto Y, Kurozumi A, Matsushita R, Enokida H, et al. MicroRNA-205 inhibits cancer cell migration and invasion via modulation of centromere protein F regulating pathways in prostate cancer. *Int J Urol* 22: 867–877, 2015.

[226] Chen W, Kong KK, Xu XK, Chen C, Li H, et al. Downregulation of miR-205 is associated with glioblastoma cell migration, invasion, and the epithelial-mesenchymal transition, by targeting ZEB1 via the Akt/mTOR signaling pathway. *Int J Oncol* 52: 485-495, 2018.

[227] Verdoodt B, Neid M, Vogt M, Kuhn V, Liffers ST, et al. MicroRNA-205, a novel regulator of the anti-apoptotic protein Bcl2, is downregulated in prostate Cancer. *Int J Oncol* 43: 307-314, 2013.

[228] Schaefer A, Jung M, Mollenkopf HJ, Wagner I, Stephan C, et al. Diagnostic and prognostic implications of microRNA profiling in prostate carcinoma. *Int J Cancer* 126: 1166–1176, 2010.

[229] Srivastava A, Goldberger H, Dimtchev A, Ramalinga M, Chijioke J, et al. MicroRNA profiling in prostate cancer–the diagnostic potential of urinary miR-205 and miR-214. *PLoS One* 8: e76994, 2013.

[230] Rajendiran S, Parwani AV, Hare RJ, Dasgupta S, Roby RK, et al. MicroRNA-940 suppresses prostate cancer migration and invasion by regulating MIEN1. *Mol Cancer* 13: 250, 2014.

[231] Sikand K, Slaibi JE, Singh R, Slane SD, Shukla GC. miR 488* inhibits androgen receptor expression in prostate carcinoma cells. *Int J Cancer* 129: 810-819, 2011.

[232] Kroiss A, Vincent S, Decaussin-Petrucci M, Meugnier E, Viallet J, et al., Androgen-regulated microRNA-135a decreases prostate cancer cell migration and invasion through downregulating ROCK1 and ROCK2. *Oncogene* 34: 2846-2855, 2015.

[233] Dang CV. Enigmatic MYC conducts an unfolding systems biology symphony. *Genes Cancer* 1: 526–531, 2010.

[234] Tao J, Zhao X, Tao J. c-MYC–miRNA circuitry: a central regulator of aggressive B-cell malignancies. *Cell Cycle* 13: 191–198, 2014.

[235] Barros-Silva D, Costa-Pinheiro P, Duarte H, Sousa EJ, Evangelista AF, et al. MicroRNA-27a-5p regulation by promoter methylation and *MYC* signaling in prostate carcinogenesis. *Cell Death Dis* 9: 167, 2018.

[236] Fletcher CE, Dart DA, Sita-Lumsden A, Cheng H, Rennie PS, Bevan CL. Androgen-regulated processing of the oncomir miR-27a, which targets Prohibitin in prostate cancer. *Hum Mol Genet* 21: 3112-3127, 2012.

[237] Wang J, Ye H, Zhang D, Hu Y, Yu X, Wang L, et al. MicroRNA-410-5p as a potential serum biomarker for the diagnosis of prostate cancer. *Cancer Cell International* 16: 12, 2016.

[238] Musumeci M, Coppola V, Addario A, Patrizii M, Maugeri-Sacca M, et al. Control of tumor and microenvironment cross-talk by miR-15a and miR-16 in prostate cancer. *Oncogene* 30: 4231-4242, 2011.

[239] Endzelins E, Melne V, Kalnina Z, Lietuvietis V, Riekstina U, et al. Diagnostic, prognostic and predictive value of cell-free miRNAs in prostate cancer: a systematic review. *Mol Cancer* 15: 41, 2016.

[240] Tian B, Huo N, Li M, Li Y, He Z. Let-7a and its target, insulin-like growth factor 1 receptor, are differentially expressed in recurrent prostate cancer. *Int J Mol Med* 36: 1409-1416, 2015.

[241] Varambally S, Cao Q, Mani RS, Shankar S, Wang X, et al. Genomic loss of microRNA-101 leads to overexpression of histone methyltransferase EZH2 in cancer. *Science* 322: 1695-1699, 2008.

[242] Majid S, Dar AA, Saini S, Shahryari V, Arora S, et al. miRNA-34b inhibits prostate cancer through demethylation, active chromatin modifications, and AKT pathways. *Clin Cancer Res* 19: 73-84, 2013.

[243] Mavridis K, Stravodimos K, Scorilas A. Downregulation and prognostic performance of microRNA 224 expression in prostate cancer. *Clin Chem* 59: 261-269, 2013.

[244] Chiu SC, Chung HY, Cho DY, Chan TM, Liu MC, et al. Therapeutic potential of microRNA let-7: tumor suppression or impeding normal stemness. *Cell Transplant* 23: 459-469, 2014.

[245] Tang G, Du R, Tang Z, Kuang Y. MiRNALet-7a mediates prostate cancer PC-3 cell invasion, migration by inducing epithelial-mesenchymal transition through CCR7/MAPK pathway. *J Cell Biochem* 119: 3725-3731, 2018.

[246] Yaman Agaoglu F, Kovancilar M, Dizdar Y, Darendeliler E, Holdenrieder S, et al. Investigation of miR-21, miR-141, and miR-221 in blood circulation of patients with prostate cancer. *Tumour Biol* 32: 583-588, 2011.

[247] Stuopelyte K, Daniunaite K, Jankevicius F, Jarmalaite S. Detection of miRNAs in urine of prostate cancer patients. *Med* (Kaunas) 52: 116-124, 2016.

[248] Wiggins JF, Ruffino L, Kelnar K, Omotola M, Patrawala L, et al. Development of a lung cancer therapeutic based on the tumor suppressor microRNA-34. *Cancer Res* 70: 5923-5930, 2010.

[249] Zheng Q, Peskoe SB, Ribas J, Rafiqi F, Kudrolli T, et al. Investigation of miR-21, miR-141, and miR-221 expression levels in prostate adenocarcinoma for associated risk of recurrence after radical prostatectomy. *Prostate* 74: 1655-1662, 2014.

[250] Lichner Z, Ding Q, Samaan Q, Saleh C, Nasser A, et al. miRNAs dysregulated in association with Gleason grade regulate extracellular matrix, cytoskeleton and androgen receptor pathways. *J Pathol* 237: 226-237, 2015.

[251] Lichner Z, Fendler A, Saleh C, Nasser AN, Boles D, et al. MicroRNA signature helps distinguish early from late biochemical failure in prostate cancer. *Clin Chem* 59: 1595-1603, 2013.

[252] Sun T, Wang X, He H, Sweeney CJ, Liu SX, et al. MiR-221 promotes the development of androgen independence in prostate cancer cells via downregulation of HECTD2 and RAB1A. *Oncogene* 33: 2790-2800, 2014.

[253] Ambs S, Prueitt RL, Yi M, Hudson RS, Howe TM, et al. Genomic profiling of microRNA and messenger RNA reveals deregulated microRNA expression in prostate cancer. *Cancer Res* 68: 6162-6170, 2008.

[254] Huang F, Tang J, Zhuang X, Zhuang Y, Cheng W, et al. MiR-196a promotes pancreatic cancer progression by targeting nuclear factor kappa-B-inhibitor alpha. *PLoS One* 9: e87897, 2014.

[255] Parikh A, Lee C, Joseph P, Marchini S, Baccarini A, et al. microRNA-181a has a critical role in ovarian cancer progression through the regulation of the epithelial-mesenchymal transition. *Nat Commun* 5: 2977, 2014.

[256] Bertoli G, Cava C, Castiglioni I. MicroRNAs as Biomarkers for Diagnosis, Prognosis and Theranostics in Prostate Cancer. *Int J Mol Sci* 17: 421, 2016.

[257] Song CJ, Chen H, Chen LZ, Ru GM, Guo JJ, Ding QN. The potential of microRNAs as human prostate cancer biomarkers: A meta-analysis of related studies. *J Cell Biochem* 119: 2763-2786, 2018.

[258] Pashaei E, Ahmady M, Ozen M, Aydin N. Meta-analysis of miRNA expression profiles for prostate cancer recurrence following radical prostatectomy. *PLoS One* 12: e0179543, 2017.

[259] Ferlay J, Shin HR, Bray F, Forman D, Mathers C, Parkin DM. Estimates of worldwide burden of cancer in 2008: GLOBOCAN 2008. *Int J Cancer* 127: 2893–2917, 2010.

[260] Thomas J, Ohtsuka M, Pichler M, Ling H. MicroRNAs: clinical relevance in colorectal cancer. *Int J Mol Sci* 12: 28063-28076, 2015.

[261] Slaby O, Svoboda M, Michalek J, Vyzula R. MicroRNAs in colorectal cancer: translation of molecular biology into clinical application. *Mol Cancer* 1: 1, 2009.

[262] Shirafkan N, Mansoori B, Mohammadi A, Shomali N, Ghasbi M, Eet al. MicroRNAs as novel biomarkers for colorectal cancer: New outlooks. *Biomed Pharmacother* 97:1319-1330, 2018.

[263] Koga Y, Yasunaga M, Takahashi A, Kuroda J, Moriya Y, et al. MicroRNA expression profiling of exfoliated colonocytes isolated from feces for colorectal cancer screening. *Cancer Prev Res* 3:1435–1442, 2010.

[264] He Y, Wang J, Wang J, Yung VYW, Hsu E, et al. MicroRNA-135b regulates apoptosis and chemoresistance in colorectal cancer by targeting large tumor suppressor kinase 2. *Am J Cancer Res* 4: 1382, 2015.

[265] Kanaan Z, Roberts H, Eichenberger MR, Ocheretner G, et al. A plasma microRNA panel for detection of colorectal adenomas: a step toward more precise screening for colorectal cancer. *Ann Surg* 258: 400–408, 2013.

[266] Feng J, Yang Y, Zhang P, Wang F, Ma Y, et al. miR-150 functions as a tumour suppressor in human colorectal cancer by targeting c-Myb. *J Cell Mol Med* 18: 2125–2134, 2014.

[267] Mogilyansky E, Rigoutsos I. The miR-17/92 cluster: a comprehensive update on its genomics, genetics, functions and increasingly important and numerous roles in health and disease. *Cell Death Differ* 20: 1603–1614, 2013.

[268] Ng EK, Chong WW, Jin H, Lam EK, Shin VY, et al. Differential expression of microRNAs in plasma of patients with colorectal cancer: a potential marker for colorectal cancer screening. *Gut* 58: 1375–1381, 2009.

[269] Dews M, Homayouni A, Yu D, Murphy D, Sevignani C, et al. Augmentation of tumor angiogenesis by a Myc-activated microRNA cluster. *Nat Genet* 38: 1060–1065, 2006.

[270] Huang Z, Huang D, Ni S, Peng Z, Sheng W. Plasma microRNAs are promising novel biomarkers for early detection of colorectal cancer. *Int J Cancer* 127: 118-126, 2010.

[271] Knudsen KN, Nielsen BS, Lindebjerg J, Hansen TF, Holst R, et al. microRNA-17 Is the Most Up-Regulated Member of the miR-17-92 Cluster during Early Colon Cancer Evolution. *PLoS One* 10: e0140503, 2015.

[272] Motoyama K, Inoue H, Takatsuno Y, Tanaka F, Mimori K, et al. microRNAs in human colorectal cancer. *Int J Oncol* 34: 1069–1075, 2009.

[273] Kara M, Yumrutas O, Ozcan O, Celik OI, Tasdemir S. Differential expressions of cancer-associated genes and their regulatory miRNAs in colorectal carcinoma. *Gene* 567: 81–86, 2015.

[274] Monzo M, Navarro A, Bandres E, Artells R, Moreno I, et al. Overlapping expression of microRNAs in human embryonic colon and colorectal cancer. *Cell Res* 18: 823–833, 2008.

[275] Garofalo M, Croce CM. MicroRNAs as therapeutic targets in chemoresistance. *Drug Resist Updates* 3: 47-59, 2013.

[276] Michael MZ, O'Connor SM, van Holst Pellekaan NG, Young GP, James RJ. Reduced accumulation of specific microRNAs in colorectal neoplasia. *Mol Cancer Res* 12: 882-891, 2003.

[277] Pichler M, Winter E, Stotz M, Eberhard K, Samonigg H, et al. Down-regulation of KRAS-interacting miRNA-143 predicts poor prognosis but not response to EGFR-targeted agents in colorectal cancer. *Br J Cancer* 11: 1826-1832, 2012.

[278] Zhang Y, Wang Z, Chen M, Peng L, Wang X, et al. MicroRNA-143 targets MACC1 to inhibit cell invasion and migration in colorectal cancer. *Mol Cancer* 1: 1, 2012.

[279] Sheng N, Tan G, You W, Chen H, Gong J, et al. MiR-145 inhibits human colorectal cancer cell migration and invasion via PAK4-dependent pathway. *Cancer Med* 6: 1331-1340, 2017.

[280] Ahmed FE, Jeffries CD, Vos PW, Flake G, Nuovo GJ, et al. Diagnostic microRNA markers for screening sporadic human colon cancer and active ulcerative colitis in stool and tissue. *Cancer Genom Proteom* 6: 281–295, 2009.

[281] Yokobori T, Yajima R, Morita H, Fujii T, Yamaguchi S. et al. MicroRNA-7 expression in colorectal cancer is associated with poor prognosis and regulates cetuximab sensitivity via EGFR regulation. *Carcinogenesis* 36: 338-345, 2015.

[282] Pichler M, Winter E, Ress AL, Bauernhofer T, Gerger A. et al. miR-181a is associated with poor clinical outcome in patients with colorectal cancer treated with EGFR inhibitör. *J Clin Pathol* 3: 198-203, 2014.

[283] Ren A, Dong Y, Tsoi H, Yu J. Detection of miRNA as non-invasive biomarkers of colorectal cancer. *Int J Mol Sci* 16: 2810–2823, 2015.

[284] Xu K, Liang X, Cui D, Wu Y, Shi W, et al. miR-1915 inhibits Bcl-2 to modulate multidrug resistance by increasing drug-sensitivity in human colorectal carcinoma cells. *Mol Carcinog* 1: 70-78, 2013.

[285] Hollis M, Nair K, Vyas A, Chaturvedi LS, Gambhir S. MicroRNAs potential utility in colon cancer: early detection, prognosis, and chemosensitivity. *World J Gastroenterol: WJG* 27: 8284, 2015.

[286] Xu F, Xu L, Wang M, An G, Feng G. The accuracy of circulating microRNA-21 in the diagnosis of colorectal cancer: a systematic review and meta-analysis. *Colorectal Dis* 17: O100-7, 2015.

[287] Toiyama Y, Takahashi M, Hur K, Nagasaka T, Tanaka K,et al. Serum miR-21 as a diagnostic and prognostic biomarker in colorectal cancer. *J Natl Cancer Inst* 105: 849–859, 2013.

[288] Ogata-Kawata H, Izumiya M, Kurioka D, Honma Y, Yamada Y., et al. Circulating exosomal microRNAs as biomarkers of colon cancer. *PLoS One* 9: e92921, 2014.

[289] Schee K, Boye K, Abrahamsen TW, Fodstad Ø, Flatmark K. Clinical relevance of microRNA miR-21, miR-31, miR-92a, miR-101, miR-106a and miR-145 in colorectal cancer. *BMC Cancer* 12: 505, 2012.

[290] Peng Q, Zhang X, Min M, Zou L, Shen P, et al. The clinical role of microRNA-21 as a promising biomarker in the diagnosis and prognosis of colorectal cancer: a systematic review and meta-analysis. *Oncotarget* 8: 44893-44909, 2017.

[291] Brunet Vega A, Pericay C, Moya I, Ferrer A, Dotor E, et al. microRNA expression profile in stage III colorectal cancer: circulating miR-18a and miR-29a as promising biomarkers. *Oncol Rep* 30: 320–326, 2013.

[292] Yamada A, Horimatsu T, Okugawa Y, Nishida N, Honjo H, et al. Serum miR-21, miR-29a, and miR-125b are promising biomarkers for the early detection of colorectal neoplasia. *Clin Cancer Res* 21: 4234–4242, 2015.

[293] Weissmann-Brenner A, Kushnir M, Lithwick Yanai G, Aharonov R, et al. Tumor microRNA-29a expression and the risk of recurrence in stage II colon cancer. *Int J Oncol* 6: 2097-2103, 2012.

[294] Basati G, Emami Razavi A, Abdi S, Mirzaei A. Elevated level of microRNA-21 in the serum of patients with colorectal cancer. *Med Oncol* 31: 205, 2014.

[295] Li L, Guo Y, Chen Y, Wang J, Zhen L, et al. The diagnostic efficacy and biological effects of microRNA-29b for colon cancer. *Technol Cancer Res Treat* 15: 772–779, 2016.

[296] Yan B, Guo Q, Fu FJ, Wang Z, Yin Z, et al. The role of miR-29b in cancer: regulation, function, and signaling. *Onco Targets Ther* 8: 539–548, 2015.

[297] Zhi ML, Liu ZJ, Yi XY, Zhang LJ, Bao YX. Diagnostic performance of microRNA-29a for colorectal cancer: a meta-analysis. *Genet Mol Res* 14: 18018-18025, 2015.

[298] Zhang X, Li X, Tan F, Yu N, Pei H. STAT1 inhibits miR-181a expression to suppress colorectal cancer cell proliferation through PTEN/Akt. *J Cell Biochem* 118: 3435-3443, 2017.

[299] Liu Y, Uzair Ur R, Guo Y, Liang H, Cheng R, et al. miR-181b functions as an oncomiR in colorectal cancer by targeting PDCD4. *Protein Cell* 7: 722-734, 2016.

[300] Ji D, Chen Z, Li M, Zhan T, Yao Y, et al. MicroRNA-181a promotes tumor growth and liver metastasis in colorectal cancer by targeting the tumor suppressor WIF-1. *Molecular Cancer* 13: 86, 2014.

[301] Toiyama Y, Hur K, Tanaka K, Inoue Y, Kusunoki M, Boland CR, et al. Serum miR-200c is a novel prognostic and metastasis-predictive biomarker in patients with colorectal cancer. *Ann Surg* 259:735–743, 2014.

[302] Cheng H, Zhang L, Cogdell DE, Zheng H, Schetter AJ, et al. Circulating plasma MiR-141 is a novel biomarker for metastatic colon cancer and predicts poor prognosis. *PLoS One* 3: e17745, 2011.

[303] Wu X, Li S, Xu X, Wu S, Chen R, et al. The potential value of miR-1 and miR-374b as biomarkers for colorectal cancer. *Int J Clin Exp Pathol* 3: 2840-2851, 2014.

[304] Zhang Y, Lu Q, Cai X. MicroRNA-106a induces multidrug resistance in gastric cancer by targeting RUNX3. *Febs Lett* 587: 3069–3075, 2013.

[305] Wang R, Li Y, Hou Y, Yang Q, Chen S, et al. The PDGF-D/miR-106a/Twist1 pathway orchestrates epithelial-mesenchymal transition in gemcitabine resistance hepatoma cells. *Oncotarget* 6: 7000–7010, 2015.

[306] Catela IT, Aralica G, Cacev T, Loncar B, Kapitanovic S. miR-106a overexpression and pRB downregulation in sporadic colorectal cancer. *Exp Mol Pathol* 94: 148–514, 2013.

[307] Diaz R, Silva J, Garcia JM, Lorenzo Y, Garcia V, et al. Deregulated expression of miR-106a predicts survival in human colon cancer patients. *Genes Chromosomes Cancer* 47: 794–802, 2008.

[308] Hao H, Liu L, Zhang D, Wang C, Xia G, et al.. Diagnostic and prognostic value of miR-106a in colorectal cancer. *Oncotarget* 8: 5038-5047, 2017.

[309] Yau TO, Wu CW, Dong Y, Tang CM, Ng SS, et al. MicroRNA-221 and microRNA-18a identification in stool as potential biomarkers for the non-invasive diagnosis of colorectal carcinoma. *Br J Cancer* 111: 1765–1771, 2014.

[310] Carter JV, Galbraith NJ, Yang D, Burton JF, Walker SP, Galandiuk S. Blood-based microRNAs as biomarkers for the diagnosis of colorectal cancer: a systematic review and meta-analysis. *Br J Cancer* 116: 762-774, 2017.

[311] Yan L, Zhao W, Yu H, Wang Y, Liu Y, Xie C. A Comprehensive Meta-Analysis of MicroRNAs for Predicting Colorectal Cancer. *Medicine* (Baltimore) 95: e2738, 2016.

[312] Strimbu K, Tavel JA. What are biomarkers? *Curr Opin HIV AIDS* 5: 463–466, 2010.

[313] Detassis S, Grasso M, Del Vescovo V, Denti MA. microRNAs Make the Call in Cancer Personalized Medicine. *Front Cell Dev Biol* 5: 86, 2017.

[314] Yanaihara N, Caplen N, Bowman E, Seike M, Kumamoto K, et al. Unique microRNA molecular profiles in lung cancer diagnosis and prognosis. *Cancer Cell* 9: 189–198, 2006.

[315] Rosenfeld N, Aharonov R, Meiri E, Rosenwald S, Spector Y, et al. MicroRNAs accurately identify cancer tissue origin. *Nat Biotechnol* 26: 462–469, 2008.

[316] Xiong DD, Lv J, Wei KL, Feng ZB, Chen JT, et al. A nine-miRNA signature as a potential diagnostic marker for breast carcinoma: an integrated study of 1,110 cases. *Oncol Rep* 37: 3297–3304, 2017.

[317] Lu M, Kong X, Wang H, Huang G, Ye C, et al. A novel microRNAs expression signature for hepatocellular carcinoma diagnosis and prognosis. *Oncotarget* 8: 8775–8784, 2017.

[318] Ochoa AE, Choi W, Su X, Siefker-Radtke A, Czerniak B, et al. Specific micro-RNA expression patterns distinguish the basal and luminal subtypes of muscle-invasive bladder cancer. *Oncotarget* 7: 80164–80174, 2016.

[319] Blenkiron C, Goldstein LD, Thorne N P, Spiteri I, Chin SF, et al. MicroRNA expression profiling of human breast cancer identifies new markers of tumor subtype. *Genome Biol* 8: R214, 2007.

[320] Rapa I, Votta A, Felice B, Righi L, Giorcelli J, et al. Identification of MicroRNAs differentially expressed in lung carcinoid subtypes and progression. *Neuroendocrinology* 101: 246–255, 2015.

[321] Zheng D, Haddadin S, Wang Y, Gu LQ, Perry MC, et al. Plasma microRNAs as novel biomarkers for early detection of lung cancer. *Int J Clin Exp Pathol* 4: 575–586, 2011.

[322] Yamada Y, Enokida H, Kojima S, Kawakami K, Chiyomaru T, et al. MiR-96 and miR-183 detection in urine serve as potential tumor markers of urothelial carcinoma: correlation with stage and grade, and comparison with urinary cytology. *Cancer Sci* 102: 522–529, 2011.

[323] Nair VS, Maeda LS, Ioannidis JP. Clinical outcome prediction by microRNAs in human cancer: a systematic review. *J Natl Cancer Inst* 104: 528–540, 2012.

[324] Hayes J, Peruzzi PP, Lawler S. MicroRNAs in cancer: biomarkers, functions and therapy. *Trends Mol Med* 20: 460-469, 2014.

[325] Wandt H, Haferlach T, Thiede C, Ehninger G. WHO classification of myeloid neoplasms and leukemia. *Blood* 115: 748–749 author reply 749–750, 2010.

[326] Garzon R, Volinia S, Liu CG, Fernandez-Cymering C, Palumbo T, et al. MicroRNA signatures associated with cytogenetics and prognosis in acute myeloid leukemia. *Blood* 111: 3183–3189, 2008.

[327] Liu C, Kelnar K, Vlassov AV, Brown D, Wang J, et alDistinct microRNA expression profiles in prostate cancer stem/progenitor cells and tumor-suppressive functions of let-7. *Cancer Res* 72: 3393–3404, 2012.

[328] Farazi TA, Ten Hoeve JJ, Brown M, Mihailovic A, Horlings HM, et al. Identification of distinct miRNA target regulation between breast cancer molecular subtypes using AGO2- PAR-CLIP and patient datasets. *Genome Biol* 15: R9, 2014.

[329] Dvinge H, Git A, Graf S, Salvon-Divon M, Curtis C, et al. The shaping and functional consequences of the microRNA landscape in breast cancer. *Nature* 497: 378–382, 2013.

[330] Pencheva N, Tavazoie SF. Control of metastatic progression by microRNA regulatory networks. *Nat Cell Biol* 15: 546–554, 2013.

[331] Png KJ, Halberg N, Yoshida M, Tavazoie SF. A microRNA regulon that mediates endothelial recruitment and metastasis by cancer cells. *Nature* 481: 190–194, 2012.

[332] Scholl V, Hassan R, Zalcberg IR. miRNA-451: A putative predictor marker of Imatinib therapy response in chronic myeloid leukemia. *Leuk Res* 36: 119-121, 2012.

[333] Pardini B, Rosa F, Barone E, Di Gaetano C, Slyskova J, et al. Variation within 30-UTRs of base excision repair genes and response to therapy in colorectal cancer

patients: a potential modulation of microRNAs binding. *Clin Cancer Res* 19: 6044–6056, 2013.

[334] Schwarzenbach H, Nishida N, Calin GA, Pantel K. Clinical relevance of circulating cell-free microRNAs in cancer. *Nat Rev Clin Oncol* 11: 145–156, 2014.

[335] Wang Y, Gu J, Roth JA, Hildebrandt MA, Lippman SM, et al. Pathway-based serum microRNA profiling and survival in patients with advanced stage non-small cell lung cancer. *Cancer Res* 73: 4801–4809, 2013.

[336] Lawrie CH, Gal S, Dunlop HM, Pushkaran B, Liggins AP, et al. Detection of elevated levels of tumour-associated microRNAs in serum of patients with diff use large B-cell lymphoma. *Br J Haematol* 141: 672–75, 2008.

[337] Roush S, Slack FJ. The let-7 family of microRNAs. *Trends Cell Biol* 18: 505–516, 2008.

[338] Boyerinas B, Park SM, Hau A, Murmann AE, Peter ME. The role of let-7 in cell differentiation and cancer. *Endocr. Relat. Cancer* 17: F19–F36, 2010.

[339] Mayr C, Hemann MT, Bartel DP. Disrupting the pairing between let-7 and Hmga2 enhances oncogenic transformation. *Science* 315: 1576–1579, 2007.

[340] Wang Y, Lu Y, Toh ST, Sung WK, Tan P, et al. Lethal-7 is down-regulated by the hepatitis B virus x protein and targets signal transducer and activator of transcription 3. *J Hepatol* 53: 57–66, 2010.

[341] He X, Duan C, Chen J, Ou-Yang X, Zhang Z, et al. Let-7a elevates p21(WAF1) levels by targeting of NIRF and suppresses the growth of A549 lung cancer cells. *FEBS Lett.* 583: 3501–3507, 2009.

[342] Sampson VB, Rong NH, Han J, Yang Q, Aris V, et al. MicroRNA let-7a down-regulates MYC and reverts MYC-induced growth in Burkitt lymphoma cells. *Cancer Res* 67: 9762–9770, 2007.

[343] Legesse-Miller A, Elemento O, Pfau SJ, Forman JJ, Tavazoie S, et al. Let-7 Overexpression leads to an increased fraction of cells in G2/M, direct down-regulation of Cdc34, and stabilization of Wee1 kinase in primary fibroblasts. *J Biol Chem* 284: 6605–6609, 2009.

[344] Heegaard NH, Schetter AJ, Welsh JA, Yoneda M, Bowman ED, et al. Circulating micro-RNA expression profiles in early stage nonsmall cell lung cancer. *Int J Cancer* 130: 1378–1386, 2012.

[345] Tsujiura M, Ichikawa D, Komatsu S, Shiozaki A, Takeshita H, et al. Circulating microRNAs in plasma of patients with gastric cancers. *Br J Cancer* 102: 1174–1179, 2010.

[346] Chung YW, Bae HS, Song JY, Lee JK, Lee NW, et al. Detection of microRNA as novel biomarkers of epithelial ovarian cancer from the serum of ovarian cancer patients. *Int J Gynecol Cancer* 23: 673–679, 2013.

[347] Hung CH, Hu TH, Lu SN, Kuo FY, Chen CH, et al. Circulating microRNAs as biomarkers for diagnosis of early hepatocellular carcinoma associated with hepatitis B virus. *Int J Cancer* 138: 714–720, 2016.

[348] Li XX, Gao SY, Wang PY, Zhou X, Li YJ, et al. Reduced expression levels of let-7c in human breast cancer patients. *Oncol Lett* 9: 1207–1212, 2015.

[349] Fayyad-Kazan H, Bitar N, Najar M, Lewalle P, Fayyad-Kazan M, et al. Circulating miR-150 and miR-342 in plasma are novel potential biomarkers for acute myeloid leukemia. *J Transl Med* 11: 31, 2013.

[350] Yu S, Liu Y, Wang J, Guo Z, Zhang Q, et al. Circulating microRNA profiles as potential biomarkers for diagnosis of papillary thyroid carcinoma. *J Clin Endocrinol Metab* 97: 2084–2092, 2012.

[351] Zuo Z, Calin GA, de Paula HM, Medeiros LJ, Fernandez MH, et al. Circulating microRNAs let-7a and miR-16 predict progression-free survival and overall survival in patients with myelodysplastic syndrome. *Blood* 118: 413–415, 2011.

[352] Kubiczkova L, Kryukov F, Slaby O, Dementyeva E, Jarkovsky J, et al. Circulating serum microRNAs as novel diagnostic and prognostic biomarkers for multiple myeloma and monoclonal gammopathy of undetermined significance. *Haematologica* 99: 511–518, 2014.

[353] Ali S, Almhanna K, Chen W, Philip PA, Sarkar FH. Differentially expressed miRNAs in the plasma may provide a molecular signature for aggressive pancreatic cancer. *Am J Transl Res* 3: 28–47, 2010.

[354] Zheng H, Zhang L, Zhao Y, Yang D, Song F, et al. Plasma miRNAs as diagnostic and prognostic biomarkers for ovarian cancer. *PLoS ONE* 8: e77853, 2013.

[355] Joosse SA, Muller V, Steinbach B, Pantel K, Schwarzenbach H. Circulating cell-free cancer-testis MAGE-A RNA, BORIS RNA, let-7b and miR-202 in the blood of patients with breast cancer and benign breast diseases. *Br J Cancer* 111: 909–917, 2014.

[356] Ma L, Teruya-Feldstein J, Weinberg RA. Tumour invasion and metastasis initiated by microRNA-10b in breast cancer. *Nature* 449: 682–688, 2007.

[357] Sun L, Yan W, Wang Y, Sun G, Luo H, et al. MicroRNA-10b induces glioma cell invasion by modulating MMP-14 and uPAR expression via HOXD10. *Brain Res* 1389: 9–18, 2011.

[358] Tian Y, Luo A, Cai Y, Su Q, Ding F, et al. MicroRNA-10b promotes migration and invasion through KLF4 in human esophageal cancer cell lines. *J Biol Chem* 285: 7986–7994, 2010.

[359] Ibrahim SA, Yip GW, Stock C, Pan JW, Neubauer C, et al. Targeting of syndecan-1 by microRNA miR-10b promotes breast cancer cell motility and invasiveness via a Rho-GTPase- and E-cadherin-dependent mechanism. *Int J Cancer* 131: E884–E896, 2012.

[360] Liao CG, Kong LM, Zhou P, Yang XL, Huang JG, et al. miR-10b is overexpressed in hepatocellular carcinoma and promotes cell proliferation, migration and invasion through RhoC, uPAR and MMPs. *J Transl Med* 12: 234, 2014.

[361] Liu Z, Zhu J, Cao H, Ren H, Fang X. miR-10b promotes cell invasion through RhoC-AKT signaling pathway by targeting HOXD10 in gastric cancer. *Int J Oncol* 40: 1553–1560, 2012.

[362] Tsukerman P, Stern-Ginossar N, Gur C, Glasner A, Nachmani D, et al. miR-10b downregulates the stress-induced cell surface molecule MICB, a critical ligand for cancer cell recognition by natural killer cells. *Cancer Res* 72: 5463–5472, 2012.

[363] Zhang L, Sun J, Wang B, Ren JC, Su W, et al. MicroRNA-10b Triggers the Epithelial-Mesenchymal Transition (EMT) of Laryngeal Carcinoma Hep-2 Cells by Directly Targeting the E-cadherin. *Appl. Biochem Biotechnol* 176: 33–44, 2015.

[364] Teplyuk NM, Uhlmann EJ, Wong AH, Karmali P, Basu M, et al. MicroRNA-10b inhibition reduces E2F1-mediated transcription and miR-15/16 activity in glioblastoma. *Oncotarget* 6: 3770–3783, 2015.

[365] Biagioni F, Bossel Ben-Moshe N, Fontemaggi G, Canu V, Mori F, et al. miR-10b*, a master inhibitor of the cell cycle, is down-regulated in human breast tumours. *EMBO Mol Med* 4: 1214–1229, 2012.

[366] Valadi H, Ekstrom K, Bossios A, Sjostrand M, Lee JJ, et al. Exosome-mediated transfer of mRNAs and microRNAs is a novel mechanism of genetic exchange between cells. *Nat Cell Biol* 9: 654–659, 2007.

[367] Ismail N, Wang Y, Dakhlallah D, Moldovan L, Agarwal K, et al. Macrophage microvesicles induce macrophage differentiation and miR-223 transfer. *Blood* 121: 984–995, 2013.

[368] Xu H, Yao Y, Meng F, Qian X, Jiang X, et al. Predictive Value of Serum miR-10b, miR-29c, and miR-205 as Promising Biomarkers in Esophageal Squamous Cell Carcinoma Screening. *Medicine* (Baltimore) 94: e1558, 2015.

[369] Teplyuk NM, Mollenhauer B, Gabriely G, Giese A, Kim E, et al. MicroRNAs in cerebrospinal fluid identify glioblastoma and metastatic brain cancers and reflect disease activity. *Neuro Oncol* 14: 689–700, 2012.

[370] Anker P, Mulcahy H, Stroun M. Circulating nucleic acids in plasma and serum as a noninvasive investigation for cancer: Time for large-scale clinical studies? *Int J Cancer* 103: 149–152, 2003.

[371] Maclellan SA, Lawson J, Baik J, Guillaud M, Poh CF, et al. Differential expression of miRNAs in the serum of patients with high-risk oral lesions. *Cancer Med* 1: 268–274, 2012.

[372] Hu Z, Dong J, Wang LE, Ma H, Liu J, et al. Serum microRNA profiling and breast cancer risk: The use of miR-484/191 as endogenous controls. *Carcinogenesis* 33: 828–834, 2012.

[373] Zhang J, Song Y, Zhang C, Zhi X, Fu H, et al. Circulating MiR-16–5p and MiR-19b-3p as Two Novel Potential Biomarkers to Indicate Progression of Gastric Cancer. *Theranostics* 5: 733–745, 2015.

[374] Wang H, Wang L, Wu Z, Sun R, Jin H, et al. Three dysregulated microRNAs in serum as novel biomarkers for gastric cancer screening. *Med Oncol* 31: 298, 2014.

[375] Li H, Zhang K, Liu LH, Ouyang Y, Guo HB, et al. MicroRNA screening identifies circulating microRNAs as potential biomarkers for osteosarcoma. *Oncol Lett* 10: 1662–1668, 2015.

[376] Ge W, Yu DC, Li QG, Chen X, Zhang CY, et al. Expression of serum miR-16, let-7f, and miR-21 in patients with hepatocellular carcinoma and their clinical significances. *Clin Lab* 60: 427–434, 2014.

[377] Li BX, Yu Q, Shi L, Li P, Fu S. Circulating microRNAs in esophageal squamous cell carcinoma: Association with locoregional staging and survival. *Int J Clin Exp Med* 8: 7241–7250, 2015.

[378] Matsumura T, Sugimachi K, Iinuma H, Takahashi Y, et al. Exosomal microRNA in serum is a novel biomarker of recurrence in human colorectal cancer. *Br J Cancer* 113: 275–281, 2015.

[379] Pfeffer SR, Grossmann KF, Cassidy PB, Yang CH, et al. Detection of Exosomal miRNAs in the Plasma of Melanoma Patients. *J Clin Med* 4: 2012–2027, 2015.

[380] Anfossi S, Giordano A, Gao H, Cohen EN, Tin S, et al. High serum miR-19a levels are associated with inflammatory breast cancer and are predictive of favorable clinical outcome in patients with metastatic HER2+ inflammatory breast cancer. *PLoS ONE* 9: e83113, 2014.

[381] Feng Y, Liu J, Kang Y, He Y, Liang B, et al. miR-19a acts as an oncogenic microRNA and is up-regulated in bladder cancer. *J Exp Clin Cancer Res* 33: 67, 2014.

[382] Hao M, Zang M, Wendlandt E, Xu Y, An G, et al. Low serum miR-19a expression as a novel poor prognostic indicator in multiple myeloma. *Int J Cancer* 136: 1835–1844, 2015.

[383] Lin Q, Chen T, Lin Q, Lin G, Lin J, et al. Serum miR-19a expression correlates with worse prognosis of patients with non-small cell lung cancer. *J Surg Oncol* 107: 767–771, 2013.

[384] Shen J, Hruby GW, McKiernan JM, Gurvich I, Lipsky MJ, et al. Dysregulation of circulating microRNAs and prediction of aggressive prostate cancer. *Prostate* 72: 1469–1477, 2012.

[385] Li Q, Liu M, Ma F, Luo Y, Cai R, et al. Circulating miR-19a and miR-205 in serum may predict the sensitivity of luminal A subtype of breast cancer patients to neoadjuvant chemotherapy with epirubicin plus paclitaxel. *PLoS ONE* 9: e104870, 2014.

[386] Hong Q, Fang J, Pang Y, Zheng J. Prognostic value of the microRNA-29 family in patients with primary osteosarcomas. *Med Oncol* 31: 37, 2014.

[387] Basati G, Razavi AE, Pakzad I, Malayeri FA. Circulating levels of the miRNAs, miR-194, and miR-29b, as clinically useful biomarkers for colorectal cancer. *Tumour Biol* 37: 1781-1788, 2016.

[388] Yang IP, Tsai HL, Huang CW, Huang MY, Hou MF, et al. The functional significance of microRNA-29c in patients with colorectal cancer: A potential circulating biomarker for predicting early relapse. *PLoS ONE* 8: e66842, 2013.

[389] Zeng X, Xiang J, Wu M, Xiong W, Tang H, et al. Circulating miR-17, miR-20a, miR-29c, and miR-223 combined as non-invasive biomarkers in nasopharyngeal carcinoma. *PLoS ONE* 7: e46367, 2012.

[390] Kota J, Chivukula RR, O'Donnell KA, Wentzel EA, Montgomery CL, et al. Therapeutic microRNA delivery suppresses tumorigenesis in a murine liver cancer model. *Cell* 137: 1005–1017, 2009.

[391] Bader AG, Brown D, Winkler M. The promise of micro-RNA replacement therapy. *Cancer Res* 70: 7027–7030, 2010.

[392] Tutar L, Tutar E, Tutar Y. MicroRNAs and cancer; an over-view. *Curr Pharm Biotechnol* 15: 430–437, 2014.

[393] Shah MY, Ferrajoli A, Sood AK, Lopez-Berestein G, Calin GA. microRNA Therapeutics in Cancer -An Emerging Concept. *EBioMed* 12: 34-42, 2016.

[394] Garzon R, Marcucci G, Croce CM. Targeting microRNAs in cancer: rationale, strategies and challenges. *Nat Rev Drug Discov* 9: 775–789, 2010.

[395] Hutvagner G, Simard MJ, Mello CC, Zamore PD. Sequence-specific inhibition of small RNA function. *PLoS Biol.* 2: E98, 2004.

[396] Krutzfeldt J, Rajewsky N, Braich R, Rajeev KG, Tuschl T, et al. Silencing of microRNAs in vivo with 'antagomirs'. *Nature* 438: 685–689, 2005.

[397] Griveau A, Bejaud J, Anthiya S, Avril S, Autret D, et al. Silencing of miR-21 by locked nucleic acid-lipid nanocapsule complexes sensitize human glioblastoma cells to radiation-induced cell death. *Int J Pharm* 454: 765–774, 2013.

[398] Ma L, Young J, Prabhala H, Pan E, Mestdagh P, et al. miR-9, a MYC/MYCN-activated microRNA, regulates E-cadherin and cancer metastasis. *Nat Cell Biol* 12: 247–256, 2010.

[399] Xiao J, Yang B, Lin H, Lu Y, Luo X, et al. Novel approaches for genespecific interference via manipulating actions of microRNAs: examination on the pacemaker channel genes HCN2 and HCN4. *J Cell Physiol* 212: 285–292, 2007.

[400] Cortez MA, Ivan C, Valdecanas D, Wang X, Peltier HJ, et al. PDL1 Regulation by p53 via miR-34. *J Natl Cancer Inst* 108, 2016.

[401] Wei JS, Song YK, Durinck S, Chen QR, Cheuk AT, et al. The MYCN oncogene is a direct target of miR-34a. *Oncogene* 27: 5204-5213, 2007.

[402] Xu X, Chen W, Miao R, Zhou Y, Wang Z, et al. miR-34a induces cellular senescence via modulation of telomerase activity in human hepatocellular carcinoma by targeting FoxM1/c-Myc pathway. *Oncotarget* 6: 3988-4004, 2015.

[403] Adams BD, Parsons C, Slack FJ. The tumor-suppressive and potential therapeutic functions of miR-34a in epithelial carcinomas. *Expert Opin Ther Targets* 20: 737-753, 2016.

[404] Bader AG. miR-34-a microRNA replacement therapy is headed to the clinic. *Front Genet* 3: 120, 2012.

[405] Misso G, Di Martino MT, De Rosa G, Farooqi AA, Lombardi A, et al. Mir-34: a new weapon against cancer? *Mol. Ther. Nucleic Acids* 3: e194, 2014.

[406] Tazawa H, Tsuchiya N, Izumiya M, Nakagama H. Tumor-suppressive miR- 34a induces senescence-like growth arrest through modulation of the E2F pathway in human colon cancer cells. *Proc Natl Acad Sci USA* 104: 15472-15477, 2007.

[407] Ivkovic TC, Voss G, Cornella H, Ceder Y. microRNAs as cancer therapeutics: A step closer to clinical application. *Cancer Letters* 407: 113-122, 2017.

[408] Swaika A, Hammond WA, Joseph RW. Current state of anti-PD-L1 and anti- PD-1 agents in cancer therapy. *Mol Immunol* 67: 4-17, 2015.

[409] https://clinicaltrials.gov/ct2/show/NCT01829971.

[410] Beg MS, Brenner AJ, Sachdev J, Borad M, Kang YK, et al. Phase I study of MRX34, a liposomal miR-34a mimic, administered twice weekly in patients with advanced solid tumors. *Investig New Drugs* 35: 180-188, 2017.

[411] Sun CY, She XM, Qin Y, Chu ZB, Chen L, et al. miR-15a and miR-16 affect the angiogenesis of multiple myeloma by targeting VEGF. *Carcinogenesis* 34: 426-435, 2014.

[412] Pothof J, Verkaik NS, van IW, Wiemer EA, Ta VT, et al. MicroRNA-mediated gene silencing modulates the UV-induced DNA-damage response. *EMBO J* 28: 2090-2099, 2009.

[413] Kasar S, Salerno E, Yuan Y, Underbayev C, Vollenweider D, et al. Systemic in vivo lentiviral delivery of miR-15a/16 reduces malignancy in the NZB de novo mouse model of chronic lymphocytic leukemia. *Genes Immun* 13: 109-119, 2012.

[414] Takeshita F, Patrawala L, Osaki M, Takahashi RU, Yamamoto Y, et al. Systemic delivery of synthetic microRNA-16 inhibits the growth of metastatic prostate tumors via downregulation of multiple cell-cycle genes. *Mol Ther* 18: 181-187, 2010.

[415] Tung YT, Huang PW, Chou YC, Lai CW, Wang HP, et al. Lung tumorigenesis induced by human vascular endothelial growth factor (hVEGF)-A165 overexpression in transgenic mice and amelioration of tumor formation by miR-16. *Oncotarget* 6: 10222-10238, 2015.

[416] Sun Z, Song X, Li X, Su T, Qi S, et al. In vivo multimodality imaging of miRNA-16 iron nanoparticle reversing drug resistance to chemotherapy in a mouse gastric cancer model. *Nanoscale* 6: 14343-14353, 2014.

[417] Dai L, Wang W, Zhang S, Jiang Q, Wang R, et al. Vector-based miR- 15a/16-1 plasmid inhibits colon cancer growth in vivo. *Cell Biol Int* 36: 765-770, 2012.

[418] Dwivedi SK, Mustafi SB, Mangala LS, Jiang D, Pradeep S, et al. Therapeutic evaluation of microRNA-15a and microRNA-16 in ovarian cancer. *Oncotarget* 7: 15093-15104, 2016.

[419] Bandi N, Vassella E. miR-34a and miR-15a/16 are co-regulated in non-small cell lung cancer and control cell cycle progression in a synergistic and Rbdependent manner. *Mol Cancer* 10: 55, 2011.

[420] Reid G, Kao SC, Pavlakis N, Brahmbhatt H, MacDiarmid J, et al. Clinical development of TargomiRs, a miRNA mimic-based treatment for patients with recurrent thoracic cancer. *Epigenomics* 8: 1079-1085, 2016.

[421] https://clinicaltrials.gov/show/NCT02369198

[422] Higgs G, Slack F, The multiple roles of microRNA-155 in oncogenesis. *J Clin Bioinforma* 3: 17, 2013.

[423] Johansson J, Berg T, Kurzejamska, E Pang MF, Tabor V, et al. MiR-155-mediated loss of C/EBPbeta shifts the TGF-beta response from growth inhibition to epithelial-mesenchymal transition, invasion and metastasis in breast cancer. *Oncogene* 32: 5614-5624, 2013.

[424] Cheng CJ, Bahal R, Babar IA, Pincus Z, Barrera F, et al. MicroRNA silencing for cancer therapy targeted to the tumour microenvironment. *Nature* 518: 107-110, 2015.

[425] Babar LA, Cheng CJ, Booth CJ, Liang X, Weidhaas JB, et al. Nanoparticle-based therapy in an in vivo microRNA-155 (miR-155)-dependent mouse model of lymphoma. *Proc Natl Acad Sci USA* 109: E1695-E1704, 2012.

[426] Zhang Y, Roccaro AM, Rombaoa C, Flores L, Obad S, et al. NA-mediated anti-miR-155 silencing in low-grade B-cell lymphomas. *Blood* 120: 1678-1686, 2012.

[427] Wang J, Yu F, Jia X, Iwanowycz S, Wang Y, et al. MicroRNA-155 deficiency enhances the recruitment and functions of myeloid-derived suppressor cells in tumor microenvironment and promotes solid tumor growth. *Int J Cancer* 136: E602-E613, 2015.

[428] Yu F, Jia X, Du F, Wang J, Wang Y, et al. miR-155-deficient bone marrow promotes tumor metastasis. *Mol Cancer Res* 11: 923-936, 2013.

[429] Cubillos-Ruiz JR, Baird JR, Tesone AJ, Rutkowski MR, Scarlett UK, et al. Reprogramming tumor-associated dendritic cells in vivo using miRNA mimetics triggers protective immunity against ovarian cancer. *Cancer Res* 72: 1683-1693, 2012.

[430] Tili E, Croce CM, Michaille JJ. miR-155: on the crosstalk between inflammation and cancer, *Int Rev Immunol* 28: 264-284, 2009.

[431] https://clinicaltrials.gov/ct2/show/NCT02580552 .

[432] Querfeld C, Pacheco T, Foss FM, Halwani AS, Porcu P, et al. Preliminary results of a phase 1 trial evaluating MRG-106, a synthetic microRNA antagonist (LNA antimiR) of microRNA-155, in patients with CTCL. *Blood* 128: 1829-1829, 2016.

In: Research on New Generation …
Editor: Yildiz Dincer
ISBN: 978-1-53614-367-6

Chapter 6

CANCER STEM CELLS CONCEPTS AND THEIR ASSOCIATION WITH DIAGNOSTIC AND THERAPEUTIC STRATEGIES

Federica Collino[1], Maria Isabel D. Rossi[2] and Rafael Soares Lindoso[1,*]
[1]Carlos Chagas Institute of Biophysics,
Federal University of Rio de Janeiro, Rio de Janeiro, Brazil
[2]Biomedical Science Institute and Clementino Fraga Filho University Hospital,
Federal University of Rio de Janeiro,
Rio de Janeiro, Brazil

ABSTRACT

The cancer stem cell (CSC) concept was proposed to explain cancer-related phenomena that were not fully explained by the stochastic model. CSCs have characteristics similar to those of stem cells in adult tissues, such as self-renewal capabilities and rapid propagation, or limited proliferative potential through asymmetric division. CSCs are also tumor initiating cells because of their capacity to generate a tumor comprising a heterogenous cell population.

The origin of CSCs is still not clear and could be associated with the accumulated mutations of adult stem cells during cell division, or de-differentiation of mutated somatic cells that start to present stem cell features. Irrespective of the hypothesis, advances in knowledge about CSCs has led to new therapeutic strategies in cancer treatments associated with specific antibodies targeting quiescent CSCs and their surrounding niche. In addition, several biomarkers for cancer diagnosis and prognosis have now emerged as strong candidates for clinical use. CSCs have been shown to

* Corresponding Author Email:lindoso@biof.ufrj.br.

secrete extracellular vesicles (EVs) containing bioactive molecules that resemble the cell of origin, including some specific proteins and microRNAs. Such EVs can be found in body fluids such as blood and urine, allowing tumor progression or response to treatment to be monitored without invasive processes. We therefore present an overview of aspects associated with the CSC hypothesis and their relevance to new therapeutic strategies and clinical applications.

Keywords: cancer stem cell, tumor niche, biomarkers, extracellular vesicles

1. Cancer Stem Cells: Origin and Role in Tumor Initiation

Tissue homeostasis, especially in tissues that are regularly renewed such as the hematopoietic system, epidermis and gut epithelium, is maintained by a population of adult stem cells. Even tissues with very slow turnover such as nervous system and striated muscle harbor such a cell population. Adult stem cells are thought to originate during development [1].

The hypothesis that tissue renewal depends upon a population of adult stem cells dates back to 1906 (see Figure 1), when the Russian scientist Alexander Maximow proposed that all hematopoietic precursor cells originate from stem cells under the local influence of the bone marrow stroma, which creates conditions for differentiation [8].

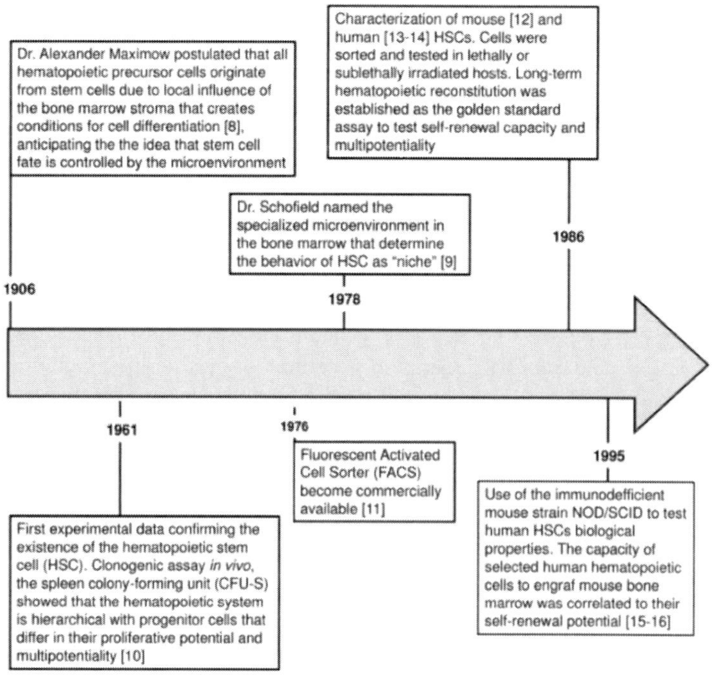

Figure 1. A timeline of important discoveries in Hematopoietic stem cell (HSC) field.

Interestingly, Dr. Maximow also anticipated the idea that stem cell fate is controlled by the microenvironment, a hypothesis that was proposed many years later by Dr. Schofield, who named the specialized microenvironment that determines stem cell behavior "*niche*" [9]. Experimental data demonstrating the biological properties of hematopoietic stem cells (HSC), the first tissue stem cell type characterized, were obtained using a clonogenic assay *in vivo*, the spleen colony-forming unit (CFU-S) [10]. The Fluorescent Activated Cell Sorter (FACS) [11] allowed mouse [12] and human [13, 14] HSCs to be characterized phenotypically (Figure 1). Selected cells were then injected into lethally or sublethally irradiated hosts to examine their capacity for long-term hematopoietic reconstitution. This approach became the gold standard assay for self-renewal potential. Human HSC candidates were assayed in immunodeficient mice, usually non-obese diabetic mice with severe combined immunodeficiency disease (NOD/SCID mice), and named SCID-repopulating cells (SRCc) [3, 6, 15, 16]. Data obtained by different groups led to the conclusion that HSCs comprise a rare population of quiescent undifferentiated cells with self-renewal capacity, generating a transit-amplifying pool of more differentiated progenitors that finally transform into mature cells [2-3, 5, 6]. Since HSCs were the first adult stem cell population identified, purified, experimentally tested, and used in clinical practice, their biological properties became paradigmatic and were used as templates for identifying other tissue stem cells. However, new and challenging facets of adult stem cell biology, including HSCs, have been uncovered. Adult stem cells are not always rare or quiescent, and a plasticity hierarchy is observed [4, 6, 17-22]. For instance, in the intestinal crypt, a population of stem cells expressing leucine rich repeat-containing G protein-coupled receptor 5 (Lgr5) cycles every 24 hours and the progeny compete for the niche. However, quiescent cells located at position +4 from the base of the crypt contribute to the renewal of the gut epithelium during acute loss of $Lgr5^+$ cells [21, 22]. Similarly, chief cells function as stem cells if the gastric corpus epithelium is injured; this epithelium is otherwise maintained by a population of adult stem cells located within the tubular glands [23, 24]. The skin epidermis and its appendages (hair follicles, and sebaceous and sweat glands) harbor stem cell populations that differ in location, quiescence and multipotentiality [4, 7].

Cancer is a multi-step disease involving dynamic changes in the genomes and biological properties of cells, including sustained proliferation, resistance to apoptosis, migration and invasion [25]. Different hypotheses have been proposed to explain the observed heterogeneity of tumors. Genetic instability of tumor cells and their clonal evolution, the cancer stem cell (CSC) model, microenvironment cues, and epigenetics have been proposed as independent mechanisms. However, recent findings suggest they are not mutually exclusive.

The CSC model presumes that tumors have a hierarchical organization similar to normal tissues, which means they are maintained by a long-lived self-renewal stem cell population, the CSCs, which generate short-term progeny. Although the model implies nothing about the origin of CSCs, it proposes a unidirectional pathway with cells that have acquired stem cell properties at the apex of the hierarchy. The CSC hypothesis challenges the way neoplasms are conventionally treated, since CSCs represent a small quiescent population of cells within tumors that are resistant to chemotherapy. Therefore, cancer chemotherapy or radiation kills the short-term progeny of CSCs, but the CSCs themselves survive, leading to relapse after treatment. Not surprisingly, interest in this field has increased during the 21st century, with attempts to characterize CSCs in different tumors and develop new therapeutic strategies [5, 6, 26-37]. However, accumulating evidence challenging the definition of adult stem cells compels us to put our knowledge on the issue in perspective.

1.1. A Historical Perspective on the Cancer Stem Cell Concept

During the 19th century, the pathologist Rudolf Virchow observed similarities between tumors and embryonic tissues. This led to the hypothesis, proposed later by Julius Cohnheim, that tumors are derived from dormant remnants of embryonic tissues by a process that alters proliferation and arrests differentiation [27, 38]. At the beginning of the 20th century, it was shown that injection of single leukemic murine cells induced leukemia in five mice among a group of 97 [39]. A limiting dilution *in vivo* assay showed that the number of leukemic lymphocytes required to induce leukemia in 50% of the injected hosts ranged from 0.7 to 3.0 [40]. These findings suggested that, in rare cases, leukemia is due to the malignant transformation of a single leukocyte and provided direct evidence that leukemic cells can colonize and proliferate in a new host. Analysis of chromosomes in rat ascites tumor cells serially transferred to new hosts showed a high frequency of definite patterns specific to the kind of tumor injected, suggesting the presence of long-lived tumor stem cells [41]. However, the hypothesis of cancer as a disease derived from normal tissue stem cells, with their capacity for proliferation and differentiation controlled at levels different from their normal counterparts, was proposed by Dr. G. Barry Pierce and colleagues, who conducted a series of landmark experiments [38, 42-45]. First, they showed that teratocarcinomas are hierarchically organized. Single undifferentiated cells isolated from a murine teratocarcinoma could generate a differentiated progeny [42]. Subsequently, they labeled proliferating cells in squamous carcinomas from rat epidermis with ^{3}H-thymidine and followed their progeny by autoradiography. Immediately after injection, the labelled cells were found exclusively in undifferentiated areas of the tumor. In contrast, at later time point, well-differentiated cells were labelled, suggesting that proliferating immature cells differentiated into a short-lived benign cell population [43]. These findings led to the concept of a tumor as a caricature of a normal tissue that develops from stem cells involved in tissue renewal and not from embryonic remnants [38, 44, 45]. *In vitro* clonogenic assays were developed to address the number and proliferative capacity of tumor stem-like cells and the results confirmed the scarcity of this cell population in cancers [46]. Although the concept is appealing, since the ability to induce the differentiation of tumorigenic stem-like cells would result in a cure [38], there has not been definitive proof to date that bona fide cancer stem cells exist, and such cells have not been characterized.

The hypothesis of a hierarchical organization of cancers was challenged by the clonal evolution model proposed by Dr. Peter C. Nowell in 1976. According to Dr. Nowell, tumors are heterogeneous because of their genetic instability, so progenies that vary in their biological properties are generated stochastically from a single transformed cell (see Figure 2). Therefore, clones with advantages in metabolism, proliferative capacity, chemoresistance, and mechanisms for escaping immunosurveillance predominate. In Dr. Nowell's own words, "biological characteristics of tumor progression (for example, morphological and metabolic loss of differentiation, invasion and metastasis, resistance to therapy) parallel the stages of genetic evolution" [47].

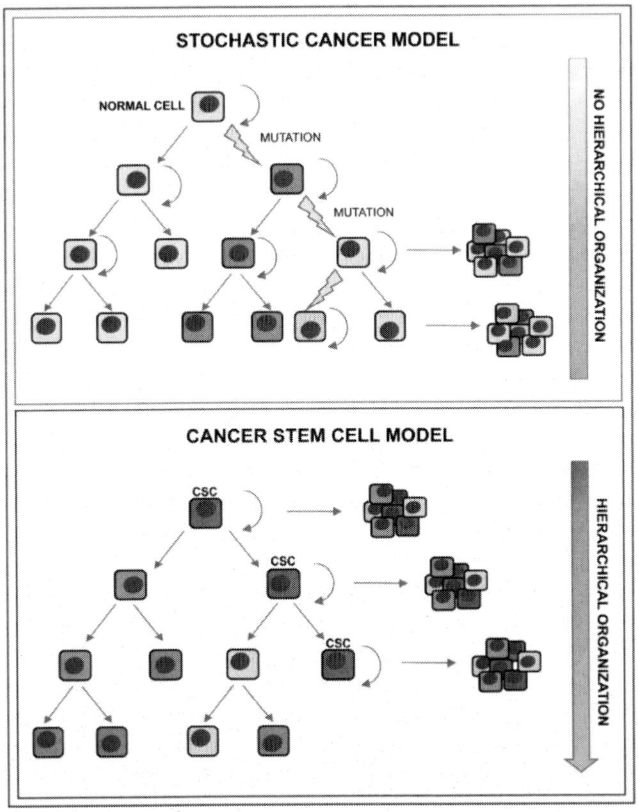

Figure 2. Cancer models. Upper panel shows stochastic cancer model, where normal cells can acquire the appropriate set of somatic mutations, producing different clones able to generate tumors. Lower panel shows cancer stem cell model that is characterized by a small population of cells with ability to self-renew, to initiate tumors and to give rise to a heterogeneous cell population, determining a hierarchical organization of tumors.

The discovery of oncogenes and tumor suppressor genes in the 20th century shifted the direction of scientific investigations into cancer. Isolation of a retrovirus as the etiological agent of a chicken sarcoma, the Rous sarcoma virus (RSV) [48], allowed a nucleotide sequence and protein, called Src, to be identified in vertebrate cells, related, respectively, to the transforming gene of Rous sarcoma virus and its product [49-52]. The existence of antioncogenes or tumor suppressor genes was first hypothesized when the tumorigenic phenotype was abrogated after some neoplastic cells were fused with normal cells. Dr. Knudson's data on human retinoblastomas also led to investigations into chromosomal abnormalities and gene mutations in this tumor [53]. Dr. Robert Weinberg and his colleagues sequenced the Retinoblastoma (Rb) gene and observed that it was inactivated in all human retinoblastomas and in some other tumors, such as small lung and bladder carcinomas [53-56].

The renaissance of the cancer stem cell concept was due to the findings of Dr. John Dick and colleagues on human acute myeloid leukemia (AML) about the end of the 20th century. Although the hypothesis of HSC transformation, at least in chronic myeloid leukemia (CML), was first raised by the observation that non-leukemic lymphocytes harbor the translocated chromosome Philadelphia (Ph) [57], there was no direct proof of it. Dr. Dick sorted leukemic cells expressing different combinations of surface markers from patients with AML and xenografted them into severe combined immunodeficient (SCID) mice. The group reported that only the cells showing the phenotype described for human HSCs ($CD45^+CD34^+CD38^-$) homed to the bone marrow and generated leukemic cells morphologically similar to those seen in the patients [58]. These findings were confirmed using the well-established NOD/SCID model and the scientists introduced the term 'SCID leukemia-initiating cells' (SL-IC) [59]. Later, the group demonstrated by serial transplantation that SL-ICs, called leukemic stem cells (LSCs), comprise a hierarchically organized population, differing in their capacity for self-renewal from normal HSCs [60]. In parallel, Dr. Connie Eaves and colleagues showed that cells with the human HSC phenotype in CML patients harbor the Ph chromosome, are multipotent, and expand in the bone marrow of more highly immunodeficient mice that lack natural killer cells (NK cells): β_2-Microglobulin deficient NOS/SCID mice (NOD/SCID-$\beta_2M^{-/-}$) [61]. Following these studies, attempts were made to characterize CSCs in solid cancers using the same approach. Tumor cells were sorted on the basis of a specific phenotype and grafted into immunocompromised mice to observe the growth of tumors similar to the original ones. The first report showed that 100 $CD44^+CD24^{-/Low}$ cells isolated from human breast cancer and xenografted into NOD/SCID mice were sufficient to induce tumors that mirrored the cellular heterogeneity of those in the cancer patients. Furthermore, serial passages of the tumorigenic population reproduced the same results, suggesting a hierarchical organization of human breast cancer with $CD44^+CD24^{-/Low}$ cells at the apex, as CSCs [62]. This seminal work was followed by characterization of putative CSCs in other human solid tumors, such as medulloblastomas and glioblastomas

[63], renal cancer [64, 65], prostatic cancer, melanoma, colon cancer, and others [6, 28, 30-32, 35]. Readers can consult other articles for additional information on the history of the CSC concept [21, 27, 33, 36, 66].

Although these studies suggest that many cancers follow the stem cell hierarchical model, they do not address the question about the origin of CSCs, or the role of non-tumorigenic cells in tumor development. Moreover, the experimental approaches used to characterize CSCs in most tumors actually reveal cells with the capacity to propagate and maintain long-term neoplastic cell populations. The terms cancer initiating cells, tumor maintaining cells, and tumor-propagating cells (TCPs) were introduced, since they reflect the uncertainty about the cell-of-origin and the readout of experimental approaches used to characterize CSCs. Although these terms seem to be more specific, they can be confusing [30, 67].

1.2. Revisiting the CSC Concept: Challenges and Caveats

In order to characterize CSCs, the tumorigenic capacities of a small number of selected cells expressing a specific combination of surface antigens were tested *in vitro* using the clonogenic sphere-forming assay described earlier [68], and *in vivo* in immunocompromised mice, the gold standard assay. This experimental strategy involved mechanical and enzymatic disruption of tissue, incubation of cells with specific antibodies, selection of a particular cell population, sphere-forming assays, and injection into new hosts with different levels of immunodeficiency. All these steps have inherent technical limitations and caveats [6, 32, 33, 69].

Xenotransplantation of human melanoma cells into a severely immunodeficient mouse strain, NOD/SCID interleukin-2 receptor gamma chain null mice (NOD/SCID $il\text{-}2rg^{-/-}$), revealed that the number of tumorigenic cells could have been underestimated. Approximately 25% of unselected melanoma cells formed tumors in the host and the addition of matrigel to the injection site increased the frequency of tumors formed in limiting dilution assays. Moreover, the tumorigenic cells were heterogeneous, suggesting that this property is not restricted to a specific phenotype but also depends upon the milieu [32, 70, 71]. Similar results were reported for skin and peripheral nerve sheath cancers, where the addition of stromal cells or laminin, respectively, enhanced the frequency of TCPs [36].

A landmark study by Dr. Robert Weinberg's group led to another challenge to the view that the CSC property is intrinsic to the cells. The group proposed that the epithelial-mesenchymal transition (EMT), a developmental program pivotal to cancer invasion and metastasis, correlates with a CSC phenotype. EMT was induced in immortalized human epithelial mammary cell lines by overexpression of the transcription factors Snail or Twist, which are downstream of transforming growth factor beta 1

(TGFβ1) signaling. The acquisition of mesenchymal traits was accompanied by expression of stem cell markers, and the capacity of these cells to form spheres *in vitro* (mammospheres) was enhanced. Furthermore, putative CSCs isolated from murine mammary gland or human breast cancers expressed EMT markers [72]; and short-term activation of Twist was sufficient to prime a subset of mammary epithelial cells to exhibit stem cell-like traits. The emergence of stable stem cell-like properties follows deactivation of Twist, which means that transient activation of an EMT program could imprint a stable phenotype [73]. These findings indicate that the surrounding microenvironment has a role in generating cancer stem-like cells by such a transient activation. However, it also raises questions about the target cells that are prone to EMT.

EMT is a complex biological program that depends on multiple signals leading to the expression of various transcription factors, suggesting that epigenetic regulatory mechanisms are involved, especially those related to chromatin remodeling [74]. Moreover, as mentioned earlier, CSCs could arise as a result of oncogenic hits targeting tissue stem cells, progenitor cells, or more differentiated cells. Depending on the cell of origin, tumors show phenotypic differences that include the degree of differentiation, hierarchical organization and aggressiveness, which indicates that activation of the EMT program also depends on cell-intrinsic properties [74-77]. In fact, oncogenic transformation of interfollicular epidermic stem cells or of hair follicle stem cells revealed a distinct chromatin organization that correlated with the EMT gene regulatory network and aggressiveness in hair follicle stem cell-derived skin carcinomas [78].

1.3. Lineage-Tracing Studies: CSCS in Untouched Tumors

As mentioned previously, selection of cancer cells and their engraftment into a new host do not address the contribution of each cell population to tumor growth or identify the cells-of-origin. Lineage-tracing assays were developed in order to answer these questions in untouched cancers. This experimental approach provided evidence that some cancers follow the CSC model and adult tissue stem cells are the cells-of-origin, while in other cases more differentiated cells acquire CSC behavior. Moreover, these studies are elucidating the plasticity of CSCs and the importance of the location of these cells within the tumor, that is, the influence of tumor microenvironment on cell behavior.

Studies using lineage-tracing approaches are revealing that tissue stem cells, progenitor cells, and differentiated cells can all become CSCs. A mosaic-genetic modified murine strain was used to study the development of gliomas induced by mutations in neural stem cells. Surprising, analysis of proliferative activity, cellular markers, cytoarchitecture, and transcriptomes all pointed to oligodendrocyte precursor cells and not neural stem cells as the cells-of-origin [79]. In contrast, skin neoplasms seem to arise from transformation of tissue stem cells. In chemical carcinogenesis, a well-established multi-step model of tumor initiation and progression, analysis of clonal expansion in a mouse strain carrying a tamoxifen-inducible specific keratin-14-cre YFP allele (*K14CreER/Rosa-YFP* mice) showed that while most of the labeled tumor cells in a benign papilloma have limited proliferative capacity and are lost after terminal differentiation, a fraction of them with stem cell characteristics persist and give rise to progeny that become conspicuous in the tumor. Interesting, in invasive squamous cell carcinoma, there was a pattern consistent with clonal expansion of a stem cell population with limited differentiation capacity [80]. Oncogenic hedgehog signaling in the skin of the same mouse strain showed that only interfollicular epidermal stem cells were able to form hierarchically organized tumors, with clonal expansion due to an increase in symmetrical division and resistance to apoptosis [81]. Intravital microscopy of a mouse model that overexpresses the polyomavirus middle T antigen (PyMT) and the Cre-inducible confetti construct made it possible to trace clones derived from cells that stochastically express one of the confetti colors during mammary cancer development. Mammary adenomas comprise a multitude of randomly-labeled cells, whereas the growth patterns of clones differ in carcinomas and large areas of the tumor express a single confetti color. Although this approach does not distinguish the cell-of-origin, it reinforces the CSC model and indicates plasticity in the CSC state [82].

In order to investigate the cell-of-origin in human colorectal cancers, scientists took advantage of genetic modifications to track $Lgr5^+$ cell lineages growing in the *in vitro* organoid model. Organoid is a three-dimensional cell culture system that allows stem cells to proliferate, differentiate, and self-assemble, forming structures that mimic the histological pattern of organs *in vivo*. Adult intestinal $Lgr5^+$ stem cells form organoids (mini-guts) that contain the cell population of the gut epithelium [83]. Lineage-tracing of $Lgr5^+$ cancer cells in organoids indicated that these cells self-renew and generate mucosecreting and absorptive cell populations [84, 85]. Moreover, ablation of $Lgr5^+$ cancer cells in organoids led to tumor regression followed by tumor regrowth after this cell population re-emerged [85]. Development of a multicolor approach to track human colon cancer cells xenografted into NOD/SCID mice revealed a decrease in the number of clones within the tumor over time owing to expansion of few clones at the leading edges. The study suggests that clonal predominance is influenced not only by cell-intrinsic properties but also by the microenvironment [86].

Melanomas can arise from activation of melanocyte stem cells [87] or from pigmented differentiated melanocytes [88], and the difference could reflect the heterogeneity of melanomas. Genetically modified mouse strains were developed to allow the lineages of melanocyte stem cells to be traced. Mice with tamoxifen-inducible Cre under the control of the melanocyte-specific tyrosinase promoter (*Tyr-CreER*) and doxycycline- or tamoxifen-regulation to drive fluorescent reporters (LSL-tdTomato or *Tre-H2B-Gfp* respectively) were further used to generate a strain carrying the oncogene

$BRAF^{V600E}$ allele (*Tyr-CreER;*^{LSL}*Braf*^{V600E}) in the presence or absence of PTEN. Ultraviolet (UVB) irradiation of the shaved skins of these mice induced regional inflammation that activated quiescent melanocyte stem cells at the hair follicle, leading to dermal melanoma formation [87]. In contrast, lineage-tracing of the melanocytic cells using the *Tyr-CreER*$^{T2/+}$ mouse strain with the fluorescent reporters *iDct-H2B-Gfp*, *LSL-tdTomato*, *LSL-YFP*, or *LSL-confetti* in the *Braf*^{V600E}; Pten/induced melanoma strain showed that interfollicular mature pigmented melanocytes expand radially at first. Subsequently, a subset of epidermal melanocytes undergoes reprogramming prior to dermal invasion and a second reprograming event takes place during the invasive phase. Amelanotic melanocytes and melanoma stem cells were resistant to transformation while in their niche, suggesting a tumor suppressor role for the surrounding microenvironment [88].

1.4. Interaction of Cancer Stem Cells with the Surrounding Niche: From Reactive Stroma to the Concept of Tumor Niche

Normally, the fate of adult stem cells is controlled by a specific microenvironment, the stem cell niche, initially characterized in the hematopoietic system. A variety of tissue-specific stem cells have been shown to occupy specialized niches composed of neighboring cells (stromal cells, hematopoietic cells, and the progeny of stem cells), vascular network, extracellular matrix (ECM) proteins, and growth factors,. creating an informative three-dimensional microenvironment. This specialized microenvironment ensures that stem cell self-renewal is maintained throughout life. The compositions of niches and the mechanisms underlying their function differ among different tissue stem cells [89, 90]. The CSC concept evolved to the idea that CSC behavior also depends on the surrounding microenvironment or CSC niche. The CSC niche is part of the tumor microenvironment and encompasses non-CSC tumor cells, stromal and immune cells, vascular network, ECM molecules, and growth factors, all modified during tumorigenesis [69, 89, 91]. It influences the stemness properties of tumor cells, implying that it is an important player on CSC plasticity, tumor heterogeneity, tumor propagating capacity, and tumor aggressiveness, as exemplified above. The hypothesis that tumor development in adult tissues is regulated by the surrounding microenvironment is not new. In 1975, Drs. Beatrice Mintz and Karl Illmensee injected malignant mouse teratoma cells with normal karyotype into blastocysts and obtained normal mosaic littermates. The authors concluded that "the origin of this tumor from a disorganized embryo suggests that malignancies of some other, more specialized, stem cells might arise comparably through tissue disorganization, leading to developmental aberrations of gene expression rather than changes in gene structure" [92]. Later, Dr. Mina Bissell's group demonstrated a pivotal role for the microenvironment in cancer pathobiology. They injected Rous sarcoma virus into chicken wing muscle and observed localized tumor growth, as expected, though the virus was found in all tissues examined. In contrast, tumor growth was observed at sites injured at the time of virus injection, suggesting that tissue lesions and inflammation promote tumor development [93]. Therefore, normal environments inhibit tumor progression, while modified ones favor signals that promote tumorigenesis and progression [94]. A reprogramming of aggressive melanoma cells in embryonic environments that reduces the invasive and tumorigenic potential of those cells corroborates the hypothesis that the microenvironment provides tumor suppressor cues [66].

Although hematopoietic cells, non-CSC cells, and ECM architecture, as part of the tumor microenvironment, are important in tumorigenesis [37, 69], the role of stromal cells in tumor progression has attracted attention since it was noticed that skin fibroblasts derived from patients with breast cancer show distinctive behavior *in vitro* [95-97]: Enhanced proliferation rate, anchorage-independent growth in soft agar [95], and aberrant migratory pattern in collagen gels [96]. The development of monoclonal antibodies recognizing α-smooth muscle actin (αSMA) allowed fibroblast-like cells expressing this molecule, called myofibroblasts, to be identified in the stroma of neoplastic lesions and in areas of inflammation and fibrosis [98-101], supporting the idea that tumor stromal cells are modified. These stromal cells, named cancer-associated fibroblasts (CAFs), are major players in tumorigenesis and tumor progression [91, 94, 102].

The origin and molecular characterization of CAFs are still a matter of debate. Besides fibroblasts in the vicinity of a tumor, cells of small and medium capillary walls, named pericytes, generate myofibroblasts. More recently, a role for local or circulating mesenchymal stromal/stem cells (MSCs) has been proposed [100, 102-106].

The mechanisms underlying CAF induction are complex and could be regarded as CSC niche-inducers. Tumors cause systemic modifications in the host. Local paracrine signaling involving cell-cell and cell-ECM interactions, secretion of chemokines, cytokines, and growth factors, and, as shown more recently, liberation of extracellular vesicles (exosomes and microvesicles) by tumor cells, can modify the surrounding environment to generate a permissive niche [102, 107]. Moreover, metastatic cancers promote the development of microenvironments in distant organs that favor the survival and growth of metastatic tumor cells. This premetastatic niche is formed, before the cancer cells arrive, by extracellular vesicles secreted by the tumor [108], corroborating the "seed and soil" hypothesis proposed by Paget in the 19th century to explain the distribution of metastases [109].

1.5. Extracellular Vesicles as Communication Mechanisms in Tumor Niches

Extracellular vesicles (EVs) are small lipid bilayer membrane vesicles involved in cell communication. They are classified into three main groups on the basis of their characteristics: microvesicles, exosomes and apoptotic bodies [110]. Microvesicles (MVs) have diameters in the 100 nm to 1 μm range. Their secretion is associated with specific stimuli such as calcium influx into the cell, which promotes alterations of phospholipid composition on the inner and outer sides of the plasma membrane along with cytoskeletal protein contraction, resulting in MV budding from the plasma membrane. The exosomes are a class of vesicles in the 10 nm to 150 nm diameter range. The biogenesis of exosomes is related to the endosome machinery; they remain stored inside multi-vesicular bodies (MVBs) until fusion with the plasma membrane releases them from the cell [110]. Apoptotic bodies are secreted only by cells undergoing apoptotic cell death. They contain organelles and have diameters of 500 nm-4 μm [110].

The mechanism involved in EV secretion is also associated with cellular stress (e.g., metabolic, hypoxic or genotoxic). The tumor environment continually confronts these phenomena, so tumor cells secrete large quantities of EVs [111-113]. EVs secreted by cancer cells are implicated in tumor cell interactions, delivering to recipient cells bioactive molecules including lipids, proteins and RNAs that induce modifications in the cell program and phenotype [114]. They are associated with increased tumor growth, drug resistance, angiogenesis, invasion and metastasis and have been the focus of several studies on tumor biology and biomarkers for clinical applications [115]. More specifically, the CSC population also secretes EVs that can modify normal surrounding cells into a pro-tumorigenic population, thus supporting immune escape, tumor growth and metastasis [116].

Exosomes isolated from glioma CSCs have been shown to contribute to cancer propagation and initiation. Such exosomes contain functionally active Cl^- intracellular channel 1 protein (CLIC1), which is associated with glioblastoma proliferation. In addition, glioma CSCs can be internalized by $CD14^+$ monocytes, which then acquire a pro-tumorigenic phenotype and support tumor immune escape [117]. Another example involves renal CSCs, which secrete EVs that alter surrounding cells such as MSCs, promoting long-period epigenetic modifications [118]. The uptake of renal CSC-EVs by MSCs resulted in upregulation of genes associated with matrix remodeling (collagen type IV alpha 3 chain, COL4A3), migration (C-X-C motif chemokine receptor 4, CXCR4 and atypical chemokine receptor 3, CXCR7), angiogenesis, and tumor growth (interleukin-8, osteopontin and myeloperoxidase). This changed the MSC phenotype into a pro-tumorigenic profile supporting the tumor. Interestingly, the cargos transferred by renal CSC-EV subpopulations seem to contain different kinds of information; exosome-enriched subpopulations induced a pro-angiogenic stimulus and microvesicle-enriched subpopulations promoted a pro-migratory response [118].

CSCs can also be modulated by EVs derived from surrounding non-tumorigenic cells. Exosomes isolated from preadipocytes regulated early-stage breast cancer (MCF10DCIS cells) by regulating stem cell renewal, cell migration, and tumor formation. Pre-treatment of the preadipocytes with shikonin, a natural antitumorigenic agent, abrogated the effects of the EVs [119]. Furthermore, Hu and co-workers showed that CAFs, which are related to tumor maintenance and growth, can confer chemoresistance on CSCs from colorectal cancer by secreting EVs [120]. Blockage of CAF exosome secretion resulted in reduced sphere-formation and tumorigenic capability of the CSCs. Taken together, these data reveal the role of EVs in tumor communication and the capacity of CSCs to modulate the surrounding cells toward a pro-tumorigenic environment.

2. Cancer Stem Cell Biomarkers and Their Potential for Clinical Application

2.1. Surface Markers

The isolation of CSC populations requires the use of markers that allow these cells to be seaparated from others present in the niche. Classically, CSC markers are surface proteins related to the phenotype and transcriptional profile of the cells. Different markers have been reported in different cancer types, and identification of specific markers can be used as tools for cancer diagnosis and prognosis [121].

CD44: An important marker, CD44, has been reported in tumors of different tissues such as pancreas, breast, kidney, ovary and colon [13]. CD44 is a transmembrane glycoprotein, a hyaluronic acid receptor and a coreceptor for growth factors and cytokines. It can be expressed as the standard molecule or as tissue-specific splice variants (CD44v1-10) and is associated with a range of cell processes [122]. In CSCs, it has a key role in interactions with the tumor niche that regulates self-renewal, tumor initiation, metastasis, and chemoradioresistance. Its importance for tumor progression was observed in a mouse model in which $CD44^{+/+}$ animals had a higher incidence of gliomas than $CD44^{+/-}$ and $CD44^{-/-}$ mice, resulting in decreased survival [123]. CD44 was also associated with tumor initiation and progression in a gastric cancer. In a mouse model where the Wnt and prostaglandin E2 pathways were activated in the gastric mucosa (K19-Wnt1/C2Me), the $CD44^{+/+}$ animals developed large gastric tumors while $CD44^{-/-}$ animals had only small tumors, indicating that CD44 is important for tumor development [124]. A study of 290 patients revealed that those with tumors expressing the CD44 marker had lower survival rates, and expression of this marker was a prognostic factor [125]. CD44v isoforms have been more directly related to CSCs. CD44v9 is associated with cancer recurrence while CD44v6 is a constitutive marker of CSCs associated with metastasis. Other isoforms have been described as possible prognostic tools for different tumors, but further studies are necessary [126, 127].

CD24: CD24 is a 27 amino acid single-chain O- and N-glycosylated protein associated with important signaling pathways in different epithelial cancer cells [128]. Its specific role is not completely clear. *In vitro* experiments with carcinoma cells (MDA-MB-435S) have shown that CD44 increases cell motility, migration and invasion through recruitment of beta1 integrin into lipid raft domains [129]. Regulation of tumor metastasis has also been linked to CD24 in an *in vivo* model of hepatocellular carcinoma [130]. However, CSCs ($CD44^+/CD24^-$) from breast and prostate cancer showed a greater capacity for migration and invasion than cancer cells that expressed CD24, indicating that this marker could be associated with specific cancer types [131].

CD133: CSC populations from brain, pancreas, lung, colon, rectum and stomach have been shown to express the CD133 marker, a five-transmembrane domain glycoprotein that regulates tumor initiation, resistance to treatment and stemness [132]. Expression of *Nestin, BMI1*, *Olig2*, and *Nanog* was upregulated in $CD133^+$ cells in association with the maintenance of the CSC population, and *CXCR4*, associated with metastasis, was also upregulated [133-135]. CD133 was originally described in hematopoietic stem cells and subsequently shown to define a subpopulation of brain tumor cells with increased capacity for tumor initiation in xenograft models [132]. In consequence, this marker has been extensively used for isolating CSCs from other solid tumors. Several signaling pathways are correlated with CD133 expression, such as the check point kinases (Chk1 and Chk2) triggered by DNA damage, Akt/PKB and Bcl2 that confer chemoresistance, or the HEDGEHOG/GLI pathway that regulates the proliferation, self-renewal and tumorigenicity of CSCs [136-138]. Interestingly, CD133 expression is modulated by oxygen levels in glioma cells. *In vitro* experiments revealed that low oxygen tension (2-3%) increases the expression of the marker, indicating the relevance of a tumor hypoxic environment in the regulation of CSCs [139].

CD105: CD105 (endoglin) is a transmembrane glycoprotein belonging to the TGFβ complex and is predominantly expressed in endothelial cells. It has been described in breast, prostate and gastric cancers, regulating cell proliferation, migration, differentiation, and angiogenesis [140]. $CD105^+$ cells isolated from clear cell renal cell carcinoma (ccRCC) showed CSC characteristics associated with the capacity to grow in a spheroid structure, initiate tumors, differentiate into epithelial or endothelial cells, generate $CD105^-$ cells and promote metastasis [64, 142]. Although CD105 has been used as a renal CSC marker, some studies have shown that other putative subpopulations of cells with CSC-like properties are $CD105^-$ [143,144]. Some authors have therefore proposed the use of CD73 (ecto-5′-nucleotidase) as a specific cell surface biomarker for ccRCC; it is associated with cancer aggressiveness and is reportedly over-expressed in several types of cancer [145]. Further studies could elucidate the relationship between CD105 and CSCs.

2.2. microRNA Biomarkers

The microRNAs (miRNAs) are non-coding RNAs important in regulating several cellular, physiological, and developmental processes [146]. Changes in the pattern of miRNA expression are associated with several diseases, including cancer [147]. Some CSC properties are regulated by specific miRNAs that can be used as biomarkers for diagnostic and therapeutic strategies [148].

The biogenesis of miRNAs is mostly associated with transcription mediated by RNA polymerase II, which generates long primary transcripts (pri-miRNAs). These pri-miRNAs are initially edited in the nucleus by RNase III Drosha to produce 70–100 nucleotide long precursors (pre-miRNAs) that are then exported to the cytoplasm, where they are further edited by a complex composed of RNase III Dicer and the transactivating response RNA-binding protein (TRBP), generating a two-strand RNA: the guide strand (mature miRNA) and the complementary strand (miRNA*) [146]. Initially, only miRNAs were thought to have a regulatory function, but miRNAs* can also have a role in the cell. In association with the RNA-induced silencing complex (RISC), mature miRNAs regulate gene expression by binding to partially complementary sequences in the 3'-untranslated regions (3'UTRs) of mRNAs, leading to degradation or inhibition of translation [146].

miRNAs have been implicated in the regulation of cancer, affecting the expression of key oncogenes and tumor suppressor genes [149]. miRNAs that regulate oncogenes are located in sites susceptible to deletions and mutations, which lead to loss of miRNA expression and consequently to overexpression of those oncogenes. Studies involving miRNAs and cancer have shown that dysregulation of miRNA expression can result in alterations in drug resistance, anti-apoptotic response and metastasis [150-152]. miRNAs are also involved in regulating different processes in CSCs such as self-renewal and tumorigenicity [153, 154]. The miRNAs associated with CSC regulation present a specific signature for each type of cancer, which has provoked great interest in the use of these molecules as biomarkers.

2.2.1. Breast Cancer Stem Cells

Studies of the miRNA profile in breast CSCs revealed a reduced expression of the let-7 family in $CD44^{+}CD24^{-/Low}$ cells. Overexpression of let-7a resulted in changes of CSC characteristics including inhibition of sphere structure formation and reduced cell proliferation and tumor initiation capacity. In addition, *in vitro* experiments revealed that let-7 antisense oligonucleotides enhanced self-renewal in non-CSCs [153]. Together with let-7, miR-30 has a synergistic effect on the regulation of self-renewal and the differentiation properties of breast CSCs. miR-30 is expressed at low levels in CSCs and directly targets ubiquitin-conjugating enzyme 9 and integrin b3, associated with the induction of apoptosis [155].

EMT, as described below, is a program often activated during tumor invasion and metastasis and is characterized by the loss of epithelial markers and the gain of mesenchymal markers. Several groups have demonstrated an inverse correlation between EMT and the miR-200 family. This family has five members divided into two clusters: miR–200a, miR–200b and miR–429 on human chromosome 1; and miR-200c and miR-141 on human chromosome 12 [156]. Overexpression of miR-200 prevents TGFβ-induced EMT by negatively regulating the expression of ZEB1 and ZEB2 (activators of EMT). The miR-200 family is downregulated in CSCs and is involved in tumor formation [48]. Members of this family target the B-lymphoma Mo-MLV insertion region 1 homolog (BMI1) and a subunit of a polycomb repressor complex (SUZ12), which are also involved in the capacity of CSCs for self-renewal [157]. miR-205 has also been reported to regulate breast CSC invasiveness and migration, affecting the incidence and development of tumors [158].

2.2.2. Prostate Cancer Stem Cells

As mentioned in the previous section (2.1, Surface markers), $CD44^+$ cells from prostate cancer have a higher metastatic and tumorigenic potential than $CD44^-$ cells [14]. Liu et al. showed that $CD44^+$ prostate cancer cells isolated from primary tumors show downregulation of miR-34a, and that this miRNA directly targets CD44 by binding to its 3'UTR [159]. miR-34a inhibits Notch and androgen receptor (AR) signaling and impedes tumor migration, metastasis and CSC self-renewal [160].

The Wnt/β-catenin pathway is directly involved in the malignancy of tumors including prostate cancer [161]. This pathway is regulated by miR-320, which directly targets β-catenin expression. Overexpression of miR-320 decreases tumorigenesis *in vivo/in vitro* and suppresses the stem cell-like properties of CSCs, while knockdown of miR-320 significantly increases tumor sphere formation, chemoresistance and protumorigenic capacities by enriching the CSC population [162].

2.2.3. Brain Cancer Stem Cells

Glioblastoma multiform (GBM) is one of the most common types of malignant brain tumor among adults and arises from glia or their precursors. miR-451 is a well-known regulator of Myc expression and is downregulated in $CD133^+$ GBM cells [163]. Transfection of CSCs with mature miR-451 dispersed neurospheres and inhibited GBM cell growth, pointing to the importance of this miRNA in CSC properties. Other miRNAs are associated with CSC regulation, for example miR-128 and miR-34a. miR-128 inhibits CSC GBM proliferation and self-renewal by directly targeting BMI1, which is responsible for maintaining neural stem cells in an undifferentiated self-renewing state [164]. miR-34a promotes the downregulation of c-MET, Notch-1/2, and cyclin dependent kinase 6 (CDK6), leading to CSC apoptosis or cell-cycle arrest [165]. miR-34a also has a significant role in regulating prostate cancer CSCs, showing that some miRNAs can act in different types of cancer.

miRNAs are also mediators of regulation in medulloblastoma, the most common malignant pediatric brain tumor. miR-199b-5p is downregulated in medulloblastoma and directly targets Hes family bHLH transcription factor 1 (HES1), a transcription factor in the Notch signaling pathway [166]. Inhibition of this pathway resulted in diminished self-renewal capacity and reduction of the CSC population. Such modulation is also mediated by miR-34a, which targets Notch-1 and Notch-2, regulating the Notch pathway [165].

2.3. Extracellular Vesicles: Body Fluid Biomarkers

Current methods for diagnosing cancers entail difficulties such as the invasiveness of the procedures, lack of sensitivity and limited information about the types and progression of tumors. As previously mentioned, tumor cells secrete EVs that are not present exclusively in their niche but are also detected in large quantities in body fluids such as blood, urine and cerebrospinal fluid. This allows EVs and the molecules they carry to be used as biomarkers for cancer.

2.3.1. Extracellular Vesicles Proteins as Biomarkers

The content of EVs reflects the cell of origin and its state. Therefore, several studies have focused on the screening of protein markers in EVs derived from tumor cells. A large proteomic profile of different cancer cell lines from different tissues was obtained with the aim of elucidating the information carried by EVs and using this information as a diagnostic tool [167]. All EVs contain common proteins derived from different types of cancer cells, mostly related to biogenesis and secretion. In addition, EVs carry proteins expressed in the respective tissue of origin [110], which can help to identify the type of cancer and its location. The protein cargo of EVs can also indicate the type of tumor cell. Proteomic analysis of EVs and the cell of origin revealed that Agrin, a heparan sulfate proteoglycan from basement membranes that is associated with focal adhesion integrity, is present in EVs derived from adherent tumor cell lines. In contrast, leukemia cancer lines grown in suspension release EVs containing the adhesion molecule ICAM3, indicating differences between the markers for adherent tumor cells and those for tumor cells in suspension [168].

A study to evaluate the prognostic potential of membrane-bound proteins on EVs in a non-selected prospective cohort of non-small cell lung cancer (NSCLC) patients in all stages revealed that NY-ESO-1, EGFR, PLAP, EpCam and Alix correlate strongly with the prediction of reduced patient survival [169]. In another study of NSCLC patients, EVs were isolated from pleural effusion fluid and the protein content was analyzed. EGFR, K-ras, basigin (EMMPRIN, CD147), carcinoembryonic antigen-related cell adhesion molecule 6 (CEACAM6), claudin1, claudin3, and RAB-family proteins were enriched in these EVs, suggesting that those molecules are potential lung cancer biomarkers [170]. A group of proteins isolated from EVs from the plasma of patients with lung adenocarcinoma, absent from the plasma itself, included potential markers for the disease. Among the 34 proteins identified as potential biomarkers were serglycin (SRGN); actin filament-binding protein tropomyosin alpha-3 chain (TPM3); adhesive glycoprotein thrombospondin-1 (THBS1); E3 ubiquitin-protein ligase (HECT); N-terminal ubiquitin-associated (UBA); and WWE domain-containing protein 1 (HUWE1), all present in a lung adenocarcinoma cell line [171].

The potential of the protein contents of EVs as biomarkers was also explored by comparing healthy individuals with patients diagnosed with cancer at different stages. Mass spectrometry revealed that the cell surface proteoglycan glypican-1 (GPC1) is enriched in cancer-cell-derived exosomes, distinguishing healthy subjects and patients with a benign pancreatic disease from those with early- and late-stage pancreatic cancers [172]. Pancreatic cancer progression is indicated by the presence of a specific CSC population characterized by surface proteins such as CD44v6, CD44, MET, Tspan8 and CD133 [173,174]. EVs secreted by these cells carry the same proteins, which are associated with the regulation of metastasis and can be used as specific CSC-associated markers for this type of tumor.

With the increased interest in diagnosis using isolated EVs, new techniques have been developed to establish procedures with greater sensitivity, lower cost and more rapid application. Advances in liquid biopy biosensors could make it possible to achieve such goals, focusing on the detection of exosomal proteins as markers for cancer diagnosis and prognosis. Immunofluorescence is commonly used to detect proteins in EVs or on their surfaces and involves specific antibodies against those proteins. Microfluid biochips based on this technique have been developed to isolate EVs from plasma and characterize their components. Comparisons between lung and ovarian cancer patients and healthy individuals, using a microfluid biosensor, showed an increase of proteins in EVs from the cancer patients (CD9, CD81, EpCAM, α-IGF-1R) [175]. Zhang and co-workers took a different approach, using a panel of antibodies that could detect markers specific to EVs and to a type of tumor, and observed the increase in EpCAM level in EVs from ovarian cancer patients using only 2μL of plasma without sample processing [176].

The Surface Plasmon Resonance (SPR) platform is a combination of label-free and real-time detection techniques developed to analyze the components of EVs. Receptor molecules are anchored to metal surfaces and, depending on the antibodies used, the population of EVs can be characterized. Grasso and co-workers analyzed the EVs from three breast cancer cell lines (MDA-MB-231, MCF-7, and BT-474) simultaneously for multiple exosomal and cancer biomarkers (CD9, CD24, CD44, CD63, EpCAM, and HER2). The results indicated a highly sensitive technique that allowed the EVs from the different cell lines to be distinguished [177]. According to these authors, EVs from healthy individuals were also characterized by the same markers using a small volume of blood plasma (0.5 mL), corroborating the sensitivity of the technique and indicating its potential in cancer diagnosis and personalized medicine.

Several others promising approaches have been developed for identifying cancer EV protein markers: colorimetric detection exploiting the oxidation of 3,3′,5,5′-tetramethylbenzidine, which is used to detect EV proteins; surface-enhanced Raman scattering (SERS), which uses plasmonics to obtain detailed chemical information about the molecules adsorbed or attached to nanostructured metallic surfaces; electrochemical sensors that through the use of biosensors provide high detection sensitivity and can define the proteins and their concentrations; micronuclear magnetic resonance (μNMR), which allows the EV concentration to be quantified simultaneously with protein expression through a biosensor that combines microfiltration and nuclear magnetic resonance [178]. The results obtained from all these approaches show greater sensitivity of the methods, allowing proteins to be detected in low amounts of sample and EVs. This suggests the possibility of detecting subpopulations of EVs such as those produced by CSCs, which could lead to more precise and earlier diagnoses.

2.3.2. Extracellular Vesicle RNAs as Biomarkers

RNAs are among the molecules carried by EVs and are also the focus of studies as biomarkers for cancer. RNAs have been detected inside EVs in different types of tissues and cells, physiologically and in pathological conditions such as cancer [179]. EVs derived from highly metastatic ovarian cancer cells support the tumor microenvironment, promoting invasion and metastasis [180, 181]. They are enriched in mRNAs coding for metalloproteinase-1 (MMP1), an enzyme associated with extracellular matrix and basement-membrane degradation during cancer progression. MMP1 in EVs can be used as a marker for the initial stages of ovarian cancer. In addition, EVs derived from glioblastoma cells surgically removed from human tumors were enriched in mRNAs associated with angiogenesis, cell proliferation, immune response, cell migration and histone modification [182]. The detection of epidermal growth factor receptor (EGFR) mRNA in EVs also aroused interest since expression of the EGFRvIII mutant/variant is specific to some tumors and defines a clinical subtype of glioma [183], suggesting the possibility of diagnosis without invasive procedures.

miRNAs inside EVs can also be used as biomarkers. Several reports have shown a correlation between specific miRNAs or groups of miRNAs and cancer, serving for early diagnosis and for distinguishing benign from malignant diseases. An important aspect to be considered is the diversity of protocols used to isolate and characterize EVs, which leads to discrepancies in the group of miRNAs correlated with the type of cancer, for example whether EVs are isolated from plasma or serum. Platelets secreted by EVs during blood coagulation represent 50% of all the EVs in serum [184], suggesting that plasma could be a better option for identifying EV miRNAs specific to the pathology.

Ovarian cancer cells secrete large quantities of exosomes into the blood stream that carry miRNAs. Comparison between EVs from ovarian cancer and benign tumor patients led to the identification of a group of miRNAs (miR-21, miR-141, miR-200a, miR-200c, miR-200b, miR-203, miR-205 and miR-214) that were highly elevated in the cancer patients [185]. Interestingly, while the miRNA profiles were different, EVs derived from patients with benign ovarian disease and ovarian cancer were both positive for EpCAM, indicating that the miRNAs can be used as biomarkers for identifying tumor stages. In addition, analysis of miRNAs in EVs from ovarian carcinoma peritoneal and pleural effusion supernatants from 86 patients revealed that high expression of miR-21 correlated with poor overall survival, while high levels of miR-21, miR-23b and miR-29a were associated with poor progression-free survival [186].

EV miRNAs have also been used for diagnosis in breast cancer studies. Analysis of miRNAs in EVs from 50 patients with breast cancer and 12 healthy donors showed higher levels of miR-101 and miR-372 in the serum exosomes of the breast cancer patients [187]. In the same study, the authors showed that miR-373 was significantly higher in triple-negative patients than luminal cancer patients or healthy controls. The miRNAs were also correlated with metastatic characteristics of the tumor; miR-105 was significantly higher in the pre- and metastatic stages in a breast tumor-bearing animal model. Furthermore, EVs isolated from patients who developed distant metastases showed significantly higher miR-105 levels than patients that did not develop metastatic disease, showing the potential of miRNAs for predicting and diagnosing breast cancer metastasis [188].

Diagnosis and prognosis of various types of cancer, including glioma, have been associated with miR-21. EVs isolated from the cerebrospinal fluid of patients with recurrent glioma had higher levels of miR-21 than healthy individuals [189]. In the same study, a correlation was found between miR-21 levels in EVs and tumor spinal/ventricle metastasis. This association between miR-21 in the EVs and glioblastoma seems to be specific to cerebrospinal fluid since miR-21 expression in serum-isolated EVs did not differ between healthy individuals and cancer patients.

CSCs isolated from renal carcinoma release EVs enriched in specific groups of miRNAs [141]. Examination of miRNAs in CSCs and non-CSCs from the tumor revealed enrichments of miR-19b, miR-29c, miR-92, miR-141, miR-151, miR-650 and miR-200c. miR-29a, miR-151 and miR-650 are associated with tumor invasion and metastasis [190-192], whereas miR-19b, miR-29c, and miR-151 are more highly expressed in renal carcinomas than normal renal tissue [193]. In addition, the authors showed an angiogenic and metastatic role for EVs secreted by CSCs. Such results indicate the specificity of miRNAs as biomarkers for detecting subpopulations of tumor cells, including CSCs.

3. Therapies Associated with the Cancer Stem Cell Hypothesis

Anti-CSC therapies began during the 1970s with the blockage of leukemic cells in an undifferentiated state [194]. Subsequently, new therapies based on CSC differentiation were developed through epigenetic inhibition of multiple hematological malignancies [195]. Investigations of CSC properties, their niche and the regulatory molecules associated with tumor progression have indicated the possibility of creating new cancer treatment strategies. Some of these new approaches are discussed in this section.

3.1. Targeting the CSC Niche

As mentioned in section 1.4, interaction with the surrounding niche is essential for maintaining the CSC population. The resistance of CSCs under survival pressure factors such as hypoxia, nutrient deprivation and even drug treatment depends on communications with the heterogeneous population of tumor components (endothelial, hematopoietic, fibroblast, perivascular and vascular cells). Furthermore, non-CSCs can be sufficiently plastic to undergo phenotypic transition under a specific stimulus and replace the lost CSC population. In this context, new strategies have been delineated targeting the CSC niche instead of the CSC population directly.

Vascular endothelial growth factor (VEGF) is a key regulator of angiogenesis in tumors that supports CSC survival and drug resistance. Treatment with bevacizumab, a monoclonal antibody directed against VEGF, prevents the growth and maintenance of tumor blood vessels and diminishes the glioblastoma stem cell population [196]. Furthermore, the tumor-initiating cell ($CD133^+$/nestin$^+$) population was significantly reduced in U87 glioma cell xenografts mice treated with bevacizumab. The microvasculature and tumor size wre also reduced [197]. Despite the promising results, the use of anti-angiogenic agents increased tumor invasiveness and metastasis owing to activation of a CSC survival program including the hypoxia-inducible factor 1α and Akt/β-catenin regulatory pathways [198]. The effectiveness of the anti-angiogenic treatment could require direct targeting of the CSC population in parallel.

The bone marrow niche has also been a target for anti-CSC therapy. Stromal cell-derived factor-1 (SDF-1) is mostly secreted by bone marrow stromal cells and can promote the homing of leukemia cells into bone marrow and inducing cell growth and chemoresistance. The use of antagonists to the SDF-1 receptor CXCR4, such as Plerixafor (AMD3100) and T14003 analogs, modified the interactions of the malignant cell with the microenvironment, altering its mobilization and drug resistance [199]. Several clinical trials are being undertaken to explore this strategy.

3.2. Targeting Stemness Pathways

The Wnt/β-catenin and Notch pathways are associated with differentiation and the maintenance of resident stem cell niches in normal tissue and are also involved in the reversion of progenitor cells into the stem cell state when CSCs are lost [200]. Overexpression of β-catenin in CSCs conferred on these cells the capacity to form mammospheres and to acquire radiation resistance [201]. Different anti-tumor approaches based on the role of Wnt have been proposed. For example, monoclonal antibody antagonists to Wnt-1 and Wnt-2 showed significant anti-tumor effects in different cancers [202]. Notch signaling is also a potential target in CSC niche therapy. In breast cancer, Notch is activated and supports CSC self-renewal. Fan and co-workers described the blockage of CSC self-renewal and reduction of medulloblastomas by targeting Notch signaling with γ-secretase inhibitors, which cleave Notch receptors and release their intracellular domains for pathway activation [203]. Induction of CSC differentiation by bone morphogenic protein (BMP) is another possible approach to cancer treatment. In colorectal cancer, CSCs stimulated with BMP4 were induced to terminal differentiation and became less chemoresistant [204]. A similar approach involved stimulation of the oncostatin M (OSM) pathway, resulting in CSC differentiation and increase of chemosensitivity in liver and breast cancers [205, 206].

CONCLUSION

The CSC concept has led to new perspectives about tumor organization and interactions. It is accepted that tumors harbor stem-like cells, the CSCs, which gain the capacity to propagate the tumor. It is becoming clear that the cell-of-origin varies among tumors and affects the degree of heterogeneity, the hierarchical organization, the cytoarchiteture, and the aggressiveness of cancers. Epigenetics and the microenvironment influence the EMT program that imprints CSC traits during cancer progression. These events also affect clone expansion and a certain plasticity of CSCs is recognized, which reconciles the stochastic model with the hierarchical model. The suppressive cues provided by the normal microenvironment are shifted in cancer towards signals that favor tumor progression and formation of CSC niches. Local and systemic mechanisms construct permissive microenvironments at the tumor site, and also at distant organs, favoring metastasis.

Therapeutic strategies based on the CSC concept have brought encouraging results in cancer treatment and several clinical trials are investigating their potential. However, further studies are necessary to elucidate the biology of CSCs and their role in tumor development and thus achieve safer and more effective therapies. Future anti-CSC therapies must address important issues such as increased specificity, targeting exclusive

CSCs, and avoiding toxicity to normal tissue and stem cells. New studies based on gene expression profiling, next generation sequencing and high content screening could yield the information required to achieve this specificity in anti-CSC treatment.

REFERENCES

[1] Slack JMW. Origin of stem cells in organogenesis. *Science*. 2008; 322:1498-501.

[2] Weissman IL. Stem cells: Units of development, units of regeneration, and units in evolution. *Cell*. 2000; 100:157-68.

[3] Doulatov S, Notta F, Laurenti E, Dick JE. Hematopoiesis: A human perspective. *Cell Stem Cell*. 2012; 10:120-36.

[4] Blanpain C, Fuchs E. Plasticity of epithelial stem cells in tissue regeneration. *Science*. 2014; 344:1242281.

[5] Clevers H. What is an adult stem cell? *Science*. 2015; 350:1319-20.

[6] Batlle E, Clevers H. Cancer stem cell revisited. *Nature Med*. 2017; 23:1124-34.

[7] Gonzales KAU, Fuchs E. Skin and its regenerative powers: An alliance between stem cells and their niche. *Dev. Cell*. 2017; 43:387-401.

[8] Konstantinov IE. In search of Alexander A. Maximow: The man behind the unitarian theory of hematopoiesis. *Perspect. Biol. Med*. 2000; 43:269-76.

[9] Schofield R. The relationship between the spleen colony-forming cell and the haematopoietic stem cell. *Blood Cells*. 1978: 4:7-25.

[10] Till JE, McCulloch EA. A direct measurement of the radiation sensitivity of normal mouse bone marrow cells. *Radiat. Res*. 1961; 14:213-22.

[11] Herzenberg LA, Parks D, Sahaf B, Perez O, Roederer M. The history and future of the fluorescence activated cell sorter and flow cytometry: a view from Stanford. *Clin. Chem*. 2002; 48:1819–27.

[12] Spangrude GJ, Heimfeld S, Weissman IL. Purification and characterization of mouse hematopoietic stem cells. *Science*. 1988; 241:58–62.

[13] Andrews RG, Singer JW, Bernstein ID. Monoclonal antibody 12-8 recognzes a 115-kd molecule present on both unipotent and multipotent hematopoietic colony-forming cells and their precursors. *Blood* 1986; 67:842-45.

[14] Sutherland HJ, Eaves CJ, Eaves AC, Dragowaska W, Lansdorp PM. Characterization and partial purification of human marrow cells capable of initiating long-term hematopoiesis *in vitro*. *Blood*. 1989; 74:156370.

[15] Bhatia M, Wang JCY, Kapp U, Bonnet D, Dick JE. Purification of primitive human hematopoietic cells capable of repopulating immune-deficient mice. *Proc. Natl Acad. Sci. USA*. 1997; 94:5320–25.

[16] Guenechea G, Gan OI, Dorrell C, Dick JE. Distinct classes of human stem cells that differ in proliferative and self-renewal potential. *Nat. Immunol*. 2001; 2:75-82.

[17] Li L, Clevers H. Coexistence of quiescent and active adult stem cells in mammals. *Science.* 2010; 327:542-45.

[18] Copley MR, Beer PA, Eaves CJ. Hematopoietic stem cell heterogeneity takes center stage. *Cell Stem Cell.* 2012; 10:690-7.

[19] Furusawa C, Kaneko K. A dynamical-systems view of stem cell biology. *Science.* 2012; 338:215-7.

[20] Muller-Sieburg CE, Sieburg HB, Bernitz JM, Cattarossi G. Stem cell heterogeneity: implications for aging and regenerative medicine. *Blood.* 2012; 119:3900-07.

[21] Clevers H. The cancer stem cell: premises, promises and challenges. *Nat. Med.* 2011; 17:313-9.

[22] Visvader JE, Clevers H. Tissue-specific designs of stem cell hierarchies. *Nat. Cell* Biol. 2016; 18:349-55.

[23] Stange DE, Koo B-K, Huch M et al. Differentiated troy$^+$ chief cells act as reserve stem cells to generate all lineages of the stomach epithelium. *Cell.* 2013; 155:357-68.

[24] Leushacke M, Tan HS, Wong A et al. Lgr5-expressing chief cells drive epithelial regeneration and cancer in the oxyntic stomach. *Nat. Cell Biol.* 2017; 19:774-86.

[25] Hanahan D, Weinberg RA. Hallmarks of cancer: The next generation. *Cell.* 2011; 144:646-74.

[26] Reya T, Morrison SJ, Clarke MF, Weissman IL. Stem cells, cancer, and cancer stem cells. *Nature.* 2001; 414: 05-11.

[27] Huntly BJP, Gilliland DG. Leukaemia stem cells and the evolution of cancer-stem-cell research. *Nat. Rev. Cancer.* 2005; 5:311-21.

[28] Aiiles LE, Weissman IL. Cancer stem cells in solid tumors. *Curr. Op. Biotechnol.* 2007; 18:460–66.

[29] Shackleton M, Quintana E, Fearon ER, Morrison SJ. Heterogeneity in cancer: Cancer stem cells versus clonal evolution. *Cell.* 2009; 136:822-29.

[30] Alison MR, Islam S, Wright NA. Stem cells in cancer: instigators and propagators? *J. Cell Sci.* 2010; 123:2357-68.

[31] Hermann PC, Bhaskar S, Cioffi M, Heeschen C. Cancer stem cells in solid tumors. *Semin. Cancer Biol.* 2010; 20:77-84.

[32] Magee J, Piskounova E, Morrison SJ. Cancer stem cells: Impact, heterogeneity, and uncertainty. *Cancer Cell.* 2012; 21:283-96.

[33] Nguyen LV, Vanner R, Dirks P, Eaves CJ. Cancer stem cells: an evolving concept. *Nat. Rev. Cancer.* 2012; 12:133-43.

[34] Visvader JE, Lindeman GJ. Cancer stem cells: current status and evolving complexities. *Cell Stem Cell.* 2012; 10:717-28.

[35] Beck B, Blanpain C. Unravelling cancer stem cell potential. *Nat. Rev. Cancer.* 2013; 13:727-38.

[36] Nassar D, Blanpain C. Cancer stem cells: Basic concepts and therapeutic implications. *Annu. Rev. Pathol. Mech. Dis*. 2016; 11:47–76.

[37] Aponte PM, Caicedo A. Stemness in cancer: stem cells, cancer stem cells, and their microenvironment. *Stem Cells Int*. 2017; 5619472. doi.org/10.1155/2017/5619472.

[38] Sell S, Pierce GB. Maturation arrest of stem cell differentiation is a common pathway for the cellular origin of teratocarcinomas and epithelial cancers. *Lab. Invest*. 1994; 70:6–22.

[39] Furth J, Khan MC. The transmission of mice with a single cell. *Am. J. Cancer*. 1937; 31:276-82.

[40] Hewitt HB. Studies of the dissemination and quantitative transplantation of a lymphocytic leukaemia of CBA mice. *Br. J. Cancer*. 1958; 12:378–401

[41] Makino S. Further evidence favoring the concept of the stem cell in ascites tumors of rats. *Ann. NY Acad. Sci*. 1956; 63:818–30.

[42] Kleinsmith LJ, Pierce GB. Multipotentiality of single embryonal carcinoma cells. *Cancer Res*. 1964; 24:1544–51.

[43] Pierce GB, Wallace C. Differentiation of malignant to benign cells. *Cancer Res*. 1971; 31:127-34.

[44] Pierce GB. Neoplasms, differentiations and mutations. *Am. J. Pathol*. 1974; 77: 103-13.

[45] Pierce GB, Speers WC. Tumors as caricatures of the process of tissue renewal: Prospects for therapy by directing differentiation. *Cancer Res*. 1988; 48: 1996-2004.

[46] Hamburger AW, Salmon SE. Primary bioassay of human tumor stem cells. *Science*. 1977; 197:461-3.

[47] Nowell PC. The clonal evolution of tumor cell populations. *Science*. 1976; 194: 23-8.

[48] Rous P. A sarcoma of the fowl transmissible by an agent separable from the tumor cells. *J. Exp. Med*. 1911; 13:397-411.

[49] Spector DH, Varmus HE, Bishop JM. Nucleotide sequences related to the transforming gene of avian sarcoma virus are present in DNA of uninfected vertebrates. *Proc. Natl. Acad. Sci. USA*. 1978; 75:4102-06.

[50] Oppermann H, Levinson AD, Varmus HE, Levintow L, Bishop JM. Uninfected vertebrate cells contain a protein that is closely related to the product of the avian sarcoma virus transforming gene (src). *Proc. Natl. Acad. Sci. USA*. 1979; 76: 1804-08.

[51] Bishop JM. Cellular oncogenes and retroviruses. *Ann. Rev. Biochem*. 1983; 52: 301-54.

[52] Bishop JM. The molecular genetics of cancer. *Science*. 1987; 235:305-11.

[53] Weinberg RA. Oncogenes, antioncogenes, and the molecular bases of multistep carcinogenesis. *Cancer Res*. 1989; 49:3713-21.

[54] Friend SH, Horowitz JM, Gerber MR et al. Deletions of a DNA sequence in retinoblastomas and mesenchymal tumors: organization of the sequence and its encoded protein. *Proc. Nati. Acad. Sci. USA*. 1987; 84:9059-63.

[55] Horowitz JM, Park S-H, Bogenmann E et al. Frequent inactivation of the retinoblastoma anti-oncogene is restricted to a subset of human tumor cells. *Proc. Nati. Acad. Sci. USA*. 1990; 87: 2775-79.

[56] Weinberg RA. The Molecular basis of oncogenes and tumor suppressor genes. *Ann. NY Acad. Sci.* 1995; 758:331-38.

[57] Fialkow PJ, Denman AM, Jacobson RJ, Lowenthal MN. Chronic myelocytic leukemia. Origin of some lymphocytes from leukemic stem cells. *J. Clin. Invest.* 1978; 62:815-23.

[58] Lapidot T, Sirard C, Vormoord J, MPaterson B, Caligluri M, Dick JE. A cell initiating human acute myeloid leukaemia after transplantation into SCID mice. *Nature.* 1994; 367:645–8.

[59] Bonnet D, Dick JE. Human acute myeloid leukemia is organized as a hierarchy that originates from a primitive hematopoietic cell. *Nat. Med.* 1997; 3:730-37.

[60] Hope KJ, Jin L, Dick JE. Acute myeloid leukemia originates from a hierarchy of leukemic stem cell classes that differ in self-renewal capacity. *Nat. Med.* 2004; 5:738-43.

[61] Holyoake T, Jiang X, Eaves C, Eaves A. isolation of a highly quiescent subpopulation of primitive leukemic cells in chronic myeloid leukemia. *Blood.* 1999; 94:2056-64.

[62] Al-Hajj M, Wicha MS, Benito-Hernandez A, Morrison SJ, Clarke MF. Prospective identification of tumorigenic breast cancer cells. *Proc. Natl. Acad. Sci. USA*. 2003; 100:3983-88.

[63] Singh SK, Hawkins C, Clarke ID et al. Identification of human brain tumour initiating cells. *Nature.* 2004; 432:396-401.

[64] Bussolati B, Bruno S, Grange C, Ferrando U, Camussi G. Identification of a tumor-initiating stem cell population in human renal carcinomas. *FASEB J.* 2008; 22:3696-705.

[65] Bussolati B, Dekel B, Azzarone B, Camussi G. Human renal cancer stem cells. *Cancer Lett*. 2013; 338:141-6.

[66] Hendrix MJC, Seftor EA, Seftor REB, Kasemeier-Kulesa J, Kulesa PM, Postovit L-M. Reprogramming metastatic tumour cells with embryonic microenvironments. *Nat. Rev. Cancer*. 2007; 7:246-55.

[67] Valent P, Bonnet D, De Maria R et al. Cancer stem cell definitions and terminology: the devil is in the details. *Nat. Rev. Cancer.* 2012; 12:767-75.

[68] Reynolds BA, Weiss S. Clonal and population analyses demonstrate that an EGF-responsive mammalian embryonic CNS precursor is a stem cell. *Dev. Biol.* 1996; 175:1-11.

[69] Plaks V, Kong N, Werb Z. The cancer stem cell niche: How essential is the niche in regulating stemness of tumor cells? *Cell Stem Cell.* 2015; 16:225-38.

[70] Quintana E, Shackleton M, Sabel MS, Fullen DR, Johnson TM, Morrison SJ. Efficient tumour formation by single human melanoma cells. *Nature.* 2008; 456:593-99.

[71] Quintana E, Shackleton M, Foster HR, Fullen DR, Sabel MS, Johnson TM, Morrison SJ. Phenotypic heterogeneity among tumorigenic melanoma cells from patients that is reversible and not hierarchically organized. *Cancer Cell.* 2010; 18:510-23.

[72] Mani SA, Guo W, Liao M-J et al. The epithelial-mesenchymal transition generates cells with properties of stem cells. *Cell.* 2008; 133:704-15.

[73] Schmidt JM, Panzilius E, Bartsch HS et al. Stem-cell-like properties and epithelial plasticity arise as stable traits after transient twist1 activation. *Cell Rep.* 2015; 10:131-39.

[74] Tam WL, Weinberg RA. The epigenetics of epithelial-mesenchymal plasticity in cancer. *Nat. Med.* 2013; 19:1438-49.

[75] Ince TA, Richardson AL, Bell GW et al. Transformation of different human breast epithelial cell types leads to distinct tumor phenotypes. *Cancer Cell.* 2007; 12: 160-70.

[76] Chaffer CL, Brueckmann I, Scheel C et al. Normal and neoplastic nonstem cells can spontaneously convert to a stem-like state. *Proc. Natl. Acad. Sci. USA.* 2011; 108:7950-55.

[77] White AC, Lowry WE. Refining the role for adult stem cells as cancer cells of origin. *Trends Cell Biol.* 2015; 25:11-20.

[78] Latil M, Nassar D, Beck B et al. Cell-type-specific chromatin states differentially prime squamous cell carcinoma tumor-initiating cells for epithelial to mesenchymal transition. *Cell Stem Cell.* 2017; 20:1-14.

[79] Liu C, Sage JC, Miller MR et al. Mosaic analysis with double markers reveals tumor cell of origin in glioma. *Cell.* 2011; 146:209-21.

[80] Driessens G, Beck B, Caauwe A, Simons BD, Blanpain C. Defining the mode of tumour growth by clonal analysis. *Nature.* 2012; 488:527-30.

[81] Sánchez-Danés A, Hannezo E, Larsimont JC et al. Defining the clonal dynamics leading to mouse skin tumour initiation. *Nature.* 2016; 536: 298-303.

[82] Zomer A, Ellenbroek SIJ, Ritsma L, Beerling E, Vrisekoop N, Van Rheenen J. Brief report: intravital imaging of cancer stem cell plasticity in mammary tumors. *Stem Cells.* 2013; 31:602-06.

[83] Sato T, Clevers H. Growing self-organizing mini-guts from a single intestinal stem cell: mechanism and applications. *Science.* 2013; 340:1190-94.

[84] Cortina C, Turon G, Stork D et al. A genome editing approach to study cancer stem cells in human tumors. *EMBO Mol. Med.* 2017; 9:869-79.

[85] Shimokawa M, Ohta Y, Nishikori S et al. Visualization and targeting of $LGR5^+$ human colon cancer stem cells. *Nature*. 2017; 545:187-94.
[86] Lamprecht S, Schmidt EM, Blaj C et al. Multicolor lineage tracing reveals clonal architecture and dynamics in colon cancer. *Nat. Commun*. 2017; 8: 1406.
[87] Moon H, Donahue LR, Choi F et al. Melanocyte stem cell activation and translocation initiate cutaneous melanoma in response to UV exposure. *Cell Stem Cell*. 2017; 21:665-78.
[88] Köhler C, Nittner D, Rambow F et al. Mouse cutaneous melanoma induced by mutant Braf arises from expansion and dedifferentiation of mature pigmented melanocytes. *Cell Stem Cell*. 2017; 21:679-93.
[89] Morrison SJ, Spradling AC. Stem cells and niches: mechanisms that promote stem cell maintenance throughout life. *Cell*. 2008; 132:596-611.
[90] Lander AD, Kimble J, Clevers H. et al. What does the concept of the stem cell niche really mean today? *BMC Biol*. 2012; 10:19.
[91] Hanahan D, Coussens LN. Accessories to the crime: functions of cells recruited to the tumor microenvironment. *Cancer Cell*. 2012; 21:309-22.
[92] Mintz B, Illmensee K. Normal genetically mosaic mice produced from malignant teratocarcinoma cells. *Proc. Nat. Acad. Sci. USA*. 1975; 72:3585-89.
[93] Dolberg DS, Hollingsworth R, Hertle M, Bissell MJ. Wounding and its role in RSV-mediated tumor formation. *Science*. 1985; 230:676-8.
[94] Bissell MJ, Hines WC. Why don't we get more cancer? A proposed role of the microenvironment in restraining cancer progression. *Nat. Med*. 2011; 17:320-9.
[95] Azzarone B, Mareel M, Billard C, Scemama P, Chaponnier C, Macieira-Coelho A. Abnormal properties of skin fibroblasts from patients with breast cancer. *Int. J. Cancer*. 1984; 33:759-64.
[96] Schor SL, Schor AM, Durning P, Rushton G. Skin fibroblasts obtained from cancer patients display foetal-like migratory behaviour on collagen gels. *J. Cell Sci*. 1985; 73:235-44.
[97] Azzarone B, Macieira-Coelho A. Further characterization of the defects of skin fibroblasts from cancer patients. *J. Cell Sci*. 1987; 87:155-62.
[98] Skalli O, Ropraz P, Trzeciak A, Benzonana G, Gillessen D, Gabbiani G. A monoclonal antibody against a-smooth muscle actin: A new probe for smooth muscle differentiation. *J. Cell. Biol*. 1986; 103:2787-96.
[99] Tsukada T, McNutt MA, Ross R, Gown AM. HHF35, α muscle actin-specific monoclonal antibody. II. Reactivity in normal, reactive, and neoplastic human tissues. *Am. J. Pathol*. 1988; 127:389-402.
[100] Rønnov-Jessen L, Petersen OW. Koteliansky VE, Bissell MJ. The origin of the myofibroblasts in breast cancer. *J. Clin. Invest*. 1995; 95:859-73.

[101] Rønnov-Jessen L, Petersen OW, Bissell MJ. Cellular changes involved in conversion of normal to malignant breast: Importance of the stromal reaction. *Physiol. Rev.* 1996; 76:69-125.

[102] McAllister SS, Weinberg RA. The tumour-induced systemic environment as a critical regulator of cancer progression and metastasis. *Nat. Cell Biol.* 2014; 15:717-27.

[103] Erez N, Truitt M, Olson P, Hanahan D. Cancer-associated fibroblasts are activated in incipient neoplasia to orchestrate tumor-promoting inflammation in an nf-kb-dependent manner. *Cancer Cell.* 2010; 17:135-47.

[104] Quante M, Tu SP, Tomita H et al. Bone marrow-derived myofibroblasts contribute to the mesenchymal stem cell niche and promote tumor growth. *Cancer Cell.* 2011; 19:257-72.

[105] De Wever O, Van Bockstal M, Mareela M, Hendrix A, Brack M. Carcinoma-associated fibroblasts provide operational flexibility in metastasis. *Semin. Cancer Biol.* 2014; 25:33-46.

[106] Öhlund D, Elyada E, Tuveson D. Fibroblast heterogeneity in the cancer wound. *J. Exp. Med.* 2014; 211:1503-23.

[107] Barcellos-Hoff MH, Lyden D, Wang TC. The evolution of the cancer niche during multistage carcinogenesis. *Nat. Rev. Cancer.* 2013; 13:511-18.

[108] Peinado H, Zhang H, Matei IR et al. Pre-metastatic niches: organ-specific homes for metástases. *Nat. Rev. Cancer.* 2017; 17:302-17.

[109] Paget S. Distribution of secondary growths in cancer of the breast. *Lancet.* 1889; 133:571-73.

[110] Raposo G, Stoorvogel W. Extracellular vesicles: exosomes, microvesicles, and friends. *J Cell Biol.* 2013;200:373-83.

[111] Park JE, Tan HS, Datta A, Lai RC, Zhang H, Meng W, et al. Hypoxic tumor cell modulates its microenvironment to enhance angiogenic and metastatic potential by secretion of proteins and exosomes. *Mol Cell Proteomics.* 2010;9:1085-99.

[112] Clayton A, Turkes A, Navabi H, Mason MD, Tabi Z. Induction of heat shock proteins in B-cell exosomes. *J Cell Sci.* 2005;118:3631-8.

[113] De Jong OG, Verhaar MC, Chen Y, Vader P, Gremmels H, Posthuma G, et al. Cellular stress conditions are reflected in the protein and RNA content of endothelial cell-derived exosomes. *J Extracell Vesicles.* 2012;1.

[114] Colombo M, Raposo G, Théry C. Biogenesis, secretion, and intercellular interactions of exosomes and other extracellular vesicles. *Annu Rev Cell Dev Biol.* 2014;30:255-89.

[115] Webber J, Yeung V, Clayton A. Extracellular vesicles as modulators of the cancer microenvironment. *Semin Cell Dev Biol.* 2015;40:27-34.

[116] Hannafon BN, Ding WQ. Cancer stem cells and exosome signaling. *Stem Cell Investig.* 2015;2:11.

[117] Domenis R, Cesselli D, Toffoletto B, Bourkoula E, Caponnetto F, Manini I, et al. Systemic T Cells Immunosuppression of Glioma Stem Cell-Derived Exosomes Is Mediated by Monocytic Myeloid-Derived Suppressor Cells. *PLoS One.* 2017;12:e0169932.

[118] Lindoso RS, Collino F, Camussi G. Extracellular vesicles derived from renal cancer stem cells induce a pro-tumorigenic phenotype in mesenchymal stromal cells. *Oncotarget.* 2015;6:7959-69.

[119] Gernapudi R, Yao Y, Zhang Y, Wolfson B, Roy S, Duru N, et al. Targeting exosomes from preadipocytes inhibits preadipocyte to cancer stem cell signaling in early-stage breast cancer. *Breast Cancer Res Treat.* 2015;150:685-95.

[120] Hu Y, Yan C, Mu L, Huang K, Li X, Tao D, et al. Fibroblast-Derived Exosomes Contribute to Chemoresistance through Priming Cancer Stem Cells in Colorectal Cancer. *PLoS One.* 2015; 10:e0125625.

[121] Mimeault M, Batra SK. Molecular biomarkers of cancer stem/progenitor cells associated with progression, metastases, and treatment resistance of aggressive cancers. *Cancer Epidemiol Biomarkers Prev.* 2014;23:234-54.

[122] Xu H, Tian Y, Yuan X, Wu H, Liu Q, Pestell RG, et al. The role of CD44 in epithelial-mesenchymal transition and cancer development. *Onco Targets Ther.* 2015;8:3783-92.

[123] Pietras A, Katz AM, Ekström EJ, Wee B, Halliday JJ, Pitter KL, et al. Osteopontin-CD44 signaling in the glioma perivascular niche enhances cancer stem cell phenotypes and promotes aggressive tumor growth. *Cell Stem* Cell. 2014;14:357-69.

[124] Ishimoto T, Nagano O, Yae T, Tamada M, Motohara T, Oshima H, et al. CD44 variant regulates redox status in cancer cells by stabilizing the xCT subunit of system xc(-) and thereby promotes tumor growth. *Cancer Cell.* 2011;19:387-400.

[125] Cao X, Cao D, Jin M, Jia Z, Kong F, Ma H, et al. CD44 but not CD24 expression is related to poor prognosis in non-cardia adenocarcinoma of the stomach. *BMC Gastroenterol.* 2014;14:157.

[126] Jiang H, Zhao W, Shao W. Prognostic value of CD44 and CD44v6 expression in patients with non-small cell lung cancer: meta-analysis. *Tumour Biol.* 2014;35:7383-9.

[127] Ozawa M, Ichikawa Y, Zheng YW, Oshima T, Miyata H, Nakazawa K, et al. Prognostic significance of CD44 variant 2 upregulation in colorectal cancer. *Br J Cancer.* 2014;111:365-74.

[128] Fang X, Zheng P, Tang J, Liu Y. CD24: from A to Z. *Cell Mol Immunol.* 2010;7:100-3.

[129] Runz S, Mierke CT, Joumaa S, Behrens J, Fabry B, Altevogt P. CD24 induces localization of beta1 integrin to lipid raft domains. *Biochem Biophys Res Commun.* 2008;365:35-41.

[130] Yang XR, Xu Y, Yu B, Zhou J, Li JC, Qiu SJ, et al. CD24 is a novel predictor for poor prognosis of hepatocellular carcinoma after surgery. *Clin Cancer Res.* 2009;15:5518-27.

[131] Croker AK, Goodale D, Chu J, Postenka C, Hedley BD, Hess DA, et al. High aldehyde dehydrogenase and expression of cancer stem cell markers selects for breast cancer cells with enhanced malignant and metastatic ability. *J Cell Mol Med.* 2009;13:2236-52.

[132] Neuzil J, Stantic M, Zobalova R, Chladova J, Wang X, Prochazka L, et al. Tumour-initiating cells vs. cancer 'stem' cells and CD133: what's in the name? *Biochem Biophys Res Commun.* 2007;355:855-9.

[133] Ma S, Chan KW, Lee TK, Tang KH, Wo JY, Zheng BJ, et al. Aldehyde dehydrogenase discriminates the CD133 liver cancer stem cell populations. *Mol Cancer Res.* 2008;6:1146-53.

[134] Miki J, Furusato B, Li H, Gu Y, Takahashi H, Egawa S, et al. Identification of putative stem cell markers, CD133 and CXCR4, in hTERT- immortalized primary nonmalignant and malignant tumor-derived human prostate epithelial cell lines and in prostate cancer specimens. *Cancer Res.* 2007;67:3153-61.

[135] Bertolini G, Roz L, Perego P, Tortoreto M, Fontanella E, Gatti L, et al. Highly tumorigenic lung cancer $CD133^+$ cells display stem-like features and are spared by cisplatin treatment. *Proc Natl Acad Sci USA.* 2009;106:16281-6.

[136] Ma S, Lee TK, Zheng BJ, Chan KW, Guan XY. $CD133^+$ HCC cancer stem cells confer chemoresistance by preferential expression of the Akt/PKB survival pathway. *Oncogene.* 2008;27:1749-58.

[137] Bao S, Wu Q, McLendon RE, Hao Y, Shi Q, Hjelmeland AB, et al. Glioma stem cells promote radioresistance by preferential activation of the DNA damage response. *Nature.* 2006;444:756-60.

[138] Clement V, Sanchez P, de Tribolet N, Radovanovic I, Ruiz i Altaba A. HEDGEHOG-GLI1 signaling regulates human glioma growth, cancer stem cell self-renewal, and tumorigenicity. *Curr Biol.* 2007;17:165-72.

[139] Bidlingmaier S, Zhu X, Liu B. The utility and limitations of glycosylated human CD133 epitopes in defining cancer stem cells. J Mol Med (Berl). 2008;86:1025-32.

[140] Fonsatti E, Maio M. Highlights on endoglin (CD105): from basic findings towards clinical applications in human cancer. *J Transl Med.* 2004;2:18.

[141] Grange C, Tapparo M, Collino F, Vitillo L, Damasco C, Deregibus MC, et al. Microvesicles released from human renal cancer stem cells stimulate angiogenesis and formation of lung premetastatic niche. *Cancer Res.* 2011;71:5346-56.

[142] Myszczyszyn A, Czarnecka AM, Matak D, Szymanski L, Lian F, Kornakiewicz A, et al. The Role of Hypoxia and Cancer Stem Cells in Renal Cell Carcinoma Pathogenesis. *Stem Cell Rev.* 2015;11:919-43.

[143] Ueda K, Ogasawara S, Akiba J, Nakayama M, Todoroki K, Ueda K, et al. Aldehyde dehydrogenase 1 identifies cells with cancer stem cell-like properties in a human renal cell carcinoma cell line. *PLoS One*. 2013;8:e75463.

[144] Gassenmaier M, Chen D, Buchner A, Henkel L, Schiemann M, Mack B, et al. CXC chemokine receptor 4 is essential for maintenance of renal cell carcinoma-initiating cells and predicts metastasis. *Stem Cells*. 2013;31:1467-76.

[145] Song L, Ye W, Cui Y, Lu J, Zhang Y, Ding N, et al. Ecto-5'-nucleotidase (CD73) is a biomarker for clear cell renal carcinoma stem-like cells. *Oncotarget*. 2017;8:31977-31992.

[146] Winter J, Jung S, Keller S, Gregory RI, Diederichs S. Many roads to maturity: microRNA biogenesis pathways and their regulation. *Nat Cell Biol*. 2009;11: 228-34.

[147] Di Leva G, Garofalo M, Croce CM. MicroRNAs in cancer. *Annu Rev Pathol*. 2014;9:287-314.

[148] Sun X, Jiao X, Pestell TG, Fan C, Qin S, et al. MicroRNAs and cancer stem cells: the sword and the shield. *Oncogene*. 2014;33:4967-77.

[149] Iorio MV, Croce CM. MicroRNA dysregulation in cancer: diagnostics, monitoring and therapeutics. A comprehensive review. *EMBO Mol Med*. 2012;4:143-59.

[150] Cimmino A, Calin GA, Fabbri M, Iorio MV, Ferracin M, Shimizu M, et al. miR-15 and miR-16 induce apoptosis by targeting BCL2. *Proc Natl Acad Sci USA*. 2005; 102:13944-9.

[151] Tavazoie SF, Alarcón C, Oskarsson T, Padua D, Wang Q, Bos PD, G, et al. Endogenous human microRNAs that suppress breast cancer metastasis. *Nature*. 2008;451:147-52.

[152] To KK, Zhan Z, Litman T, Bates SE. Regulation of ABCG2 expression at the 3' untranslated region of its mRNA through modulation of transcript stability and protein translation by a putative microRNA in the S1 colon cancer cell line. *Mol Cell Biol*. 2008;28:5147-61.

[153] Yu F, Yao H, Zhu P, Zhang X, Pan Q, Gong C, et al. let-7 regulates self renewal and tumorigenicity of breast cancer cells. *Cell*. 2007;131:1109-23.

[154] Song SJ, Poliseno L, Song MS, Ala U, Webster K, Ng C, et al. MicroRNA-antagonism regulates breast cancer stemness and metastasis via TET-family-dependent chromatin remodeling. *Cell*. 2013;154:311-324.

[155] Yu F, Deng H, Yao H, Liu Q, Su F, Song E. Mir-30 reduction maintains self-renewal and inhibits apoptosis in breast tumor-initiating cells. *Oncogene*. 2010;29:4194-204.

[156] Gregory PA, Bert AG, Paterson EL, Barry SC, Tsykin A, Farshid G, et al. The miR-200 family and miR-205 regulate epithelial to mesenchymal transition by targeting ZEB1 and SIP1. *Nat Cell Biol*. 2008;10:593-601.

[157] Sun L, Yao Y, Liu B, Lin Z, Lin L, Yang M, et al. MiR-200b and miR-15b regulate chemotherapy-induced epithelial-mesenchymal transition in human tongue cancer cells by targeting BMI1. *Oncogene.* 2012;31:432-45.

[158] Zhang J, Wei B, Hu H, Liu F, Tu Y, He F. The association between differentially expressed micro RNAs in breast cancer cell lines and the micro RNA-205 gene polymorphism in breast cancer tissue. *Oncol Lett.* 2018;15:2139-2146.

[159] Liu C, Kelnar K, Liu B, Chen X, Calhoun-Davis T, Li H, et al. The microRNA miR-34a inhibits prostate cancer stem cells and metastasis by directly repressing CD44. *Nat Med.* 2011;17:211-5.

[160] Kashat M, Azzouz L, Sarkar SH, Kong D, Li Y, Sarkar FH. Inactivation of AR and Notch-1 signaling by miR-34a attenuates prostate cancer aggressiveness. *Am J Transl Res.* 2012;4:432-42.

[161] Tycko B, Li CM, Buttyan R. The Wnt/beta-catenin pathway in Wilms tumors and prostate cancers. *Curr Mol Med.* 2007;7:479-89.

[162] Hsieh IS, Chang KC, Tsai YT, Ke JY, Lu PJ, Lee KH, et al. MicroRNA-320 suppresses the stem cell-like characteristics of prostate cancer cells by downregulating the Wnt/beta-catenin signaling pathway. *Carcinogenesis.* 2013;34:530-8.

[163] Gal H, Pandi G, Kanner AA, Ram Z, Lithwick-Yanai G, Amariglio N, et al. MIR-451 and Imatinib mesylate inhibit tumor growth of Glioblastoma stem cells. *Biochem Biophys Res Commun.* 2008;376:86-90.

[164] Godlewski J, Nowicki MO, Bronisz A, Williams S, Otsuki A, Nuovo G, et al. Targeting of the Bmi-1 oncogene/stem cell renewal factor by microRNA-128 inhibits glioma proliferation and self-renewal. *Cancer Res.* 2008;68:9125-30.

[165] Li Y, Guessous F, Zhang Y, Dipierro C, Kefas B, Johnson E, et al. MicroRNA-34a inhibits glioblastoma growth by targeting multiple oncogenes. *Cancer Res.* 2009;69:7569-76.

[166] Garzia L, Andolfo I, Cusanelli E, Marino N, Petrosino G, De Martino D, et al. MicroRNA-199b-5p impairs cancer stem cells through negative regulation of HES1 in medulloblastoma. *PLoS One.* 2009;4:e4998.

[167] Fontana S, Saieva L, Taverna S, Alessandro R. Contribution of proteomics to understanding the role of tumor-derived exosomes in cancer progression: state of the art and new perspectives. *Proteomics.* 2013;13:1581-94.61.

[168] Hurwitz SN, Rider MA, Bundy JL, Liu X, Singh RK, Meckes DG Jr. Proteomic profiling of NCI-60 extracellular vesicles uncovers common protein cargo and cancer type-specific biomarkers. *Oncotarget.* 2016;7:86999-87015.

[169] Sandfeld-Paulsen B, Aggerholm-Pedersen N, Bæk R, Jakobsen KR, Meldgaard P, Folkersen BH, Rasmussen TR, et al. Exosomal proteins as prognostic biomarkers in non-small cell lung cancer. *Mol Oncol.* 2016;10:1595-1602.

[170] Park JO, Choi DY, Choi DS, Kim HJ, Kang JW, Jung JH, et al. Identification and characterization of proteins isolated from microvesicles derived from human lung cancer pleural effusions. *Proteomics*. 20131;13:2125-34.
[171] Vykoukal J, Sun N, Aguilar-Bonavides C, Katayama H, Tanaka I, et al. Plasma-derived extracellular vesicle proteins as a source of biomarkers for lung adenocarcinoma. *Oncotarget*. 2017;8:95466-95480.
[172] Melo SA, Luecke LB, Kahlert C, Fernandez AF, Gammon ST, Kaye J, et al. Glypican-1 identifies cancer exosomes and detects early pancreatic cancer. *Nature*. 2015; 523:177-82.
[173] Wang H, Rana S, Giese N, Büchler MW, Zöller M. Tspan8, CD44v6 and alpha6beta4 are biomarkers of migrating pancreatic cancer-initiating cells. *Int J Cancer*. 2013;133:416-26.
[174] Jung T, Castellana D, Klingbeil P, Cuesta Hernández I, Vitacolonna M, Orlicky DJ, et al. CD44v6 dependence of premetastatic niche preparation by exosomes. *Neoplasia*. 2009;11:1093-105.
[175] He M, Crow J, Roth M, Zeng Y, Godwin AK. Integrated immunoisolation and protein analysis of circulating exosomes using microfluidic technology. *Lab Chip*. 2014;14:3773-80.
[176] Zhang P, He M, Zeng Y. Ultrasensitive microfluidic analysis of circulating exosomes using a nanostructured graphene oxide/polydopamine coating. *Lab Chip*. 2016;16:3033-42.
[177] Grasso L, Wyss R, Weidenauer L, Thampi A, Demurtas D, Prudent M, et al. Molecular screening of cancer-derived exosomes by surface plasmon resonance spectroscopy. *Anal Bioanal Chem*. 2015;407:5425-32.
[178] Liu C, Yang Y, Wu Y. Recent Advances in Exosomal Protein Detection Via Liquid Biopsy Biosensors for Cancer Screening, Diagnosis, and Prognosis. *AAPS J*. 2018;20:41.
[179] Redzic JS, Balaj L, van der Vos KE, Breakefield XO. Extracellular RNA mediates and marks cancer progression. *Semin Cancer Biol*. 2014;28:14-23.
[180] Peinado H, Alečković M, Lavotshkin S, Matei I, Costa-Silva B, Moreno-Bueno G, et al. Melanoma exosomes educate bone marrow progenitor cells toward a pro-metastatic phenotype through MET. *Nat Med*. 2012;18:883-91.
[181] Yang C, Robbins PD. The roles of tumor-derived exosomes in cancer pathogenesis. *Clin Dev Immunol*. 2011;2011:842849.
[182] Skog J, Würdinger T, van Rijn S, Meijer DH, Gainche L, Sena-Esteves M, Curry WT Jr, et al. Glioblastoma microvesicles transport RNA and proteins that promote tumour growth and provide diagnostic biomarkers. *Nat Cell Biol*. 2008;10:1470-6.
[183] Pelloski CE, Ballman KV, Furth AF, Zhang L, Lin E, Sulman EP, et al. Epidermal growth factor receptor variant III status defines clinically distinct subtypes of glioblastoma. *J Clin Oncol*. 2007;25:2288-94.

[184] Gemmell CH, Sefton MV, Yeo EL. Platelet-derived microparticle formation involves glycoprotein IIb-IIIa. Inhibition by RGDS and a Glanzmann's thrombasthenia defect. *J Biol Chem*. 1993;268:14586-9.

[185] Taylor DD, Gercel-Taylor C. MicroRNA signatures of tumor-derived exosomes as diagnostic biomarkers of ovarian cancer. *Gynecol Oncol*. 2008;110:13-21.

[186] Vaksman O, Tropé C, Davidson B, Reich R. Exosome-derived miRNAs and ovarian carcinoma progression. *Carcinogenesis*. 2014;35:2113-20.

[187] Joyce DP, Kerin MJ, Dwyer RM. Exosome-encapsulated microRNAs as circulating biomarkers for breast cancer. *Int J Cancer*. 2016;139:1443-8.

[188] Zhou W, Fong MY, Min Y, Somlo G, Liu L, Palomares MR, et al. Cancer-secreted miR-105 destroys vascular endothelial barriers to promote metastasis. *Cancer Cell*. 2014;25:501-15.

[189] Shi R, Wang PY, Li XY, Chen JX, Li Y, Zhang XZ, et al. Exosomal levels of miRNA-21 from cerebrospinal fluids associated with poor prognosis and tumor recurrence of glioma patients. *Oncotarget*. 2015;6:26971-81.

[190] Gebeshuber CA, Zatloukal K, Martinez J. miR-29a suppresses tristetraprolin, which is a regulator of epithelial polarity and metastasis. *EMBO Rep*. 2009;10:400-5.

[191] Zhang X, Zhu W, Zhang J, Huo S, Zhou L, Gu Z, et al. MicroRNA-650 targets ING4 to promote gastric cancer tumorigenicity. *Biochem Biophys Res Commun*. 2010;395:275-80.

[192] Luedde T. MicroRNA-151 and its hosting gene FAK (focal adhesion kinase) regulate tumor cell migration and spreading of hepatocellular carcinoma. *Hepatology*. 2010;52:1164-6.

[193] Chow TF, Youssef YM, Lianidou E, Romaschin AD, Honey RJ, Stewart R, et al. Differential expression profiling of microRNAs and their potential involvement in renal cell carcinoma pathogenesis. *Clin Biochem*. 2010;43:150-8.

[194] Nowak D, Stewart D, Koeffler HP. Differentiation therapy of leukemia: 3 decades of development. *Blood*. 2009;113:3655-65.

[195] Stahl M, Kohrman N, Gore SD, Kim TK, Zeidan AM, Prebet T. Epigenetics in Cancer: A Hematological Perspective. *PLoS Genet*. 2016;12:e1006193.

[196] Shin DH, Lee SJ, Kim JS, Ryu JH, Kim JS. Synergistic Effect of Immunoliposomal Gemcitabine and Bevacizumab in Glioblastoma Stem Cell-Targeted Therapy. *J Biomed Nanotechnol*. 2015;11:1989-2002.

[197] Calabrese C, Poppleton H, Kocak M, Hogg TL, Fuller C, Hamner B, et al. A perivascular niche for brain tumor stem cells. *Cancer Cell*. 2007;11:69-82.

[198] Conley SJ, Gheordunescu E, Kakarala P, Newman B, Korkaya H, Heath AN, et al. Antiangiogenic agents increase breast cancer stem cells via the generation of tumor hypoxia. *Proc Natl Acad Sci USA*. 2012;109:2784-9.

[199] 199 Domanska UM, Kruizinga RC, Nagengast WB, Timmer-Bosscha H, Huls G, de Vries EG, et al. A review on CXCR4/CXCL12 axis in oncology: no place to hide. *Eur J Cancer*. 2013;49:219-30.

[200] Yu Z, Pestell TG, Lisanti MP, Pestell RG. Cancer stem cells. *Int J Biochem Cell Biol.* 2012;44:2144-51.

[201] Woodward WA, Chen MS, Behbod F, Alfaro MP, Buchholz TA, Rosen JM. WNT/beta-catenin mediates radiation resistance of mouse mammary progenitor cells. *Proc Natl Acad Sci USA*. 2007;104:618-23.

[202] O'Flaherty JD, Barr M, Fennell D, Richard D, Reynolds J, O'Leary J, et al. The cancer stem-cell hypothesis: its emerging role in lung cancer biology and its relevance for future therapy. *J Thorac Oncol*. 2012;7:1880-90.

[203] Fan X, Matsui W, Khaki L, Stearns D, Chun J, Li YM, Eberhart CG. Notch pathway inhibition depletes stem-like cells and blocks engraftment in embryonal brain tumors. *Cancer Res*. 2006;66:7445-52.

[204] Lombardo Y, Scopelliti A, Cammareri P, Todaro M, Iovino F, Ricci-Vitiani L, et al. Bone morphogenetic protein 4 induces differentiation of colorectal cancer stem cells and increases their response to chemotherapy in mice. *Gastroenterology*. 2011;140:297-309.

[205] Yamashita T, Honda M, Nio K, Nakamoto Y, Yamashita T, Takamura H, et al. Oncostatin m renders epithelial cell adhesion molecule-positive liver cancer stem cells sensitive to 5-Fluorouracil by inducing hepatocytic differentiation. *Cancer Res*. 2010;70:4687-97.

[206] West NR, Murray JI, Watson PH. Oncostatin-M promotes phenotypic changes associated with mesenchymal and stem cell-like differentiation in breast cancer. *Oncogene*. 2014;33:1485-94.

In: Research on New Generation …
Editor: Yildiz Dincer
ISBN: 978-1-53614-367-6

Chapter 7

tRNA and Their Fragments as Diagnostic and Prognostic Biomarkers in Cancer

Nurten Bahtiyar[1], PhD and Ilhan Onaran[2,*], PhD
[1]Istanbul University, Cerrahpasa Medical Faculty,
Department of Biophysics, Istanbul, Turkey
[2]Istanbul University, Cerrahpasa Medical Faculty,
Department of Medical Biology, Istanbul, Turkey

Abstract

Increasing knowledge about non-coding RNAs (ncRNA) led to modifications and, ultimately, expansions of the original concept regarding cancer. These RNAs are able to distinguish cancer types and subtypes and have the potential to be biomarkers in cancers. ncRNAs are deregulated in a widespread manner in almost every type of human cancer, and the signature of RNA expression in tumors can be potentially used as biomarkers for tumor characterization and cancer prognosis. Although the evidence on this subject come mainly from mRNA and miRNA experiments, recent studies have also reported the occurrence of circulating transfer RNA (tRNA) halves in serum and plasma samples. Furthermore, their fragments were found to be deregulated in several human cancers. In addition to differential tRNA expression in cancer cells, research has shown that there are tissue-specific differences in the expression of tRNA species and coordinated expression among tRNAs decoding amino acids with similar chemical properties in distinct tissue types. Tumor cells generally express higher levels of nuclear-encoded tRNA than every tissue examined. In particularly, tRNA fragments are in blood components can be candidate markers for the diagnosis and prognosis of different types of malignancies. However, there is a need to address the level of correlation that exists between the tRNA deregulation identified in tumor tissue and circulating tRNAs. This chapter may provide an inside look at cancer diagnosis and follow-up from tRNAs perspective.

* Corresponding Author Email: ilonaran@istanbul.edu.tr.

Keywords: tRNA, tRNA fragments, cancer, biomarker

INTRODUCTION

In recent years, there has been an intensive effort towards the development of efficient cancer biomarkers in human biological fluids and tissues for the detection of multiple cancers. Analyses of the findings from the recent standardization of high throughput approaches showed that genomic networks are modified and rewired during tumorigenesis. These studies revealed that changes in gene expression networks in tumors include non-coding RNAs (ncRNAs) as well as coding RNA. Among ncRNAs, microRNAs (miRNAs) have been most intensely studied and experimental evidence supports their potential as cancer biomarkers. In addition to miRNAs, there have been recent discoveries on cell type-specific transfer RNA (tRNA) regulation and unexpected tRNA functions. There has been also been studies demonstrating that various cancer types are correlated with tRNA deregulation, such as changes in the expression of tRNAs and of the modification enzymes, and variations in the modification level. In further studies, some tRNA fragments, in particularly tRNA halves, have been detected in plasma, serum, and other bodily fluids from cancer patients. All these have increased the interest for discovering candidate biomarkers in tRNAs as well as miRNAs which, so far have been the main focus of the investigation.

In this chapter, we present a summary of the potentially most relevant findings on this subject. We first give an overview on tRNA, its complexity, and the tissue- and time-specific expression of tRNAs. We then discuss how the modulated expression of specific tRNAs, tRNA fragments, and the changes in tRNA processing, or, in tRNA modifications, have been linked to tumorigenesis. Finally, we try to summarize the current status of tRNA–related biomarker candidates and the potential for their future development.

TRNAS: WHAT, WHERE AND WHY?

tRNA is universal to all domains of life and have an important functional role in protein synthesis. They are best known as adapter molecules in the translation of mRNA to protein, and the most abundant and extensively modified RNA among all cellular RNAs in a human cell with tens of millions of transcripts and on an average 13 modifications per molecule, respectively. Each tRNA is charged with a corresponding amino acid and delivered into the ribosome during protein biosynthesis. Additionally, a variety of evidence implicates that tRNAs can further serve as a source of tRNA fragments small functional RNAs that affect many cellular processes. Although cytoplasmic mature tRNAs molecules of human cells range in size from 73 to 93 nucleotides (about 80 nucleotides), mitochondrial tRNAs are shorter than their cytosolic counterparts, ranging in size from 59 to 75 nucleotides.

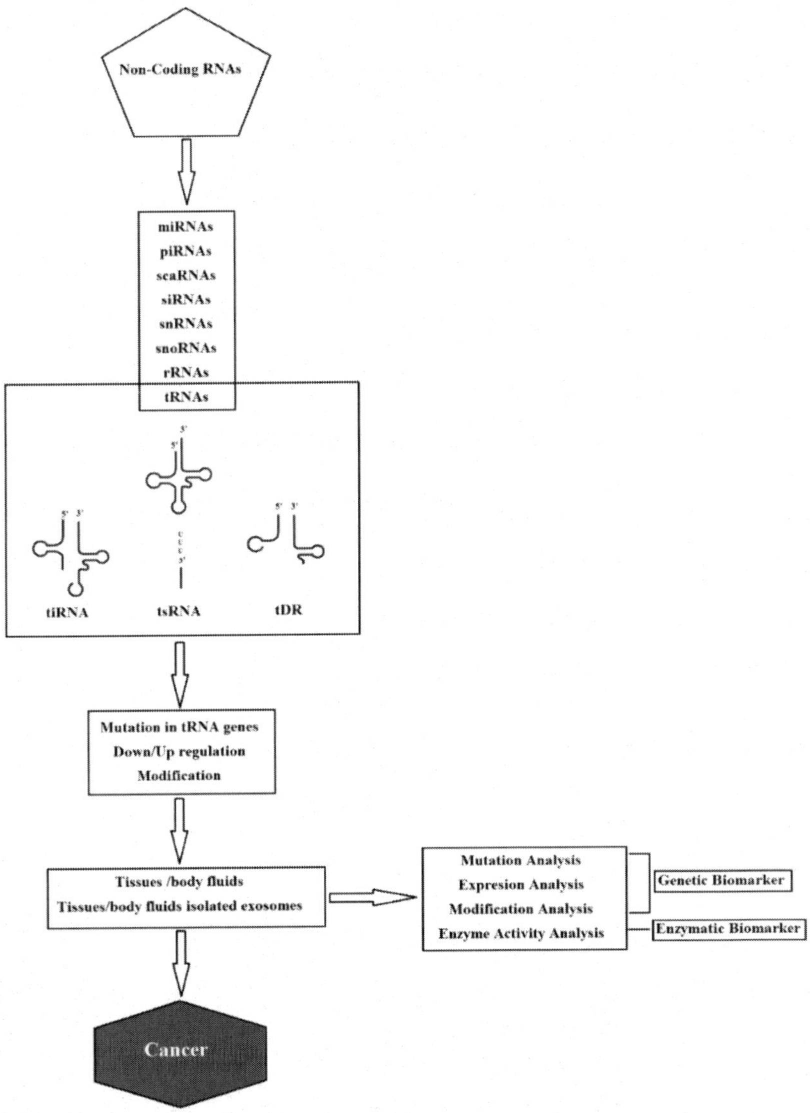

Figure 1. Schematic workflow for potential cancer tRNA biomarkers.

In humans, tRNAs are transcribed by RNA polymerase III as pre-tRNAs in the nucleus and undergo extensive modifications in the form of endonucleolytic cleavages, exonucleolytic trimming, non-templated nucleotide additions, and multiple single base modification events. After the various processing steps occur in the nucleus and the cytoplasm, the highly modified and aminoacylated tRNA may then be used within the translation. tRNAs are not always end products but are processed further into smaller ncRNAs, many of which are known to have different functional roles apart from protein translation. Processing of precursor and mature forms of tRNA result in the production of three distinct classes of tRNA-derived ncRNAs, which are heterogeneous in nucleotide composition, size, biogenesis and function: tRNA-derived small RNA (tsRNAs), tRNA-derived fragments (tRFs or tDRs), and tRNA halves (also known as stress-dependent tRFs or tiRNAs). tRFs and tiRNA are generated in the cytoplasm from mature tRNA, while tsRNAs are generated in the nucleus as a consequence of the pre-tRNA 3'-end cleavage. Currently, there is evidence that some tRNA fragments could play a critical role in cancer by interfering with gene expression regulation at different levels [1-3]. In the next sections, we will discuss of each of these tRNA-derived ncRNAs.

Human tRNA Complexity

To discover new biomarkers, it is important to understand molecular complexity and diversity of putative novel biomarker candidate(s). Therefore, we refer to several recent reviews on this topic. The actual number of nuclear-encoded tRNA genes varies considerably among organisms, even among closely related species, and generally increases with the complexity of the organism. Given the reference human genome, approximately 450 tRNA genes are dispersed throughout the human genome and are located on all chromosomes except chromosome Y (http://lowelab.ucsc.edu/GtRNAdb/Hsapi/). The duplications for some tRNA genes have been also annotated in the human genome. Additionally, 22 tRNA genes are present in human mitochondrial GENCODE website (https://www.gencodegenes.org/), the number of predicted tRNA genes is 631 for 49 isoacceptors. There is at least one tRNA species for each of the 20 standard proteinogenic amino acids.

Once an amino acid is stably attached to its corresponding tRNA, its position in a growing polypeptide chain is determined by the codon–anticodon interactions on the mRNA and not by the nature of the amino acid. Because of the degeneracy of the genetic code, there is more than one codon for most amino acids. tRNAs with distinct sequences, including the sequence of the anticodon but incorporating the same amino acid in protein synthesis are termed "isoacceptors". tRNA isoacceptors can be encoded by multiple nuclear genes, many of which are completely identical in sequence, which are distributed in different genomic locations. It has been suggested that translation rates, which also can have an effect on co-translational protein folding can be affected by the identity of the specific tRNA isoacceptor that carries the amino acid.

Human tRNAs are grouped into 49 isoacceptor families to decode the 21 amino acids (including selenocysteine) specified by the genetic code. Thus, the presence of many tRNA-encoding genes and the code's degeneracy are the first step of the complexity of the tRNA world. Furthermore, tRNA gene copy numbers show inter-individual variability in humans and can even differ among individuals. *Additional diversity* arises as results of base modifications in tRNAs, by the existence of a plethora of tRNA interacting partners. In addition to isoacceptors to the tRNA repertoire, there are also isodecoders-distinct tRNAs with the same anticodon but different bodies with sequence variations in sequence outside the anticodon.

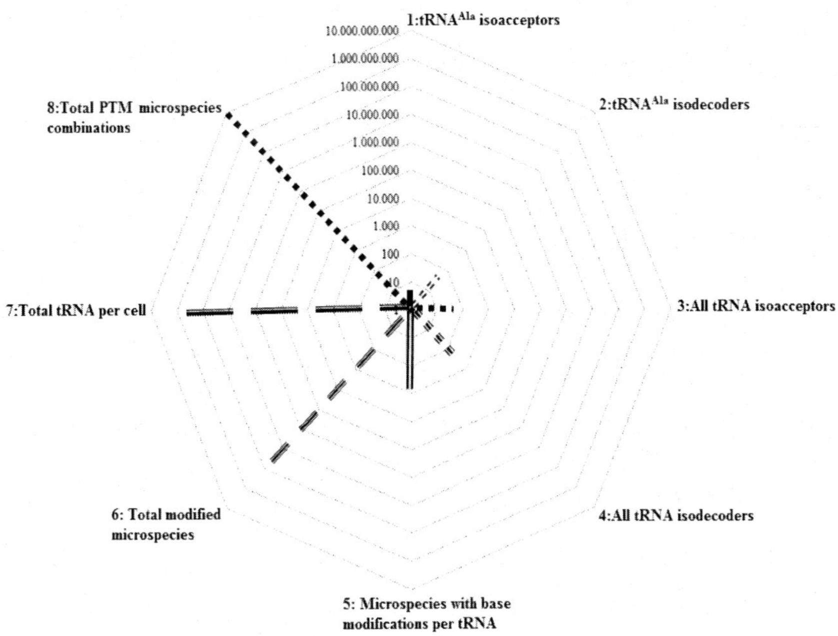

Figure 2. The emerging complexity of the tRNA world: tRNA complexity starts and progresses from a single tRNA to all combinations of the different possible mixtures of tRNA species.

Human genome is very rich in tRNA isodecoder genes. For example, there are roughly 30 genes and 24 isodecoders for $tRNA^{Gly-GCA}$ with GCA anticodon. Each tRNA isodecoder in cells is aminoacylated by its cognate aminoacyl-tRNA synthetase with the correct amino acid. Preventive and identity elements, which are found in the body sequence, have an important role in the attachment of the correct amino acid. Therefore, any change in a tRNA sequence that produces a new isodecoder may reduce or eliminate a preventive element for mischarging. The findings from a mice study showed that a genomic mutation reduced the expression of one isodecoder (Tr-20) resulted in neuro-degeneration [4]. It has been suggested that it may cause increased ribosome pausing specifically at AGA codons, and protein misfolding or degradation. Some isoacceptors within the families of isoacceptors are also isodecoders. Every human cell does not express all isodecoder genes and isodecoder expression vary widely among normal (non-diseased) human tissues [5-7].

As a rudimentary knowledge of biology, one important property of tRNA is the presence of RNA modifications. Among all types of RNA, tRNA has highest density of modifications with an average of 13 modifications per molecule. While all primary tRNA transcripts are composed of the nucleotides with A, C, G and U bases, many of these nucleotides are modified. The known modification in human tRNAs are 93 and these modifications can alter the properties of tRNAs in very different ways. The principal modifications taking place in a typical tRNA molecule during pre-tRNA processing include methylation of bases and sugars and creation of unusual bases such as dihydrouracil, ribothymidine, pseudouridine, and inosine. In addition to decoding role, tRNA modifications also help to localization, stability, ribosome binding, and translational dynamics. However, the variety of their functions is not fully understood. Taking into account only a single modification level from the modifications at 13 sites and the presence and absence of this, the combination number for similar tRNA sequences may be calculated as $2^{13} = 8192$. Thus, even if only modification signatures account, this can create further complexity which is an important factor in the identification and characterization of diagnostic and prognostic biomarkers in complex diseases such as cancers [4, 7, 8].

Dynamics of tRNA Modifications

Analysis of tRNA modifications indicates that the modification level and type of an individual tRNAs can result in marked differences in a cell-specific manner. For example, while HeLA cells have three modifications for a specific tRNA, similar tRNA from the placenta have roughly 17 modifications. In addition, an analysis of several specific tRNA base modifications on human HEK293T cell line revealed that the modification levels may differ between 10% and >80% in context to the type of modification and its location [4, 5, 7]. Moreover, tRNA modifications are believed to be highly dynamic according to the metabolic state of the cell. On the other hand, the modification status of any tRNA can have different effects on its activities [9]. Findings indicate that environmental conditions may also affect translation by differential charging of different tRNA species carrying the specific amino acid. An evidence of such an effect has been shown through starvation of a certain amino acid [10]. It is therefore of high biological interest to measure the extents of tRNA modification at individual sites as a result of altered cancer metabolism for identification of candidate cancer biomarker. All possible combinations from the above mentioned sources of potential complexity pose a significant challenge and opportunity to biological researchers (Figure 2).

tRNA Expression in Different Tissues

tRNA content is not entirely constant in multicellular organisms. There also exist intercellular and tissue variations in tRNA abundance. Comparative tRNA expression studies on various tissues showed that tRNA concentrations vary by as much as 10-fold among the tissues. For example, brain tissue contains decreased levels of tRNAs charged with hydrophobic amino acids and increased levels of tRNAs for the charged group compared with thymus, lymph node, and spleen tissue. Additionally, different tissues and cellular processes require spatiotemporal-specific expression of tRNAs, the mechanisms are currently poorly understood. It has been shown that tRNA content in a specific cell or tissue changes dramatically during cell growth, differentiation, and development. The differential tissue-specific expression of tRNAs mirrors the codon bias of tissue-specific proteins, indicating that tRNA expression is tightly arranged with the translational needs of a specific cell type. Codon usage bias affects protein translation because tRNAs that recognize synonymous codons differ in their abundance. Besides tRNA content and abundance, its other properties- modification, and charging- may play an important role in regulating translation according to cell states and conditions. In addition to immense tRNA complexity from all combinations of the different possible mixtures of tRNA species and tissue- and the existence of tissue-specific expression of tRNA species, considering the evidence listed below, tRNAs can be a good biomarker candidate for detecting in clinical samples associated with cancer. Here, it is important to bear in mind that large tRNA gene diversity and tissue-specific differences in the expression of individual tRNA species may increase both sensitivity and limitations for the discovery of new biomarkers. While the tissue-specific differences may serve as a fingerprint of the tumor source and marker of disease aggressiveness, dynamic changes in the expression of individual tRNA can be difficult to measure [11, 12].

Hallmarks of Cancer: Adding tRNAs

At this part of the chapter, how the regulation of tRNA function may be a control point over mRNA translation and cell growth is briefly introduced with the linked molecular events allowing the coordination of normal growth and oncogenic transformation. As known, a dichotomous choice for cells of many multicellular species is between proliferation and differentiation. Generally, differentiated cells are less proliferative, and proliferating cells are not terminally differentiated. Cellular proliferation and differentiation which are often distinct cellular states need changes in gene-expression programs in a carefully regulated way. Deregulated cell growth and proliferation are the important features of almost all tumors. One of the important characteristics of transformed and tumor cells is increased proliferation that requires elevated protein synthesis. In mammalian systems, tRNA concentration varies between proliferating and differentiating cells. Findings in recent years show that the direct action of tRNAs as well as amino acids on translation plays an important role in the rewiring of protein synthesis. The analyses of cancerous programs show that tRNA, which is an ncRNA may be involved in carcinogenesis. Cancer transcriptomes studies demonstrate that cancer cells than non-cancerous cells differ not only in the mRNA profiles but also in the tRNA profiles. Oncogene and tumor suppressor pathways may control the expression levels of all components of the translation machinery, which affect protein expression levels in a collaboration [1, 13, 14]. In the section that follows, it will be argued that how changes in tRNA expression play a major role in the translational control of specific mRNAs.

tRNAs Are Overabundant in Cancer Cells

Growing evidence indicates that tRNAs are implicated in a range of disorders, including cancer. Based on few studies using RNA-seq data from mostly human cell lines, it has been suggested that the alteration of single tRNA levels may cause the specific phenotypic consequences, such as the transformation from normal to the cancer cell. For example, the increased $tRNA^{Glu\text{-}UUC}$ and $tRNA^{Arg\text{-}CCG}$ levels in breast cancer cells are correlated with enhanced translation efficiency of core target transcripts regarding higher metastatic capacity and invasiveness. It has been suggested that overexpression of these tRNAs contribute the metastatic progression, enhancing the translational efficiency of metastasis-related genes in a codon-dependent manner. Additionally, with few studies using breast cancer cell lines and breast tumors, it has been found that overexpression of the initiator tRNA ($tRNAi^{Met}$) is correlated with increased cell proliferation. Besides, the same study also showed that the breast cancer had up to a 10-fold increase in nuclear- and mitochondrial- encoded tRNAs. This study noted that tRNAs carrying polar amino acids such as Thr, Ser, and Tyr are also more upregulated than tRNAs carrying small amino acids (Ala, Cys, Gly, Pro). Although it is not clear whether tRNA overexpression is the cause or consequence of carcinogenesis, increased levels of $tRNAi^{Met}$ are thought to be the key element in oncogenic transformation for several carcinomas [15-17].

Since the availability of the tRNA species has an important role on decoding times of mRNA codons, changes in the intracellular tRNA pool can have profound effects on protein expression by regulating the rate of translation for a subset of endogenous proteins. Thus, tRNA abundance have an important role on cell growth and–cancer progression. There is an evidence that RNA isoacceptor over-expression is associated with an increase of the translational efficiency of genes relevant to cancer development and progression. Several studies on yeast emphasize that variations in the abundance of tRNA isoacceptors as relate to affect the rate of translation [18].

Some opinions from broadening studies to organisms other than yeast have been proposed on how would changes in tRNAs such as global tRNA over-expression and differential tRNA isoacceptor expression affect translation? Firstly, global tRNA over-expression could enhance the rate of translation by increasing the tRNA pool reachable for incorporation into the A-site of translating ribosomes. Secondly, in addition to an absolute increase in total levels, enhancing tRNA synthesis might have an effect on the translation by changing the relative levels of tRNAs within the usual pool. The findings from several studies on mammalian cells suggest that simply increasing tRNA levels, and levels of $tRNAi^{Met}$ specifically, is sufficient to increase mRNA translation and enhance cell growth. Additionally, the findings from several bioinformatic and experimental studies showed that certain codons are used more commonly than other, "rare" codons and a specific tissue uses a specific codon to encode for an amino acid as due to translational demand. The usage of "rare" codons in a coding sequence generally decreases its translation rate, since transient pauses may occur at rare codons. Although the encoded protein sequence is same, more common codons may enhance translation speed and fidelity as selective pressures [14, 19, 20].

A mechanism allowing the use of certain codons may affect the rate of translation and protein dynamics such as protein folding and translocation. Consequently, changes in tRNA expression can trigger important phenotype changes with the related cancer-cell, such as the acquisition of invasiveness by altering the kinetics of translation. On the other hand, it is hypothesized that observed coordination at the mRNA and tRNA levels might be crucial as a means to ensure high expression of certain genes, and it is important to constitute the proliferation and differentiation programs. In this case, any increase or decrease in the expression of a given tRNA may affect the translation of all genes in a similar direction. As a result, these kinds of modulation may modify the expression of promoters of cancer progression by speeding up translation process. However, potential direct roles of specific tRNA levels in gene expression control and carcinogenesis remain poorly characterized. However, we note that the unknown mechanism is not a barrier to future biomarker discovery efforts.

tRNA Modification Changes Associated with Human Malignancies

As previously noted, variations in the modification level of tRNAs also occur in differentiating cells and malignant cells. As soon as transcribed, tRNAs undergo some of the modifications, so that they can be spliced and cleaved, and can have their bases modified. About 15 years ago, it has been detected that certain tumors are uracil more rapidly than nonmalignant tissue. In following years, a large number of tRNA modification have been identified with the advent of new method, Although the function of many of them remains unknown, available data show that tRNA modification degree is one of the sources of variability in the action of the various tRNA species and it may affect the rate and fidelity of translation, tRNA stability, cellular stress responses, and cell growth. Additionally, certain tRNA modifications or lack of correct modification may influence the translation of just a defined subset of transcripts enriched in certain types of codons, which could all be linked to a typical cell pathway that modify the regulation of complex cellular processes [8, 9, 21, 22].

Growing evidence reveals that tRNA modifications may play important roles in complex pathologies. Although tRNA modifications are mostly linked to various diseases, several studies have also reported a clear link between their modifications and different types of cancer such as lymphomas, leukemias, and carcinomas, including breast, bladder, colorectal and skin. However, much of the current literature on the relation of tRNA modification and cancer pays particular attention to the aberrant expression of the modification enzymes. There are several examples of gene expression changes that exhibit increased or decreased levels of the modification for several tRNAs in cancer cell lines and tumor tissues in context of several modification enzymes. However, several investigations have demonstrated that tRNA hypomodification is more than the hypermodification in cancer cells. For example, neoplastic cells derived from biopsies of human tumors exhibit a decreased queuosine content of tRNA. Decreased Q is further associated with the metastatic potential of tumor cells, including the brain, colon, lung, leukemia, and lymphoma. Additionally, using a quantitative HPLC/MS/MS assay, a proliferation-related modification regulation are shown for the fast proliferating cancer cells [13, 23-25].

tRNA Modification Genes That Are Linked to Cancer

Recent studies have implicated tRNA-isopentenyltransferase (tRNA-IPT), which catalyzes the addition of i6A on position 37 in tRNA and the human tRNA methyltransferase 9-like protein (hTRM9L), as tumor growth suppressor [26, 27]. There is much evidence showing that tRNA-IPT and hTRM9L transcripts are decreased in many tumor types such as lung adenocarcinomas, colorectal, breast, and bladder cancer and that their re-expression decreased tumor sizes in tumor models.

There are also several other tRNA modification enzymes linked to cancer. The transcript of Human RNA (guanine-9-)methyltransferase domain containing 2 (HRG9MTD2), which is responsible for methylation of guanine at position 9 (m^1G9) of several tRNAs, have elicited significant interest as one of seven transcripts found to display significant differences between early-onset and late-onset colorectal cancer patients [24, 28]. In the future, direct RNA sequencing combined with direct detection of tRNA modifications may be a new tool for cancer diagnosis.

Table 1. Representative tRNA alterations associated with various cancer types

Cancer	Nuclear tRNA/ Mitochondrial tRNA/ tRNA modification/tRF	tRNA modification	Samples
Skin	tRNA modification	m^5C	Tissue
Breast	tRNA modification	m^5C	Tissue
	tRF	SHOT-RNA$^{Asp\text{-}GUC}$	Tissue
		SHOT-RNA$^{His\text{-}GUG}$	Tissue
		SHOT-RNA$^{Lys\text{-}CUU}$	Tissue
		tRF$^{Asp\text{-}GTC}$	Tissue
Colorectal	Nuclear tRNA	tRNA^{Asn}	Tissue
	tRNA modification	m^1G	Tissue
	tRF	tRF/miR-1280	Tissue
Colon	Mitochondrial tRNA	tRNA^{His}	Tissue
Gastric	Mitochondrial tRNA	tRNA^{Phe}	Tissue
Ovarian	Nuclear tRNA	tRNA^{Arg}	Tissue
		tRNA^{Tyr}	Tissue
Kidney	Mitochondrial tRNA	tRNA^{His}	Tissue
Lung	tRF	tRF$^{Leu\text{-}CAG}$	Tissue and serum
	Mitochondrial tRNA	tRNA^{Arg}	Tissue
		tRNA^{Cys}	Tissue
Prostate	tRF	SHOT-RNA$^{Asp\text{-}GUC}$	Tissue
		SHOT-RNA$^{His\text{-}GUG}$	Tissue
		SHOT-RNA$^{Lys\text{-}CUU}$	Tissue
		tRNA$^{Lys\text{-}CTT}$	Tissue
		tRNA$^{Phe\text{-}GAA}$	Tissue
		tRF-544	Tissue
		tRF-315	Tissue
Renal cell carcinoma	tRF	tRF$^{Val\text{-}AAC}$	Tissue and serum

Other the hand, it has been known that peripheral leukocyte and platelet RNAs, including tRNAs, are dynamically affected by pathological conditions, and these blood cells can sequester RNAs and proteins released by cancer cells [29, 30]. In blood specimens, molecular assessment of tRNA modifications and the enzymes catalyzing such modifications linked to cancer may also provide the opportunity to identify new candidate biomarkers to detect cancer. It should be emphasized that, up until now, this has been neglected.

tRNA Fragments and Their Roles in Tumorigenesis: New Potential Biomarkers

As stated previously, the development of next generation (deep) sequencing technologies advanced the discovery of derivatives of small-noncoding RNAs, and the knowledge of their biological role. RNA-Seq data analysis showed that tRNAs also serve as a major source of small non-coding RNAs that act as regulators of key cellular processes. Studies in last five years showed that there may be a specific association between certain pathophysiologic states and the length, start and end points, and relative abundance of human tRNA fragments.

Accumulated evidence has also shown that tRNA fragments are frequently dysregulated in several cancers and they play important roles in the carcinogenesis by the regulation of cell proliferation, invasion, metastasis, apoptosis and gene expression. Some tRNA fragments, in particularly tRNA halves, have been detected in plasma, serum, and other bodily fluids as well as cell lines and tissues from cancer patients.

Mirroring recent understandings on miRNA detection, tRNA fragments may potentially serve as a promising marker for cancer detection and monitoring. The following section discusses the most relevant findings on the relation between tRNA fragments and cancer [29, 31].

tsRNA Signatures in Cancer

In recent years, deep sequencing data reveals that tsRNAs are one of the main populations of tRNA fragments found in cancer cell lines. Additionally, there is experimental evidence showing different expression patterns of these molecules between normal and cancerous cells. Studies revealed that several tsRNAs are frequently mutated and down-regulated in chronic lymphocytic leukemia and lung cancer samples from patients. It has been shown that ts-53 and ts-101 are often mutated in both cancer types. The finding in the enlarged study of the same group indicates that a cancer-specific signature of 31 tsRNAs able to distinguish among different cancers and tissues types. In addition, current evidence suggests that ts-53, ts-46 and ts-47 have tumor suppressor characteristics and they can represent a new diagnostic tool in cancer management [29].

It was suggested that a new developed an integrated web-based computing system, tRF2Cancer will help researchers to investigate the features and functions of tRFs from deep-sequencing data in multiple cancers and discover potential cancer biomarkers [32]. Nevertheless, there is still much to reveal their cancer-related changes.

tRFs

tRFs are divided into two subclasses as tRF-5s and tRF-3s, derived from 5'-end region of tRNA and containing the A box of the RNA polymerase III promoter, and derived from the 3'-end region of the tRNA and containing the B box, respectively. The expression of several tRF subtypes differs greatly in various cancers, although the factors that influence the expression of these fragments are quite unclear. As for tRF-3, Maute et al. [33] demonstrated that tRF-3027, also known as *CU1276* is down-regulated in human malignant B cells and lymphoma biopsies. The other an example is tRF^{Glu}, tRF^{Asp}, and tRF^{Gly} derived from $tRNA^{Glu}$, $tRNA^{Asp}$, and $tRNA^{Gly}$. They are found in lower levels in metastatic breast cancer relative to those in non-metastatic cancers and associated with YB-1 oncoprotein, which is up-regulated in various cancer types [34]. Overexpression of these tRFs decreases cell proliferation rate under serum-starvation. Thus, it has been suggested that they have a tumor suppressive role in response to oncogenic stress.

A database of the prostate cancer tRF transcriptome indicated that the expression of tRNA-derived fragments was significantly correlated with the histologic grade of prostate cancer. The ratio of tRF-544 (derived from $tRNA^{Phe\text{-}GAA}$)/tRF-315 (derived from $tRNA^{Lys\text{-}CTT}$) promises to be a useful clinical biomarker of disease progression in prostate cancer [34].

Using a database of small RNA sequencing, including a repository of nuclear and mitochondrial tRFs which was recently built and is currently available [35, 36], may provide valuable clues to delve further into the potential of tRFs as cancer biomarkers for different cancers. On the other hand, it has been reported that tRF expression changes not only in response to environmental stress conditions and in different disease states, but also also in ethnic groups and with gender and age as a confounding factor [37]. Therefore, when considering tRNA or tRNA fragments' deregulation in cancer with regard to biomarker discovery, all confounding factors may also have to be accounted for.

tRNA Halves

In addition to tsRNAs and tRFs, tRNA halves (also known as stress-dependent tRFs or tiRNAs) are another important type of small RNAs derived from tRNAs, that are typically greater than 32 nt in length. As several example studies below will show, there is a growing body of evidence that associates changes in the abundance of circulating 5′ tRNA halves with cancer, although their functions are not yet known. In addition, investigating the differential regulation of tRNA halves present a significant potential as biomarkers in detecting and monitoring various cancers.

In breast cancer patients, it has been shown that deregulation of specific 5′ tRNA halves in their serum correlate with various conditions: a) while several 5' stress tRFs from tRNA-Arg, −Asn,−Gln, −Gly, −Leu, −Lys, −Ser, −Trp, and –Val isoacceptors are up-regulated, tRFs derived from tRNA-Asp and –Lys isoacceptors are down-regulated. b) Deregulation of specific 5′ tRNA halves also correlate with clinicopathological characteristics of breast cancer, such as estrogen receptor (ER), progesterone receptor (PR), and human epidermal growth factor receptor 2 (HER2) status. In inflammatory breast cancer, five specific 5′ tRNA halves were found to be up-regulated, whereas, in relapsed patients, 18 tRFs were up-regulated [31]. c) Up-regulated miRNA-like tRFs were found in extra-cellular vesicles of breast adenocarcinoma cell line MCF7 [17].

Deregulation of stress derived tRFs have also been explored in cancer other than breast cancer: Firstly, stress derived tRFs were found up-regulated in serum from head and neck squamous cell carcinoma patients [38]. Secondly, an expression study also indicated that 5′-tRNA$^{Arg\text{-}CCT}$, 5′-tRNA$^{Glu\text{-}CTC}$, and 5′-tRNA$^{Lys\text{-}TTT}$ halves are expressed at lower levels in serum of patients with clear cell renal cell carcinoma. This study suggests that the describing 5′-tRNA halves may act in a tumor suppressive manner. In fact, several studies confirm the association of specific tRNA-derived fragments with tumor suppression [39]. Thirdly, 5′ tRNA halves are abundant in seminal fluid and are elevated in prostate cancer relative to noncancer patients [40].

Lastly, it has been shown that tRNA- derived small RNA termed Sex Hormone-dependent tRNA-derived RNAs (SHOT-RNAs), which are the tRNA halves, are specifically expressed in sex hormone-dependent cancers [41, 42]. There are few reports which showed specific and abundant expression SHOT-RNA accumulation in tissues from patients with breast cancer as well as in cultured cell line. Currently, several scientists working on biomarker discovery suggest that tRNA halves may have the potential to identify cancer at an early stage.

Future Perspectives, Outlook and Conclusions

Clinical Utility of tRNA and Its Fragments: Strengths, Weaknesses, Opportunities and Threats

Cancer can be arguably defined as one of the most complex diseases described to date because several molecular and cellular control mechanisms during carcinogenesis modify in a time-dependent process. Despite recent advances in the understanding of cancer pathophysiology, this characteristic of the disease is a major challenge for the discovery of new biomarkers with sufficient sensitivity and specificity to facilitate the detection of cancer in its early stages. Additionally, complications and variations of genetic, epigenetic, and environment factors in the initiation and progression of disease limit the performance of diagnostic and prognostic biomarkers. Therefore, the introduction of new cancer biomarkers into routine clinical use is slow and painful, and very few have been successfully put into practice so far [43].

Application of omics technologies with sophisticated, high-throughput and large-scale analytical methods is expected to play an increasingly important role in biomarker discovery. Clinical discovery strategies using these technologies have long concentrated on the analysis of molecular biomarkers including genes, proteins, and metabolites. This is because many oncogenes and tumor suppressors stimulate or inhibit RNA and protein synthesis, and these products can be greatly altered during disease [4, 44].

Molecular biomarker candidates may have several properties that make them attractive or (or unattractive) to the discovery of new biomarkers. From the perspective of the advantages and disadvantages, each biomolecule such as DNA, RNA, protein, peptide, as well as biomolecule chemical modification regarding biomarker provides different challenges in the diagnosis and management of cancer.

RNA biomarkers offer the advantage that their detection and quantification, even with very low abundance is easier than protein biomarkers. However, they have a handicap for stability in alkaline conditions. As RNA sequences at trace levels are amplified by PCR, the low concentration analytes may be captured with high sensitivity. Additionally, RNA biomarkers have a cost advantage oven protein biomarker, since it does not require a specific antibody for detecting each protein. The advantages of RNA oven DNA are a) by producing multiple copies, there is more RNA present in a cell than DNA b) due to their structures, they have the potential to be present stably in blood. c) they are also more informative in providing dynamic insights into cellular and regulatory processes [45].

The high-throughput technologies allow for detection of different types of ncRNAs (e.g., miRNA, small nuclear RNA, etc.) as well as the detection of protein-coding RNAs (i.e., mRNAs). They also offer new opportunities to develop new biomarkers and potential applications by providing rich information to characterize genotypes and phenotypes of diseases in a global and also dynamic manner [43].

Until recently, the large number of modifications and rigid structure of tRNAs made it difficult for efficient and quantitative high-throughput sequencing. Many tRNA modifications cause pauses or stops during reverse transcription, a key step in most sequencing protocols. Therefore, tRNA and tRNA fragments can escape from sequencing-based detection when they contain nucleoside modifications.

These obstacles have now been overcome through the use of engineered demethylases to remove base methylations (m^1G, m^1A, m^3C) in tRNA and the use of a thermostable reverse transcriptase to overcome the stable secondary structure [9]. Additionally, recently developed AlkB-facilitated RNA methylation sequencing also provides tools to profile the modified tRNA fragments [46]. These recent developments introduced to RNA-seq protocols may offer new opportunities for the discovery of cancer biomarkers or biomarker combinations. However, before developing blood tRNA as a clinical biomarker, the techniques of extraction, quantification, data analysis and a normalization strategy needs to be standardized.

Strengths of tRNAs as Biomarkers

Analyses of the findings from the recent standardization of these techniques show that the intact tRNA and its fragments also present a significant potential for detection and monitoring cancer. The studies mentioned in this chapter clearly indicate that several aspects of tRNAs and their fragments such as the changes that arise due to tRNA gene mutations, alterations in tRNAs levels, and aberrations in tRNA processing or modifications may provide the following new possibilities and challenges:

a. Utilizing a signature of altered tRNA may help to differentiate between cancer and normal tissue.
b. Use of tRNA based classifiers may identify the tissue origin of an unknown primary site of cancer.
c. Using a panel of tRNAs and its fragments that show differential expression within one cancer type may also differentiate various tumor subtypes.
d. Using a panel from tRNAs-related changes in circulating blood or in tumor-derived exosomes may provide a non-invasive method for early detection of cancers.
e. A combined evaluation of tRNA-related changes and selective mRNAs, as well as protein markers, may help in developing a more complete classifier [1847].

Unfortunately, some of these potentials have not been tested or characterized sufficiently to draw any conclusions as regarding biomarker discovery. Furthermore, clinically useful tumor biomarkers regarding tRNAs as well as miRNAs have not yet been approved by FDA, although there are several nucleic acid-based tumor biomarkers cleared or approved by the Center for Devices and Radiological Health (FDA) as according to the list obtained from http://www.fda.gov/MedicalDevices/ProductsandMedicalProcedures/InVitroDiagnostics/ucm330711.htm (accessed March 2018).

Weaknesses of tRNAS as Biomarkers

When viewed objectively, studies on tRNA present various difficulties yet to be overcome. Considering that the miRNA-related studies presented in the literature regarding biomarker discovery are at least 20 times more than that of tRNA and its fragments, we can say that simply discovering the presence of a cancer-specific alteration in some examples in Table does not make it a useful biomarker for the stratification or diagnosis of disease. Therefore, we must keep in mind that biomarker development is a long process that typically consists of at least four phases, such as initial discovery, assay development, assessment of clinical validity and utility. Even before carrying out these processes, there are still important things to be done. Most of the problems and uncertainties are not unique to ncRNAs including tRNA, but rather universal challenges in biomarker development.

Firstly, there are still crucial questions about the biology of circulating tRNA as well as other ncRNA. Furthermore, there is a need to address the level of correlation that exists between the tRNA deregulation identified in tumor tissue and circulating tRNA fragments. Additionally, there is also much uncertainty about the crosstalk between circulating RNA and tumor cells with their microenvironment. Answering these questions is necessary to discover successful biomarkers.

Secondly, there are several problems that make the transitions between the phases, such as improper choice of initial premises for biomarker discovery study, flaws in the study design, unclear guidelines and conditions for transition into clinical stage, and obscure clinical utility and complexity of interpretation. These require further scientific investigation.

Thirdly, there is still a major difficulty for the precise determination of tRNA population and its variation during tumorogenesis as well tRNA modification detection. Although the recent standardization high-throughput RNA sequencing with several procedure variations opened new possibilities for tRNA quantification, currently, there is still a major difficulty for the precise determination of tRNA population and its variation during tumorogenesis as well tRNA modification detection.

Lifestyle and Disease May Cause Inter-Individual Variation in tRNA-Related Changes

In omics fields, more studies have verified that physiological factors (e.g., age, race, and BMI) and environmental factors (e.g., lifestyle, diet, presumably circadian rhythm and geographical diversity) often contribute high levels of variability in characterizing the status and progression of many diseases. Although very little is known about the confounding factors as regarding tRNAs, it has been shown that in particularly, the expression level of tRF, which is one of three tRNA fragments, varies not only in different disease states but also in ethnic groups and with gender and age. In addition, there is evidence that the overall content of ncRNAs in the circulation is lowered in patients with chronic kidney disease. Further studies establishing tRNA expression patterns in relation to patient's age, demography, common health conditions and lifestyles are needed. Therefore, such variations and possible confounding factors should be taken into consideration when evaluating the deregulation of tRNA and its fragments in cancer with regard to biomarker discovery [1, 4, 29, 48, 49].

Taken together, as was the case of miRNAs, we can anticipate optimistically that research in the field of tRNA and its fragments are likely to increase rapidly and contribute significantly to disease diagnosis and treatment [50-78].

REFERENCES

[1] Samadder, A., Mitra, S., Chakraborty, B., et al. (2016). tRN-A-RS Acts As Biomarker for Cancer and Other Diseases. *J Mol Biomark Diagn;* S2: 019.

[2] Kumar, P., Kuscu, C., and Dutta, A. (2016). Biogenesis and function of transfer RNA-related fragments (tRFs). *Trends Biochem* Sci; 41: 679-689.

[3] Martens-Uzunova, E.S., Olvedy, M., and Jenster, G. (2013). Beyond microRNA– novel RNAs derived from small non-coding RNA and their implication in cancer. *Cancer Lett.* 340: 201-211.

[4] Pan, T. (2018). Modifications and functional genomics of human transfer RNA. *Cell Res*; 28: 395–404.

[5] Schimmel, P. (2018). The emerging complexity of the tRNA world: Mammalian tRNAs beyond protein synthesis. *Nat Rev Mol Cell Biol;* 19(1):45-58.

[6] Fujishima, K., and Kanai, A. (2014). tRNA gene diversity in the three domains of life. *Front Genet*; 5: 142.

[7] Kirchner, S., and Ignatova, Z. (2015). Emerging roles of tRNA in adaptive translation, signalling dynamics and disease. *Nat Rev Genet*; 16(2):98-112.

[8] Lorenz, C., Lünse, C.E., and Mör,l M. (2017). tRNA modifications: Impact on structure and thermal adaptation. *Biomolecules;* 7: 35.

[9] Marín, M., Fernández-Calero, T., and Ehrlich, R. (2017). Protein folding and tRNA biology. *Biophys Rev;* 9: 573-588.

[10] Novoa, E.M., and de Pouplana, L.R. (2012). Speeding with control: codon usage, tRNAs, and ribosomes. *Trends in Genet*; 28: 574-581.

[11] Dittmar, K.A., Goodenbour, J.M., and Pan, T. (2006). Tissue-specific differences in human transfer RNA expression. *PLoS Genet;* 2: e221.

[12] Sagi, D., Rak, R., Gingold, H., et al. (2016). Tissue-and time-specific expression of otherwise identical tRNA genes. *PLoS Genet;* 12: e1006264.

[13] Close, P., Bose, D., Chariot, A., et al. (2018). Dynamic Regulation of tRNA Modifications in Cancer. *Cancer and Noncoding RNA;* 163-186.

[14] Goodarzi, H., Nguyen, H.C., Zhang, S., et al. (2016). Modulated expression of specific tRNAs drives gene expression and cancer progression. *Cell*; 165: 1416-1427.

[15] Pavon-Eternod, M., Gomes, S., Geslain, R., et al. (2009). tRNA over-expression in breast cancer and functional consequences. *Nucleic Acids Res;* 37: 7268-7280.

[16] Gingold, H., Tehler, D., Christoffersen, N.R., et al. (2014). A dual program for translation regulation in cellular proliferation and differentiation. *Cell;* 158: 1281-1292.

[17] Soares, A.R., and Santos, M. (2017). Discovery and function of transfer RNA-derived fragments and their role in disease. *Wiley Interdiscip Rev RNA;* 8.

[18] Geslain, R., and Pan, T. (2010). Functional analysis of human tRNA isodecoders. *J Mol Biol;* 396: 821-831.

[19] Grewal, S.S. (2015). Why should cancer biologists care about tRNAs? tRNA synthesis, mRNA translation and the control of growth. *Biochim Biophys Acta;* 1849: 898-907.

[20] Hori, H., Tomikawa, C., Hirata, A., et al. (2014). Transfer RNA synthesis and regulation. *eLS.* https://onlinelibrary.wiley.com/doi/abs/10.1002/9780470015902.a0000529.pub3

[21] Phizicky, E.M., and Alfonzo, J.D. (2010). Do all modifications benefit all tRNAs? *FEBS Lett;* 584: 265-271.

[22] Roundtree, I.A., Evans, M.E., Pan, T., et al. (2017). Dynamic RNA modifications in gene expression regulation. *Cell;* 169: 1187-1200.

[23] Dirheimer, G., Baranowski, W., and Keith, G. (1995). Variations in tRNA modifications, particularly of their queuine content in higher eukaryotes. Its relation to malignancy grading. *Biochim;* 77: 99-103.

[24] Torres, A.G., Batlle, E., and de Pouplana, L.R. (2014). Role of tRNA modifications in human diseases. *Trends Mol Med;* 20: 306-314.

[25] Dong, C., Niu, L., Song, W., et al. (2016). tRNA modification profiles of the fast-proliferating cancer cells. *Biochem Biophys Res Commun;* 476: 340-345.

[26] Endres, L., Dedon, P.C., and Begley, T.J. (2015). Codon-biased translation can be regulated by wobble-base tRNA modification systems during cellular stress responses. *RNA Biol;* 12: 603-614.

[27] Gu, C., Begley, T.J., and Dedon, P.C. (2014). tRNA modifications regulate translation during cellular stress. *FEBS Lett;* 588: 4287-4296.

[28] Berg, M., Ågesen, T.H., Thiis-Evensen, E., et al. (2010). Distinct high resolution genome profiles of early onset and late onset colorectal cancer integrated with gene expression data identify candidate susceptibility loci. *Mol Cancer;* 9: 100.

[29] Balatti, V., Pekarsky, Y., and Croce, C.M. (2017). Role of the TRNA-derived Small Rnas in Cancer: New Potential Biomarkers and Target for Therapy. *Advances in cancer research;* 173-187.

[30] Best, M., Vancura, A. and Wurdinger, T. (2017). Platelet RNA as a circulating biomarker trove for cancer diagnostics. *J Thromb Haemost;* 15: 1295-1306.

[31] Dhahbi, J.M., Spindler, S.R., Atamna, H., et al. (2014). Deep sequencing of serum small RNAs identifies patterns of 5′ tRNA half and YRNA fragment expression associated with breast cancer. *Biomark Cancer;* 6: BIC. S20764.

[32] Zhang, L., Wan, S., Jiang, Y., et al. (2017). Molecular elucidation of disease biomarkers at the interface of chemistry and biology. *J Am Chem Soc*; 139: 2532-2540.

[33] Maute, R.L., Schneider, C., Sumazin, P., et al. (2013). tRNA-derived microRNA modulates proliferation and the DNA damage response and is down-regulated in B cell lymphoma. *Proc Natl Acad Sci U S A;* 110: 1404-1409.

[34] Sun, C., Fu, Z., Wang, S., et al. (2018). Roles of tRNA-derived fragments in human cancers. *Cancer Lett;* 414: 16-25.

[35] Zheng, L-L., Xu, W-L., Liu, S., et al. (2016). tRF2Cancer: A web server to detect tRNA-derived small RNA fragments (tRFs) and their expression in multiple cancers. *Nucleic Acids Res;* 44: W185-W193.

[36] Pliatsika, V., Loher, P., Telonis, A.G., et al. (2016). MINTbase: a framework for the interactive exploration of mitochondrial and nuclear tRNA fragments. *Bioinformatics;* 32: 2481-2489.

[37] Lambertz, U., Ovando, M.E.O., Vasconcelos, E.J., et al. (2015). Small RNAs derived from tRNAs and rRNAs are highly enriched in exosomes from both old and new world Leishmania providing evidence for conserved exosomal RNA Packaging. *BMC genomics;* 16: 151.

[38] Martinez, B.V., Dhahbi, J.M., Lopez, Y.O.N., et al. (2015). Circulating small non coding RNA signature in head and neck squamous cell carcinoma. *Oncotarget;* 6: 19246.

[39] Nientiedt, M., Deng, M., Schmidt, D., et al. (2016). Identification of aberrant tRNA-halves expression patterns in clear cell renal cell carcinoma. *Sci Rep;* 6: 37158.

[40] Dhahbi, J.M., Spindler, S.R., Atamna, H., et al. (2013). 5′ tRNA halves are present as abundant complexes in serum, concentrated in blood cells, and modulated by aging and calorie restriction. *BMC genomics;* 14: 298.

[41] Honda, S., and Kirino, Y. (2016). SHOT-RNAs: A novel class of tRNA-derived functional RNAs expressed in hormone-dependent cancers. *Mol Cell Oncol;* 3: e1079672.

[42] Honda, S., Loher, P., Shigematsu, M., et al. (2015). Sex hormone-dependent tRNA halves enhance cell proliferation in breast and prostate cancers. *Proc Natl Acad Sci U S A;* 112: E3816-E3825.

[43] Liu, R., Wang, X., Aihara, K., et al. (2014). Early diagnosis of complex diseases by molecular biomarkers, network biomarkers, and dynamical network biomarkers. *Med Res Rev;* 34: 455-478.

[44] Zhu, L., Liu, X., Pu, W., et al. (2018). tRNA-derived small non-coding RNAs in human disease. *Cancer Lett*; 419:1-7

[45] Xi, X., Li, T., Huang, Y., et al. (2017). RNA biomarkers: frontier of precision medicine for cancer. *Non-coding RNA;* 3: 9.

[46] Cozen, A.E., Quartley, E., Holmes, A.D., et al. (2015). ARM-seq: AlkB-facilitated RNA methylation sequencing reveals a complex landscape of modified tRNA fragments. *Nat Methods;* 12: 879.

[47] Zaporozhchenko, I.A., Ponomaryova, A.A., Rykova, E.Y., et al. (2018). The potential of circulating cell-free RNA as a cancer biomarker: challenges and opportunities. *Expert Rev Mol Diagn*; 18(2):133-145.

[48] Balatti, V., Nigita, G., Veneziano, D., et al. (2017). tsRNA signatures in cancer. *Proc Natl Acad Sci U S A;* 114(30):8071-8076.

[49] Patrick, E., Schramm, S-J., Ormerod, J.T., et al. (2017). A multi-step classifier addressing cohort heterogeneity improves performance of prognostic biomarkers in three cancer types. *Oncotarget*; 8(2):2807-2815.

Readings

[50] Bennike, T., Birkelund, S., Stensballe, A., et al. (2014). Biomarkers in inflammatory bowel diseases: current status and proteomics identification strategies. *World J Gastroenterol: WJG;* 20(12):3231-44.

[51] Breschi, A., Gingeras, T.R., and Guigó, R. (2017). Comparative transcriptomics in human and mouse. *Nat Rev Genet;* 18: 425.

[52] Clark, W. (2016). Dynamic tRNA Modification Landscape. *University of Chicago.*

[53] Devaux, Y. (2017). Transcriptome of blood cells as a reservoir of cardiovascular biomarkers. *Biochim Biophys Acta;* 1864: 209-216.

[54] Evans, M.E, Clark, W.C., Zheng, G., et al. (2017). Determination of tRNA aminoacylation levels by high-throughput sequencing. *Nucleic Acids Res;* 45: e133-e133.

[55] Feng, S., Zheng, Z., Xu, Y., et al. (2017). A noninvasive cancer detection strategy based on gold nanoparticle surface-enhanced raman spectroscopy of urinary modified nucleosides isolated by affinity chromatography. *Biosens and Bioelectron;* 91: 616-622.

[56] Fredrick, K., and Ibba, M. (2010). How the sequence of a gene can tune its translation. *Cell;* 141: 227-229.

[57] Geslain, R., and Eriani, G. (2014). Regulation of translation dynamic and neoplastic conversion by tRNA and their pieces. *Translation;* 2: e28586.

[58] Guzman, N., Agarwal, K., Asthagiri, D., et al. (2015). Breast cancer–specific miR signature unique to extracellular vesicles includes "microRNA-like" tRNA fragments. *Mol Cancer Res*; 13: 891-901.

[59] Haussecker, D., Huang, Y., Lau, A., et al. (2010). Human tRNA-derived small RNAs in the global regulation of RNA silencing. *Rna;* 16: 673-695.

[60] Heyns, M., and Kovalchuk, O. (2015). Non-coding RNAs including miRNAs, piRNAs, and tRNAs in human cancer. *Oncotarget;* 6: 23055.

[61] Kim, E.Y., Jung, J.Y., Kim, A., et al. (2017). Methionyl-tRNA synthetase overexpression is associated with poor clinical outcomes in non-small cell lung cancer. BMC cancer; 17: 467.

[62] Mayeux, R. (2004). Biomarkers: potential uses and limitations. *NeuroRx;* 1: 182-188.

[63] Motorin, Y., and Helm, M. (2010). tRNA stabilization by modified nucleotides. *Biochemistry;* 49: 4934-4944.

[64] Olvedy, M., Scaravilli, M., Hoogstrate, Y., et al. (2016). A comprehensive repertoire of tRNA-derived fragments in prostate cancer. *Oncotarget;* 7: 24766.

[65] Pan, T., and Rosner, M. (2010). tRNAs as Biomarkers and Regulators for Breast Cancer. *Chicago Univ Il.*

[66] Parisien, M., Wang, X., and Pan, T. (2013). Diversity of human tRNA genes from the 1000-genomes project. *RNA Biol;* 10: 1853-1867.

[67] Roche, S., Gabelle, A., and Lehmann, S. (2008). Clinical proteomics of the cerebrospinal fluid: towards the discovery of new biomarkers. *Proteomics Clin Appl;* 2: 428-436.

[68] Rogers, H.H., and Griffiths-Jones, S. (2014). tRNA anticodon shifts in eukaryotic genomes. *RNA;* 20: 269-281.

[69] Rostami, R., Aghasi, M., Mohammadi, A., et al. (2013). Enhanced oxidative stress in Hashimoto's thyroiditis: inter-relationships to biomarkers of thyroid function. *Clin Biochem;* 46: 308-312.

[70] San Koh, C., and Sarin, L.P. (2018). Transfer RNA modification and infection–Implications for pathogenicity and host responses. *Biochim Biophys Acta;* 1861(4):419-432.

[71] Sarin, L.P., and Leidel, S.A. (2014). Modify or die?-RNA modification defects in metazoans. *RNA Biol;* 11: 1555-1567.

[72] Stoecklin, G., and Diederichs, S. (2014). tRNAs: new tricks from old dogs. *EMBO J;* 33: 1981-1983.

[73] Tatusov, R.L., Koonin, E.V. and Lipman, D.J. (1997). A genomic perspective on protein families. *Science;* 278: 631-637.

[74] Tiberio, P., Callari, M., Angeloni, V., et al. (2015). Challenges in using circulating miRNAs as cancer biomarkers. *Biomed Res Int;* 2015. https://www.hindawi.com/journals/bmri/2015/731479/ .

[75] Topisirovic, I., and Sonenberg, N. (2014). Distinctive tRNA repertoires in proliferating versus differentiating cells. *Cell;* 158: 1238-1239.

[76] Touzot, M., Dahirel, A., Cappuccio, A., et al. (2015). Using Transcriptional Signatures to Assess Immune Cell Function: From Basic Mechanisms to Immune-Related Disease. *J Mol Biol;* 427: 3356-3367.

[77] Dhahbi, J.M., Atamna, H., and Selth, L.A. (2018). Data Mining of Small RNA-Seq Suggests an Association Between Prostate Cancer and Altered Abundance of 5′ Transfer RNA Halves in Seminal Fluid and Prostatic Tissues. *Biomark Cancer;* 10: 1179299X18759545.

[78] Garcia-Silva, M.R., Cabrera-Cabrera, F., Güida, M.C., et al. (2013). Novel aspects of tRNA-derived small RNAs with potential impact in infectious diseases. *Adv Biosci Biotechnol;* 4: 17.

In: Research on New Generation …
Editor: Yildiz Dincer

ISBN: 978-1-53614-367-6

Chapter 8

LONG NON-CODING RNAS: POTENTIAL DIAGNOSTIC BIOMARKERS

Burcu Bayoglu*
Istanbul University Cerrahpasa Medical Faculty,
Department of Medical Biology, Istanbul, Turkey

ABSTRACT

Recently a novel class of non-coding RNAs called long non-coding RNAs (lncRNAs) have been shown to be important players in the regulation of gene expression. LncRNAs are a large class of RNA transcripts >200 nt in length, capped and polyadenylated. LncRNAs are classified according to their localizations in the genome as intronic, intergenic, antisense, overlapping, bidirectional and processed. Thousands of lncRNAs are identified and they have distinct biogenesis. They have functions in the epigenetic control of gene expression by serving as scaffold, signal, decoy, enhancer and guide. They are mostly found in the nucleus of the cell, however, extracellular lncRNAs are also circulating in body fluids such as whole blood, saliva, urine, and plasma. LncRNAs are reported to promote tumor progression through altering various signaling pathways. Moreover, polymorphic variants on lncRNA genes may affect their expression levels in various tissues and body fluids which may trigger tumor formation, development and progression. Thus, the dysregulation in the lncRNA levels and their complex relationship with other kinds of non-coding RNAs may tend to various cancer types. Recent studies are focused on lncRNAs as candidate biomarkers for some types of cancers. Gain and loss of function experiments have shown that up-regulation of specific oncogenic lncRNAs can lead to augmented proliferation and diminished cell cycle arrest and apoptosis which promotes tumorigenesis. Certain lncRNAs are also reported to function as a tumor suppressor gene in some cancers. These functional tumor suppressor lncRNAs are shown to play a negative regulatory role in tumor growth and metastasis. Transcriptome analysis of lncRNAs will reveal the functional roles of these novel

* Corresponding Author Email: burcu.bayoglu@istanbul.edu.tr, burcubayoglu@yahoo.com.

oncogenic and tumor suppressor non-coding RNA types. In future, various promising antisense drugs for lncRNA targets may be developed and used in clinical applications and disease management.

Keywords: long non-coding RNAs, tumor biomarkers, non-coding transcripts, epigenetic regulation

INTRODUCTION

Cancer is the second leading cause of death in developed countries [1]. The initiation and the progression of cancer include multiple molecular mechanisms. Alterations in the genome sequence of a cancer cell is one of the mechanisms in cancer initiation. With the developments in the sequencing technologies, nowadays cancer genomes can totally be sequenced and changes between the normal and diseased genomes can be distinguished. During cancer progression, both genetic and epigenetic events occur. Besides the genetic events like, mutations of oncogenes and tumor suppressor genes, gene amplifications, deletions, and gene fusions, epigenetic events including dysregulation of histon modifiers, DNA methylation and demethylation take place in the cancer cell genome. As a result of these mechanisms, regulation of gene expression alters leading to cancer. Non-coding RNAs (ncRNAs) take place in the epigenetic events of the regulation of gene expression [2]. Long non-coding RNAs (lncRNAs) play an important role in gene silencing either by recruiting chromatin remodeling complexes which induce histon methylation [3] or by the recruitment of RNA-binding proteins which control histon deacetylation [4, 5]. RNA has high flexibility leading to easy activation, modification, transportation and degradation [6].

Recent international projects like ENCODE (encyclopedia of DNA elements) have shown that vast majority of the human genome (~80%) encode for functional RNA transcripts [7]. However, only less than 2% of the human genome transcribe into proteins. So in recent years the remaining non-protein coding genome are investigated for having an association with some types of cancer. Cancer is an important cause of death in all over the world [8, 9]. The dysregulation in the expression of cancer-related genes has been associated with protein-coding genes in past approaches. The molecular etiopathogenesis of cancer has to be known thoroughly to perform efficient pharmacotherapies.

Long non-coding RNAs are a large class of non-protein coding RNA transcripts >200 nt in length. They have diverse localizations in the genome. They have various functions including signal, scaffold, decoy, guide, enhancer RNAs and short peptides [10, 11]. LncRNAs have a coding potential of <100 amino acids if they are longer than 2 kb. However, they have no protein coding potential if they have the length of >200 nt [12-

14]. Nowadays, the number of the identified lncRNAs are found to be more than protein-coding genes [15-17]. Non-coding RNAs revealed a main source of biomarkers [18-22], molecular therapeutic targets [20-23], and unexplained intronic GWAS variants [24].

Various lncRNAs are associated with some types of cancer [25] including glioblastoma [26-28], colorectal cancer [26, 29], liver cancer [30, 31], breast cancer [32], and leukemia [26, 33]. Dysregulation of lncRNA expression may lead to various cellular functional anomalies like cell proliferation, inhibition of apoptosis, triggering angiogenesis, promotion of metastasis and dysfunction of tumor suppressor proteins [25, 34].

Long Non-Coding RNAs - General Characteristics

LncRNAs are abundant in the human genome [13, 35]. They usually do not have open reading frames (ORFs). However, some lncRNAs have bifunctional features comprising both protein-coding and non-coding functions [36, 37]. This feature gives rise to thought that many protein-coding genes also have non-coding activities [38]. Their genomic locations can be within 10 kb of protein-coding genes [39], long intergenic (lincRNAs) [40], enhancer RNAs (eRNAs) in the distal sequence of the gene [41], and antisense to coding genes [42] (Figure 1).

LncRNAs appear as slight amounts in body fluids and tissues [43]. It is reported that the number of lncRNA genes are more than protein-coding genes [43] and 90% of these lncRNAs have no protein product [44, 45]. RNA-sequencing studies demonstrated that most of the lncRNAs are not conserved in DNA sequence and are specific to primate species. However, 3% of lncRNAs are highly conserved and observed from *Xenopus* to human [46]. Although no conserved sequence was displayed within lncRNAs, identical biological functions were conserved [47]. LncRNAs are well-regulated and limited to specific types of cells [48]. Their function, structure and microhomology regions are conserved in evolution although they have low sequence similarity [47, 49, 50]. RNA molecules may not need sequence conservation as in proteins to accomplish their function. Moreover, when compared with protein-coding gene promoters, lncRNA promoters have more conserved sequence [13]. The function of most of the transcripts are not identified to date. Sequence effects on function are not well-understood in lncRNA biology. However, lncRNAs have some features similar to protein-coding genes, like 5′-cap and alternative splicing [43]. They may possess several exons and some of them have poly (A) tails [51]. Since they achieve many functions, they appear in many tissues with diverse expression [52]. As they are measured in different cellular compartments like cytoplasm and nucleus [43], they also are observed in body fluids [53]. While chromatin interactions, regulation of transcription and RNA processing issues are regulated by nuclear lncRNAs, on the other side, cellular lncRNAs modulate mRNA stability and translation and also affect cellular signals [54]. They usually exist in small amounts [55]. They have half-lives ranged between less than 2 hours to over 12 hours [56].

LncRNA expression is mostly tissue-specific since they have different regulatory mechanisms from protein-coding promoters [57]. LncRNA promoters have high degree of A/T nucleotides and low CpG islands. They also possess diverse transcription factor binding sequences and chromatin mark patterns when compared with protein-coding genes [58]. LncRNAs have more methylations at transcriptional start sites than protein-coding genes without any connection to their expression levels [59]. Furthermore, lncRNA genes have high degree of Myc binding sites and they are more sensitive to Dicer-miRNA-myc circuit than protein-coding genes. Myc interacts with lncRNA genes to activate their transcriptional initiation and elongation [60]. Dicer-miRNA pathway and myc oncogene regulate mRNA and lncRNA expression [61].

Developing technology functionally give account of cancer-related lncRNAs. Therefore, in cancer phenotypes, lncRNAs are provided to be identified as biomarkers and further studies will reveal their roles in the cellular mechanisms of cancer pathophysiology.

Long Non-Coding RNAs- Mechanisms of Action in Cancer

In recent years, it was showed that mutations in the non-coding genome cause human diseases [62]. Epigenetic changes [63], somatic non-coding mutations [64], and somatic copy number variations [65] are reported in various cancer types.

LncRNAs use diverse mechanisms in cancer development and progression including chromatin interactions, chromatin remodeling, competing endogenous RNAs, and natural antisense transcripts (NATs). Several functional roles of lncRNAs are to act as signals and scaffolds in chromatin remodeling process. Also performing as enhancers, they make chromatin loops stable and maintain the situation [41, 66-68]. Enhancers located on the distal part of gene of interest, interact with the promoter region with the help of chromatin looping [69-71]. Some lncRNAs called as competing enhancer RNAs accumulate regulatory factors like transcription factors, enzymes, chromatin modifying proteins and miRNAs to control transcription [72-75].

Epigenetic alterations including chromatin remodeling mechanism is one of the regulation types in transcription and lncRNAs have emerging roles in this process. An antisense lncRNA transcript KCNQ1OT1 (potassium voltage-gated channel subfamily KT member 1 opposite strand/antisense transcript 1) is an example of this mechanism. Increased expression of this antisense transcript is reported in colon cancer [76]. The antisense transcript KCNQ1OT1, as a signal molecule, recruits G9a histon methyltransferases and polycomb repressive complex 2 (PRC2) which suppresses gene expression by the demethylation of lysine 9 and 27 on histon 3 [77, 78]. Within the chromatin remodeling process, KCNQ1OT1, suppresses gene expression through *cis* and *trans* acting ways [38].

Another antisense transcript named ANRIL (CDKN2B-AS1) in the INK4 locus of chromosome 9, acting as a scaffold, performs in transcriptional silencing by recruiting polycomb repressive complex 1 (PRC1) and PRC2 [79, 80]. PRC1 and PRC2 function to repress gene expression [81]. PRC1 mediates the methylation of H3K27me3 and PRC2 mediates both H3K27me3 and H2AK119ub1. PRC2 has multiple subunits including EED, SUZ12 and EZH1 and EZH2 exhibiting the methyltransferase activity [82-85]. It has essential roles in the control of development and cellular differentiation. Dysregulation of PRC2 due to the mutations or overexpression may lead to numerous cancers [86-90]. LncRNAs make interaction with the chromatin remodeling complex, PRC2. Some cancer types were related with abnormal levels of PRC2 complex or its subunits [91]. Thus, the association of lncRNA levels with tumor formation may occur through PRC2 complex [38].

LncRNAs can also function as enhancers to regulate transcription on chromatin loops. In experimental studies, diminished expression levels of these elncRNAs lead to corrupt chromatin loops and thus cause a change in target gene expression. One example of this relation is estrogen receptor α (ERα) [66], a predicted biomarker in breast cancer. Another example is prostate specific antigen (PSA) which is accepted to be a biomarker in prostate cancer. Studies have revealed an association between eRNA levels expressed from the *PSA* upstream enhancers and *PSA* gene expression which proposes a relation between chromatin interaction and eRNAs [92]. In a study, it was shown that the tumor suppressor transcription factor p53, which has important tasks in cell cycle, bind some enhancers and expressed RNAs to interact with some adjacent genes. Inhibiting these eRNAs cause reduced adjacent gene expression which decreases the function of p53 [93]. Also, in recent years, integrator which is a complex related with RNA polymerase II, has been shown to have a role in eRNA biogenesis. The inhibition of the integrator lowers eRNA induction and prevents chromatin looping needed for gene transcription [94].

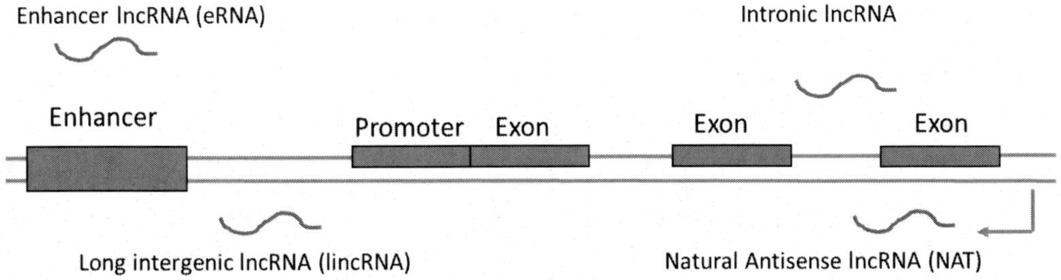

Figure 1. Genomic locations of lncRNAs [modified from 38].

One of the significant mechanisms in cancer initiation is the differential expression of lncRNAs. Regulatory DNA mutations influence transcription by changing the promoter/enhancer activity or chromatin states and this affects lncRNA expression. There are extremely conserved non-coding regions in the genome and they are mutated and dysregulated in cancer [95].

LncRNAs play a role as the drivers of cancer phenotypes and they contribute to the major characteristics of cancer; proliferation, growth suppression, immortality, motility, angiogenesis and viability of cancer phenotypes [60] (Figure 2).

One of the examples of the lncRNAs taking part in proliferation mechanism is LUNAR1. In T cell acute lymphoblastic leukemia, *Notch1* oncogene induces LUNAR1 and drives growth. As a result, upregulation of insulin-like growth factor 1 receptor expression and signaling occurs [96].

Also, recently some lncRNAs have been shown to play a role in influencing tumor suppressors and growth arrest in cell cycle. The antisense transcript p15-AS promotes to silence $p15^{INK4b}$ [97] and increased p15-AS expression was related with diminished $p15^{INK4b}$ expression in leukemic cells [60].

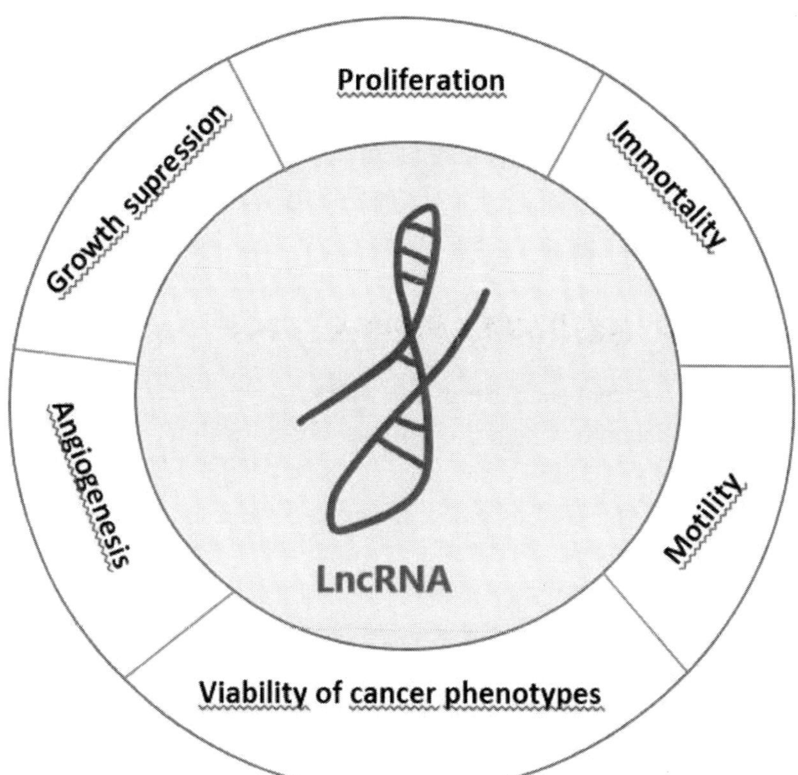

Figure 2. The roles of lncRNAs in the development of cancer mechanisms [modified from 60].

In the viability circuits in cancer, it was observed that tumor cells were immortal with telomere maintenance. TERC, a telomere RNA component, has SNPs on its locus associated with telomere length and increased risk of developing high-grade glioma [98]. Furthermore, MALAT1, a nuclear lncRNA depletion was reported to impair cellular motility *in vitro* and metastasis in mice lung carcinoma cells [99].

LncRNA recruitment also occurs on the distant promoters and enhancers in cancer transcription. It was recently reported that the lncRNA Pauper which is restricted to central nervous system, interacts with various promoters for cell cycle regulation and has a role in neuroblastoma [100].

LONG NON-CODING RNA BIOMARKERS IN CANCER

Cancer is basically a genetic disease changing the information in the cell and leading to disrupt cellular homeostasis. Along with the investigation of the mutations in the protein-coding genes that drive cancer, some targeted therapies are developed for the malignancies. Human tumors are quite complex and subtypes of these tumors exhibit different characteristics of molecular features. Since, less than 2% of the genome is protein coding, carrying a mutation in the non-coding regions became more important for tumor formation. Nowadays, functional mutations in the non-coding section of the genome are detected by genome-wide mutation analyses. These sections of the genome have significant effects on the transcription of lncRNAs. Transcription of lncRNAs are regulated critically and can help to understand the malignant transformation signals [60].

Recent years, next generation sequencing (NGS) technique revealed to identify various lncRNAs in tumors and dysregulation of these transcripts have been demonstrated in common malignancies [2]. In a recent study, 60.000 lncRNA genes have been found out in various tumors, cell lines and normal tissues [101]. There are distinct mechanisms that lncRNAs involve in cancer development. They commonly guide chromatin modifying complexes through epigenetic mechanisms. 38% of lncRNAs were reported to bind PRC2 or chromatin modifying proteins RCOR1 (COREST) and KDM5C (SMCX) and the rest of them bind to trithorax chromatin activating complexes and activated chromatin [102]. Previous studies have identified some lncRNAs and their mechanism of actions. For instance, ANRIL (CDKN2B-AS1), HOTAIR, XIST, and KCNQ1OT1 are well-characterized lncRNAs which have the ability to recruit chromatin modifiers to a specific locus to change the state of the chromatin [102].

Epigenetic gene modifications can be used for helpful biomarker investigations. A good example for this is especially seen in prostate cancer. Approximately, in 85% of prostate cancer patients hypermethylation of glutathione-S-transferase (*GSTP1*) gene is examined rather than benign hyperplasias [103] indicating for being a valuable biomarker. A prostate specific lncRNA, PCAT, has been shown to have a role in tumor progression [104]. Yet another lncRNA, SCHLAP1 (second chromosome locus associated with prostate-1), also called LINC00913, expression was identified as predicting lethal prostate cancer. Its molecular mechanism acts through the blockage of the regulatory function and localization of SWI/SNF complex in prostate cancer [105, 106]. Also, some lncRNAs including TP53COR1/Linc-p21, HOTTIP, MALAT1, HOTAIR, PCAT1, interact with some miRNAs and their stability were decreased by that interaction [107, 108]. SChLAP1 overexpression was also linked to increased risk of metastatic progression in men treated with radical prostatectomy [109]. In prostate cancer, another lncRNA, differential display 3, also called PCA3, was also identified as a specific biomarker [53]. PCA3 was also confirmed by Food and Drug Administration (FDA) as a diagnostic marker for prostate cancer. PCA3 as a biomarker in the urine, is the first lncRNA approved by FDA. It is a non-invasive beneficial test for prostate cancer. It has also improved specificity and is advantageous when compared with serum prostate specific antigen testing [110]. Identifying thousands of differentially expressed lncRNAs between cancer and normal tissues are now possible with genome-wide scanning approaches [34, 101, 112]. These approaches give rise to find new biomarkers among lncRNAs for specific types of cancer. Expression profiles of lncRNAs can also determine subtypes of cancer to reveal tumor behavior and progression [112]. Further studies have reported other lncRNAs with specificity for prostate cancer including tumor grade and lymph node metastasis relationship [113].

Various histon modifications such as acetylation, deacetylation, ubiquitylation, phosphorylation, and methylation play a significant role in the initiation process of tumor formation. These epigenetic changes during the early stages of the clonal expansion may lead to specify the genetic changes and thus promote malignancy [114]. Single nucleotide polymorphisms (SNPs), copy number variations and mutations in the non-coding regions can change the lncRNA expression leading to cancer. Some potential lncRNA biomarkers related with specific cancer types are given in Table 1 (Table 1).

Some potential lncRNA biomarkers were identified on 8q24 locus containing many cancer-related SNPs. Some of these SNPs contribute to express cancer-related lncRNAs. For instance, lncRNAs CCAT2 and PCAT-1 in this locus may serve as biomarkers in colorectal and prostate cancers respectively [141].

Another example of lncRNA which can serve as a biomarker is HOTAIR. Overexpression of HOTAIR in early stage breast cancer is reported to be profoundly predictive to metastatic progression and complete survival [32]. It was reported that the dysregulation of HOTAIR leads to 26 kinds of human tumors [142]. Also, HOTAIR was associated with the sensitivity of ovarian cancer patients to a kind of platinum chemotherapy [122]. The association of HOTAIR expression and treatment with carboplatin was related with poor survival. However, same group has improved responses to cisplatin relative to HOTAIR negative patients [122]. With these observations, using HOTAIR as a biomarker may help to choose the right platinum containing chemotherapeutic agents in patients with ovarian cancer.

Yet another identified lncRNA biomarker is AA174084 which has the ability of making a distinction between gastric cancer and benign alterations of gastric epithelium in gastric secretions [139].

Table 1. Potential lncRNA biomarkers in various cancer types [38, 60]

LncRNA	Cancer type	Status	Reference
PTCSC3	Tyroid cancer	Downregulated	[115]
HULC	Hepatocellular carcinoma	Upregulated, rs7763881	[116-118]
XIST	Various cancers	Dysregulated	[33, 119]
GAPLINC	Gastric cancer	Poor prognosis	[120]
MALAT1	Liver, lung, colorectal	Poor prognosis and metastasis	[121]
HOTAIR	Colorectal, liver, pancreatic, breast, gastric	Metastasis, can be used in platinum chemotherapies in ovarian cancer	[32, 122]
ANRIL	Prostate, various	Upregulated, 9p21.3 SNPs	[123]
KCNQ1OT1	Colorectal	Upregulated	[76]
PRNCR1	Prostate	Upregulated	[124]
H19	Hepatocellular carcinoma, gastric	High expression, bioavailable in blood	[125]
PVT1	Colorectal	8q24 amplification	[126]
PCAT-1	Prostate	8q24 SNPs	[104, 127]
CCAT2	Colorectal	8q24 SNPs	[128, 129]
PTCSC3	Tyroid	rs944289	[130]
TERC	Oral cavity	3q26 amplification	[131]
GAS5	Hepatocellular carcinoma	rs145204276	[132]
FAL1	Ovarian	Poor prognosis	[133]
HOTTIP	Hepatocellular carcinoma	Increased risk of progression	[134]
MEG3	Meningioma	Tumor grade/progression risk	[135]
NBAT1	Neuroblastoma	Good prognosis	[136]
NKILA	Breast	Decreased risk of metastasis	[137]
SCHLAP1	Prostate	Increased risk of metastasis	[109]
CCAT1	Colorectal	Can be used as BET inhibitor	[138]
AA174084	Gastric	Bioavailable in gastric secretions	[139]
PCA3	Prostate	Bioavailable in urine	[53]
SPRY4-IT1	Melanoma	Tumor formation	[140]

Future Directions and Therapies

Non-coding RNAs and the non-coding regions in the genome play a critical role in cancer initiation and progression. High-throughput technologies including next generation sequencing (NGS) and gene expression profiling make various tumors molecularly well-characterized. These techniques make it possible to identify cancer driving mutations, amplifications, deletions, gene fusions, etc. in cancer diagnosis. DNA and histon modifications are also observed in various tumors and these epigenetic alterations are seen both in cancer initiation and progression.

LncRNAs interact with other RNAs, DNA and proteins to drive many cancer types. Thus, it is very important to determine cancer-related transcripts in order to use them for therapeutic applications. Future approaches should focus both on identifying new lncRNAs by high-throughput sequencing technologies and also determining functional characterization of these lncRNAs to develop new therapies. Also, recent genome editing technologies may elaborate the mechanisms of cancer initiation and progression to advance therapeutic targeting.

References

[1] Siegel, R. L., Miller, K. D. & Jemal, A. (2016). Cancer statistics, 2016. *CA Cancer J Clin*, *66*, 7–30.

[2] Chakravarthi, B. V., Nepal, S. & Varambally, S. (2016). Genomic and Epigenomic Alterations in Cancer. *Am J Pathol*, *186* (7), 1724-1735.

[3] Rinn, J. L., Kertesz, M., Wang, J. K., Squazzo, S. L., Xu, X., Brugmann, S. A., et al. (2007). Functional demarcation of active and silent –chromatin domains in human HOX loci by noncoding RNAs. *Cell*, *129*, 1311–1323.

[4] Mercer, T. R., Dinger, M. E. & Mattick, J. S. (2009). Long non-coding RNAs: insights into functions. *Nat Rev Genet* 10, 155–159.

[5] Nagano, T. & Fraser, P. (2009). Emerging similarities in epigenetic gene silencing by long noncoding RNAs. *Mamm Genome*, *20*, 557–562.

[6] Cao, J. (2014). The functional role of long non-coding RNAs and epigenetics. *Biol Proced Online*, *16*, 11.

[7] ENCODE Project Consortium. (2004). The ENCODE (ENCyclopedia Of DNA Elements) Project, *Science*, *306*, 636–640.

[8] Chen, W., Zheng, R., Baade, P. D., Zhang, S., Zeng, H., Bray, F., et al. (2016). Cancer statistics in China, 2015. *CA Cancer J Clin*, *66*(2), 115–132.

[9] Simard, E. P., Ward, E. M., Siegel, R. & Jemal, A. (2012). Cancers with increasing incidence trends in the United States: 1999 through 2008. *CA Cancer J Clin*, *62*, 118–128.

[10] Li, X. L., Wu, Z. Q., Fu, X. B. & Han, W. D. (2014). LncRNAs: insights into their function and mechanics in underlying disorders. *Mutat Res Rev Mut Res*, *762*, 1–21.

[11] Moran, V. A., Perera, R. J. & Khalil, A. M. (2012). Emerging functional and mechanistic paradigms of mammalian long non-coding RNAs. *Nucleic Acids Res*, *40*, 6391–6400.

[12] Dinger, M. E., Pang, K. C., Mercer, T. R. & Mattick, J. S. (2008). Differentiating protein-coding and noncoding RNA: challenges and ambiguities. *PLoS Comput Biol*, *4*, e1000176.

[13] Carninci, P., Kasukawa, T., Katayama, S., Gough, J., Frith, M. C., Maeda, N., et al. (2005). The transcriptional landscape of the mammalian genome. *Science*, *309*, 1559–1563.

[14] Perkel, J. M. (2013). Visiting "noncodarnia". *Biotechniques*, *54*(301), 303–304.

[15] ENCODE Project Consortium, Birney, E., Stamatoyannopoulos, J. A., Dutta, A., Guigó, R., Gingeras, T. R., et al. (2007). Identification and analysis of functional elements in 1% of the human genome by the ENCODE pilot project. *Nature*, *447*, 799–816.

[16] Carninci, P. & Hayashizaki, Y. (2007). Noncoding RNA transcription beyond annotated genes. *Curr Opin Genet Dev*, *17*, 139–144.

[17] Kapranov, P., Cheng, J., Dike, S., Nix, D. A., Duttagupta, R., Willingham, A. T., et al. (2007). RNA maps reveal new RNA classes and a possible function for pervasive transcription. *Science*, *316*, 1484–1488.

[18] Clark, B. S. & Blackshaw, S. (2014). Long non-coding RNA-dependent transcriptional regulation in neuronal development and disease. *Front Genet*, *5*, 164.

[19] Gibb, E. A., Brown, C. J. & Lam, W. L. (2011). The functional role of long non-coding RNA in human carcinomas. *Mol Cancer* 10, 38.

[20] Qureshi, I. A. & Mehler, M. F. (2013). Long non-coding RNAs: novel targets for nervous system disease diagnosis and therapy. *Neurotherapeutics: the journal of the American Society for Experimental NeuroTherapeutics*, *10*, 632–646.

[21] Reis, E. M. & Verjovski-Almeida, S. (2012). Perspectives of Long Non-Coding RNAs in Cancer Diagnostics. *Front Genet*, *3*, 32.

[22] Vergara, I. A., Erho, N., Triche, T. J., Ghadessi, M., Crisan, A., Sierocinski, T., et al. (2012). Genomic "Dark Matter" in Prostate Cancer: Exploring the Clinical Utility of ncRNA as Biomarkers. *Front Genet*, *3*, 23.

[23] Wahlestedt, C. (2013). Targeting long non-coding RNA to therapeutically upregulate gene expression. *Nat Rev Drug Discov*, *12*, 433–446.

[24] St Laurent, G., Vyatkin, Y. & Kapranov, P. (2014). Dark matter RNA illuminates the puzzle of genome-wide association studies. *BMC Med*, *12*, 97.

[25] Gutschner, T. & Diederichs, S. (2012). The hallmarks of cancer: a long noncoding RNA point of view. *RNA Biol*, *9*, 703–719.

[26] Ellis, B. C., Molloy, P. L. & Graham, L. D. (2012). CRNDE: a long non-coding RNA involved in cancer, neurobiology, and development. *Front Genet*, *3*, 270.

[27] Wang, P., Ren, Z. & Sun, P. (2012). Overexpression of the long non-coding RNA MEG3 impairs in vitro glioma cell proliferation. *J Cell Biochem*, *113*, 1868–1874.

[28] Yao, Y., Ma, J., Xue, Y., Wang, P., Li, Z., Liu, J., et al. (2015). Knockdown of long non-coding RNA XIST exerts tumor-suppressive functions in human glioblastoma stem cells by up-regulating miR-152. *Cancer Lett*, *359*, 75–86.

[29] Kogo, R., Shimamura, T., Mimori, K., Kawahara, K., Imoto, S., Sudo, T., et al. (2011). Long noncoding RNA HOTAIR regulates polycomb–dependent chromatin modification and is associated with poor prognosis in colorectal cancers. *Cancer Res* 71, 6320–6326.

[30] Yang Z., Zhou L., Wu L. M., Lai M. C., Xie H. Y., Zhang F., et al. (2011). Overexpression of long non-coding RNA HOTAIR predicts tumor recurrence in hepatocellular carcinoma patients following liver transplantation. *Ann Surg Oncol*, *18*, 1243–1250.

[31] Quagliata L., Matter M., Piscuoglio S., Makowska Z., Heim M., Tornillo L., et al. (2013). Hoxa13 and Hottip expression levels predict patients' survival and metastasis formation in hepatocellular carcinoma. *J Hepatol*, *58*, S39–40.

[32] Gupta R. A., Shah N., Wang K. C., Kim J., Horlings H. M., Wong D. J., et al. (2010). Long non-coding RNA HOTAIR reprograms chromatin state to promote cancer metastasis. *Nature*, *464*, 1071–1076.

[33] Yildirim E., Kirby J. E., Brown D. E., Mercier F. E., Sadreyev R. I., Scadden D. T., et al. (2013). Xist RNA is a potent suppressor of hematologic cancer in mice. *Cell*, *152*, 727–742.

[34] Brunner A. L., Beck A. H., Edris B., Sweeney R. T., Zhu S. X., Li R., et al. (2012). Transcriptional profiling of long non-coding RNAs and novel transcribed regions across a diverse panel of archived human cancers. *Genome Biol*, *13*, R75.

[35] Consortium E. P., Birney E., Stamatoyannopoulos J. A., Dutta A., Guigo R., Gingeras T. R., et al. (2007). Identification and analysis of functional elements in 1% of the human genome by the ENCODE pilot project. *Nature*, *447*, 799–816.

[36] Chooniedass-Kothari S., Emberley E., Hamedani M. K., Troup S., Wang X., Czosnek A., et al. (2004). The steroid receptor RNA activator is the first functional RNA encoding a protein. *FEBS Lett*, *566*, 43–47.

[37] Warden, C. D., Kim, S. H. & Yi, S. V. (2008). Predicted functional RNAs within coding regions constrain evolutionary rates of yeast proteins. *PLoS One*, *3*, e1559.

[38] Fang Y. & Fullwood M. J. (2016). Roles, Functions, and Mechanisms of Long Non-coding RNAs in Cancer. *Genomics Proteomics Bioinformatics*, *14*(1), 42-54.

[39] Jia, H., Osak, M., Bogu, G. K., Stanton, L. W., Johnson, R. & Lipovich, L. (2010). Genome-wide computational identification and manual annotation of human long noncoding RNA genes. *RNA*, *16*, 1478–1487.

[40] Ulitsky, I. & Bartel, D. P. (2013). LincRNAs: genomics, evolution, and mechanisms. *Cell*, *154*, 26–46.

[41] Lam, M. T, Li, W., Rosenfeld, M. G. & Glass, C. K. (2014). Enhancer RNAs and regulated transcriptional programs. *Trends Biochem Sci*, *39*, 170–182.

[42] Rinn, J. L. & Chang, H. Y. (2012). Genome regulation by long noncoding RNAs. *Annu Rev Biochem*, *81*, 145–166.

[43] Derrien, T., Johnson, R., Bussotti, G., Tanzer, A., Djebali, S., Tilgner, H., et al. (2012). The GENCODE v7 catalog of human long noncoding RNAs: analysis of their gene structure, evolution, and expression. *Genome Res*, *22*, 1775–1789.

[44] Banfai, B., Jia, H., Khatun, J., Wood, E., Risk, B., Gundling, W. E., et al. (2012). Long noncoding RNAs are rarely translated in two human cell lines. *Genome Res*, *22*, 1646–1657.

[45] Guttman, M., Russell, P., Ingolia, N. T., Weissman, J. S. & Lander, E. S. (2013). Ribosome profiling provides evidence that large noncoding RNAs do not encode proteins. *Cell*, *154*, 240–251.

[46] Necsulea A., Soumillon M., Warnefors M., Liechti A., Daish T., Zeller U., et al. (2014). The evolution of lncRNA repertoires and expression patterns in tetrapods. *Nature*, *505*, 635–640.

[47] Ulitsky, I., Shkumatava, A., Jan, C. H., Sive, H. & Bartel, D. P. (2011). Conserved function of lincRNAs in vertebrate embryonic development despite rapid sequence evolution. *Cell*, *147*, 1537–1550.

[48] Cabili, M. N., Trapnell, C., Goff, L., Koziol, M., Tazon-Vega, B., Regev, A., et al. (2011). Integrative annotation of human large intergenic noncoding RNAs reveals global properties and specific subclasses. *Genes Dev*, *25*, 1915–1927.

[49] Hezroni, H., Koppstein, D., Schwartz, M. G., Avrutin, A., Bartel, D. P. & Ulitsky, I. (2015). Principles of long noncoding RNA evolution derived from direct comparison of transcriptomes in 17 species. *Cell Rep*, *11*, 1110–1122.

[50] Quinn, J. J., Zhang, Q. C., Georgiev, P., Ilik, I. A., Akhtar, A. & Chang, H. Y. (2016). Rapid evolutionary turnover underlies conserved lncRNA-genome interactions. *Genes Dev*, *30*, 191–207.

[51] Cheng, J., Kapranov, P., Drenkow, J., Dike, S., Brubaker, S., Patel, S., et al. (2005). Transcriptional maps of 10 human chromosomes at 5- nucleotide resolution. *Science*, *308*, 1149–1154.

[52] Ravasi, T., Suzuki, H., Pang, K. C., Katayama, S., Furuno, M., Okunishi, R., et al. (2006). Experimental validation of the regulated expression of large numbers of non-coding RNAs from the mouse genome. *Genome Res*, *16*, 11–19.

[53] Bussemakers, M. J., van Bokhoven, A., Verhaegh, G. W., Smit, F. P., Karthaus, H. F., Schalken, J. A., et al. (1999). DD3: a new prostate-specific gene, highly overexpressed in prostate cancer. *Cancer Res*, *59*, 5975–5979.

[54] Batista, P. J. & Chang, H. Y. (2013). Long noncoding RNAs: cellular address codes in development and disease. *Cell*, *152*, 1298–1307.

[55] Preker, P., Nielsen, J., Kammler, S., Lykke-Andersen, S., Christensen, M. S., Mapendano, C. K., et al. (2008). RNA exosome depletion reveals transcription upstream of active human promoters. *Science*, *322*, 1851–1854.

[56] Clark, M. B., Johnston, R. L., Inostroza-Ponta, M., Fox, A. H., Fortini, E., Moscato, P., et al. (2012). Genome-wide analysis of long noncoding RNA stability. *Genome Res*, *22*, 885–898.

[57] Quinn, J. J. & Chang, H. Y. (2015). Unique features of long non-coding RNA biogenesis and function. *Nat. Rev. Genet*, *17*, 47–62.

[58] Alam, T., Medvedeva, Y. A., Jia, H., Brown, J. B., Lipovich, L., and Bajic, V. B. (2014). Promoter analysis reveals globally differential regulation of human long non-coding RNA and protein-coding genes. *PLoS One*, *9*, e109443.

[59] Sati, S., Ghosh, S., Jain, V., Scaria, V. & Sengupta, S. (2012). Genome-wide analysis reveals distinct patterns of epigenetic features in long non-coding RNA loci. *Nucleic Acids Res*, *40*, 10018–10031.

[60] Schmitt, A. M. & Chang, H. Y. (2016). Long Noncoding RNAs in Cancer Pathways. *Cancer Cell*, *29*(4), 452–463.

[61] Zheng, G. X., Do, B. T., Webster, D. E., Khavari, P. A. & Chang, H. Y. (2014). Dicer-microRNA-Myc circuit promotes transcription of hundreds of long noncoding RNAs. *Nat. Struct. Mol. Biol*, *21*, 585–590.

[62] Maurano, M. T., Humbert, R., Rynes, E., Thurman, R. E., Haugen, E., Wang, H., et al. (2012). Systematic localization of common disease-associated variation in regulatory DNA. *Science*, *337*, 1190–1195.

[63] Roadmap Epigenomics, C., Kundaje, A., Meuleman, W., Ernst, J., Bilenky, M., Yen, A., et al. (2015). Integrative analysis of 111 reference human epigenomes. *Nature*, *518*, 317–330.

[64] Melton, C., Reuter, J. A., Spacek, D. V. & Snyder, M. (2015). Recurrent somatic mutations in regulatory regions of human cancer genomes. *Nat. Genet*, *47*, 710–716.

[65] Beroukhim, R., Mermel, C. H., Porter, D., Wei, G., Raychaudhuri, S., Donovan, J., et al. (2010). The landscape of somatic copy-number alteration across human cancers. *Nature*, *463*, 899–905.

[66] Li, W., Notani, D., Ma, Q., Tanasa, B., Nunez, E., Chen, A. Y., et al. (2013). Functional roles of enhancer RNAs for oestrogen-dependent transcriptional activation. *Nature*, *498*, 516–520.

[67] Orom, U. A. & Shiekhattar, R. (2013). Long noncoding RNAs usher in a new era in the biology of enhancers. *Cell*, *154*, 1190–1193.

[68] Ounzain, S. & Pedrazzini, T. (2015). The promise of enhancer-associated long noncoding RNAs in cardiac regeneration. *Trends Cardiovasc Med*, *25*, 592–602.

[69] Sanyal, A., Lajoie, B. R., Jain, G., Dekker, J. (2012). The long-range interaction landscape of gene promoters. *Nature*, *489*, 109–113.

[70] Li, G., Ruan, X., Auerbach, R. K., Sandhu, K. S., Zheng, M., Wang, P., et al. (2012). Extensive promoter-centered chromatin interactions provide a topological basis for transcription regulation. *Cell*, *148*, 84–98.

[71] Visel, A., Rubin, E. M. & Pennacchio, L. A. (2009). Genomic views of distantacting enhancers. *Nature*, *461*, 199–205.

[72] Kallen, A. N., Zhou, X. B., Xu, J., Qiao, C., Ma, J., Yan, L., et al. (2013). The imprinted H19 lncRNA antagonizes let-7 microRNAs. *Mol Cell*, *52*, 101–112.

[73] Karreth, F. A., Tay, Y., Perna, D., Ala, U., Tan, S. M., Rust, A. G., et al. (2011). In vivo identification of tumor- suppressive PTEN ceRNAs in an oncogenic BRAF-induced mouse model of melanoma. *Cell*, *147*, 382–395.

[74] Tay, Y., Kats, L., Salmena, L., Weiss, D., Tan, S. M., Ala, U., et al. (2011). Coding-independent regulation of the tumor suppressor PTEN by competing endogenous mRNAs. *Cell*, *147*, 344–357.

[75] Salmena, L., Poliseno, L., Tay, Y., Kats, L. & Pandolfi, P. P. (2011). A ceRNA hypothesis: the Rosetta Stone of a hidden RNA language? *Cell*, *146*, 353–358.

[76] Nakano, S., Murakami, K., Meguro, M., Soejima, H., Higashimoto, K., Urano, T., et al. (2006). Expression profile of LIT1/KCNQ1OT1 and epigenetic status at the KvDMR1 in colorectal cancers. *Cancer Sci*, *97*, 1147–1154.

[77] Pandey, R. R., Mondal, T., Mohammad, F., Enroth, S., Redrup, L., Komorowski, J., et al. (2008). Kcnq1ot1 antisense noncoding RNA mediates lineage-specific transcriptional silencing through chromatin- level regulation. *Mol Cell*, *32*, 232–246.

[78] Czermin, B., Melfi, R., McCabe, D., Seitz, V., Imhof, A. & Pirrotta, V. (2002). Drosophila enhancer of Zeste/ESC complexes have a histone H3 methyltransferase activity that marks chromosomal polycomb sites. *Cell*, *111*, 185–196.

[79] Aguilo, F., Zhou, M. M. & Walsh, M. J. (2011). Long noncoding RNA, polycomb, and the ghosts haunting INK4b-ARF-INK4a expression. *Cancer Res*, *71*, 5365–5369.

[80] Kotake, Y., Nakagawa, T., Kitagawa, K., Suzuki, S., Liu, N., Kitagawa, M., et al. (2011). Long non-coding RNA ANRIL is required for the PRC2 recruitment to and silencing of p15(INK4B) tumor suppressor gene. *Oncogene*, *30*, 1956–1962.

[81] Holoch, D. & Margueron, R. (2017). Mechanisms Regulating PRC2 Recruitment and Enzymatic Activity. *Trends Biochem Sci*, *42*, 531–542.
[82] Cao, R., Wang, L., Wang, H., Xia, L., Erdjument-Bromage, H., Tempst, P., et al. (2002). Role of histone H3 lysine 27 methylation in Polycomb-group silencing. *Science*, *298*, 1039–1043.
[83] Herz, H. M., Garruss, A. & Shilatifard, A. (2013). SET for life: biochemical activities and biological functions of SET domain-containing proteins. *Trends Biochem Sci*, *38*, 621–639.
[84] Pasini, D., Bracken, A. P., Jensen, M. R., Lazzerini Denchi, E. & Helin, K. (2004). Suz12 is essential for mouse development and for EZH2 histone methyltransferase activity. *EMBO J*, *23*, 4061–4071.
[85] Cao, R. & Zhang, Y. (2004). SUZ12 is required for both the histone methyltransferase activity and the silencing function of the EED-EZH2 complex. *Mol Cell*, *15*, 57–67.
[86] Moritz, L. E. & Trievel, R. C. (2017). Structure, mechanism, and regulation of polycomb repressive complex 2. *J Biol Chem*, pii: jbc.R117.800367. doi: 10.1074/jbc.R117.800367.
[87] Kim, K. H. & Roberts, C. W. (2016). Targeting EZH2 in cancer. *Nat Med*, *22*, 128–134.
[88] Comet, I., Riising, E. M., Leblanc, B. & Helin, K. (2016). Maintaining cell identity: PRC2-mediated regulation of transcription and cancer. *Nat Rev Cancer*, *16*, 803–810.
[89] Chang, C. J. & Hung, M. C. (2012). The role of EZH2 in tumour progression. *Br J Cancer*, *106*, 243–247.
[90] Wen, Y., Cai, J., Hou, Y., Huang, Z. & Wang, Z. (2017). Role of EZH2 in cancer stem cells, from biological insight to a therapeutic target. *Oncotarget*, *8*, 37974–37990.
[91] Volkel, P., Dupret, B., Le Bourhis, X. & Angrand, P. O. (2015). Diverse involvement of EZH2 in cancer epigenetics. *Am J Transl Res*, *7*, 175–93.
[92] Hsieh, C. L., Fei, T., Chen, Y., Li, T., Gao, Y., Wang, X., et al. (2014). Enhancer RNAs participate in androgen receptor-driven looping that selectively enhances gene activation. *Proc Natl Acad Sci U S A*, *111*, 7319–7324.
[93] Melo, C. A., Drost, J., Wijchers, P. J., van de Werken, H., de Wit, E., Oude Vrielink, J. A., et al. (2013). ERNAs are required for p53-dependent enhancer activity and gene transcription. *Mol Cell*, *49*, 524–535.
[94] Lai, F., Gardini, A., Zhang, A. & Shiekhattar, R. (2015). Integrator mediates the biogenesis of enhancer RNAs. *Nature*, *525*, 399–403.
[95] Calin, G. A., Liu, C. G., Ferracin, M., Hyslop, T., Spizzo, R., Sevignani, C., et al. (2007). Ultraconserved regions encoding ncRNAs are altered in human leukemias and carcinomas. *Cancer Cell*, *12*, 215–229.

[96] Trimarchi, T., Bilal, E., Ntziachristos, P., Fabbri, G., Dalla-Favera, R., Tsirigos, A., et al. (2014). Genome-wide mapping and characterization of Notch-regulated long noncoding RNAs in acute leukemia. *Cell*, *158*, 593–606.

[97] Yu, W., Gius, D., Onyango, P., Muldoon-Jacobs, K., Karp, J., Feinberg, A. P., et al. (2008). Epigenetic silencing of tumour suppressor gene p15 by its antisense RNA. *Nature*, *451*, 202–206.

[98] Walsh, K. M., Codd, V., Smirnov, I. V., Rice, T., Decker, P. A., Hansen, H. M., et al. (2014). Variants near TERT and TERC influencing telomere length are associated with highgrade glioma risk. *Nat. Genet*, *46*, 731–735.

[99] Gutschner, T., Hammerle, M., Eissmann, M., Hsu, J., Kim, Y., Hung, G., et al. (2013). The noncoding RNA MALAT1 is a critical regulator of the metastasis phenotype of lung cancer cells. *Cancer Res*, *73*, 1180–1189.

[100] Vance, K. W., Sansom, S. N., Lee, S., Chalei, V., Kong, L., Cooper, S. E., et al. (2014). The long non-coding RNA Paupar regulates the expression of both local and distal genes. *EMBO J*, *33*, 296–311.

[101] Iyer, M. K., Niknafs, Y. S., Malik, R., Singhal, U., Sahu, A., Hosono, Y., et al. (2015). The landscape of long noncoding RNAs in the human transcriptome. *Nat Genet*, *47*, 199-208.

[102] Cheetham, S. W., Gruhl, F., Mattick, J. S. & Dinger, M. E. (2013). Long noncoding RNAs and the genetics of cancer. *Br J Cancer*, *108*, 2419–2425.

[103] Gonzalgo, M. L., Nakayama, M., Lee, S. M., De Marzo, A. M. & Nelson, W. G. (2004). Detection of GSTP1 methylation in prostatic secretions using combinatorial MSP analysis. *Urology*, *63*, 414–418.

[104] Prensner, J. R., Iyer, M. K., Balbin, O. A., Dhanasekaran, S. M., Cao, Q., Brenner, J. C., et al. (2011). Transcriptome sequencing across a prostate cancer cohort identifies PCAT-1, an unannotated lincRNA implicated in disease progression. *Nat Biotechnol*, *29*(8), 742–749.

[105] Mehra, R., Udager, A. M., Ahearn, T. U., Cao, X., Feng, F. Y., Loda, M., et al. (2016). Overexpression of the Long Non-coding RNA SChLAP1 Independently Predicts Lethal Prostate Cancer. *Eur Urol*, *70*(4), 549–552.

[106] Prensner, J. R., Iyer, M. K., Sahu, A., Asangani, I. A., Cao, Q., Patel, L., et al. (2013). The long noncoding RNA SChLAP1 promotes aggressive prostate cancer and antagonizes the SWI/SNF complex. *Nat Genet*, *45*, 1392–1398.

[107] Liz, J. & Esteller, M. (2016). lncRNAs and microRNAs with a role in cancer development. *Biochim Biophys Acta*, *1859*, 169–176.

[108] Prensner, J. R., Chen, W., Han, S., Iyer, M. K., Cao, Q., Kothari, V., et al. (2014). The long non-coding RNA PCAT-1 promotes prostate cancer cell proliferation through cMyc. *Neoplasia*, *16*, 900–908.

[109] Prensner, J. R., Zhao, S., Erho, N., Schipper, M., Iyer, M. K., Dhanasekaran, S. M., et al. (2014). RNA biomarkers associated with metastatic progression in prostate

cancer: a multi-institutional high-throughput analysis of SChLAP1. *Lancet Oncol*, *15*, 1469–1480.

[110] Wei, J. T., Feng, Z., Partin, A. W., Brown, E., Thompson, I., Sokoll, L., et al. (2014). Can urinary PCA3 supplement PSA in the early detection of prostate cancer? *J. Clin. Oncol*, *32*, 4066–4072.

[111] Yan, X., Hu, Z., Feng, Y., Hu, X., Yuan, J., Zhao, S. D., et al. (2015). Comprehensive genomic characterization of long non-coding RNAs across human cancers. *Cancer Cell*, *28*, 529–540.

[112] Du, Z., Fei, T., Verhaak, R. G., Su, Z., Zhang, Y., Brown, M., et al. (2013). Integrative genomic analyses reveal clinically relevant long noncoding RNAs in human cancer. *Nat. Struct. Mol. Biol*, *20*, 908–913.

[113] Bottcher, R., Hoogland, A. M., Dits, N., Verhoef, E. I., Kweldam, C., Waranecki, P., et al. (2015). Novel long noncoding RNAs are specific diagnostic and prognostic markers for prostate cancer. *Oncotarget*, *6*, 4036–4050.

[114] Baylin, S. B. & Ohm, J. E. (2006). Epigenetic gene silencing in cancer: a mechanism for early oncogenic pathway addiction? *Nat Rev Cancer*, *6*, 107–116.

[115] Fan, M., Li, X. Y., Jiang, W., Huang, Y., Li, J. D. & Wang, Z. M. (2013). A long non-coding RNA, PTCSC3, as a tumor suppressor and a target of miRNAs in thyroid cancer cells. *Exp Ther Med*, *5*, 1143–1146.

[116] Wang, J., Liu, X., Wu, H., Ni, P., Gu, Z., Qiao, Y., et al. (2010). CREB upregulates long non-coding RNA, HULC expression through interaction with microRNA-372 in liver cancer. *Nucleic Acids Res*, *38*, 5366–5383.

[117] Du, Y., Kong, G., You, X., Zhang, S., Zhang, T., Gao, Y., et al. (2012). Elevation of highly up-regulated in liver cancer (HULC) by hepatitis B virus X protein promotes hepatoma cell proliferation via down-regulating p18. *J Biol Chem*, *287*, 26302–26311.

[118] Liu, Y., Pan, S., Liu, L., Zhai, X., Liu, J., Wen, J., et al. (2012). A genetic variant in long non-coding RNA HULC contributes to risk of HBV-related hepatocellular carcinoma in a Chinese population. *PLoS One*, *7*, e35145.

[119] Sirchia, S. M., Tabano, S., Monti, L., Recalcati, M. P., Gariboldi, M., Grati, F. R., et al. (2009). Misbehaviour of XIST RNA in breast cancer cells. *PLoS One*, *4*, e5559.

[120] Hu, Y., Wang, J., Qian, J., Kong, X., Tang, J., Wang, Y., et al. (2014). Long noncoding RNA GAPLINC regulates CD44-dependent cell invasiveness and associates with poor prognosis of gastric cancer. *Cancer Res*, *74*, 6890–6902.

[121] Lai, M. C., Yang, Z., Zhou, L., Zhu, Q. Q., Xie, H. Y., Zhang, F., et al. (2012). Long non-coding RNA MALAT-1 overexpression predicts tumor recurrence of hepatocellular carcinoma after liver transplantation. *Med Oncol*, *29*, 1810–1816.

[122] Teschendorff, A. E., Lee, S. H., Jones, A., Fiegl, H., Kalwa, M., Wagner, W., et al. (2015). HOTAIR and its surrogate DNA methylation signature indicate carboplatin resistance in ovarian cancer. *Genome Med*, *7*, 108.

[123] Yap, K. L., Li, S. D., Munoz-Cabello, A. M., Raguz, S., Zeng, L., Mujtaba, S., et al. (2010). Molecular interplay of the noncoding RNA ANRIL and methylated histone H3 Lysine 27 by polycomb CBX7 in transcriptional silencing of INK4a. *Mol Cell*, *38*, 662–674.

[124] Chung, S. Y., Nakagawa, H., Uemura, M., Piao, L., Ashikawa, K., Hosono, N., et al. (2011). Association of a novel long non-coding RNA in 8q24 with prostate cancer susceptibility. *Cancer Sci*, *102*, 245–252.

[125] Matouk, I. J., DeGroot, N., Mezan, S., Ayesh, S., Abu-Iail, R., Hochberg, A., et al. (2007). The H19 non-coding RNA is essential for human tumor growth. *PLoS One*, *2*, e845.

[126] Tseng, Y. Y., Moriarity, B. S., Gong, W., Akiyama, R., Tiwari, A., Kawakami, H., et al. (2014). PVT1 dependence in cancer with MYC copy-number increase. *Nature*, *512*, 82–86.

[127] Eeles, R. A., Kote-Jarai, Z., Giles, G. G., Olama, A. A., Guy, M., Jugurnauth, S. K., et al. (2008). Multiple newly identified loci associated with prostate cancer susceptibility. *Nat. Genet*, *40*, 316–321.

[128] Ling, H., Spizzo, R., Atlasi, Y., Nicoloso, M., Shimizu, M., Redis, R. S., et al. (2013). CCAT2, a novel noncoding RNA mapping to 8q24, underlies metastatic progression and chromosomal instability in colon cancer. *Genome Res*, *23*, 1446–1461.

[129] Tomlinson, I., Webb, E., Carvajal-Carmona, L., Broderick, P., Kemp, Z., Spain, S., et al. (2007). A genomewide association scan of tag SNPs identifies a susceptibility variant for colorectal cancer at 8q24.21. *Nat. Genet*, *39*, 984–988.

[130] Jendrzejewski, J., He, H., Radomska, H. S., Li, W., Tomsic, J., Liyanarachchi, S., et al. (2012). The polymorphism rs944289 predisposes to papillary thyroid carcinoma through a large intergenic noncoding RNA gene of tumor suppressor type. *Proc. Natl. Acad. Sci. USA*, *109*, 8646–8651.

[131] Dorji, T., Monti, V., Fellegara, G., Gabba, S., Grazioli, V., Repetti, E., et al. (2015). Gain of hTERC: a genetic marker of malignancy in oral potentially malignant lesions. *Hum. Pathol* 46, 1275–1281.

[132] Tao, R., Hu, S., Wang, S., Zhou, X., Zhang, Q., Wang, C., et al. (2015). Association between indel polymorphism in the promoter region of lncRNA GAS5 and the risk of hepatocellular carcinoma. *Carcinogenesis*, *36*, 1136–1143.

[133] Hu, X., Feng, Y., Zhang, D., Zhao, S. D., Hu, Z., Greshock, J., et al. (2014a). A functional genomic approach identifies FAL1 as an oncogenic long noncoding RNA that associates with BMI1 and represses p21 expression in cancer. *Cancer Cell*, *26*, 344–357.

[134] Quagliata, L., Matter, M. S., Piscuoglio, S., Arabi, L., Ruiz, C., Procino, A., et al. (2014). Long noncoding RNA HOTTIP/HOXA13 expression is associated with disease progression and predicts outcome in hepatocellular carcinoma patients. *Hepatology*, *59*, 911–923.

[135] Zhang, X., Gejman, R., Mahta, A., Zhong, Y., Rice, K. A., Zhou, Y., et al. (2010). Maternally expressed gene 3, an imprinted noncoding RNA gene, is associated with meningioma pathogenesis and progression. *Cancer Res*, *70*, 2350–2358.

[136] Pandey, G. K., Mitra, S., Subhash, S., Hertwig, F., Kanduri, M., Mishra, K., et al. (2014). The riskassociated long noncoding RNA NBAT-1 controls neuroblastoma progression by regulating cell proliferation and neuronal differentiation. *Cancer Cell*, *26*, 722–737.

[137] Liu, B., Sun, L., Liu, Q., Gong, C., Yao, Y., Lv, X., et al. (2015). A cytoplasmic NF-kappaB interacting long noncoding RNA blocks IkappaB phosphorylation and suppresses breast cancer metastasis. *Cancer Cell* 27, 370–381.

[138] McCleland, M. L., Mesh, K., Lorenzana, E., Chopra, V. S., Segal, E., Watanabe, C., et al. (2016). CCAT1 is an enhancer-templated RNA that predicts BET sensitivity in colorectal cancer. *J. Clin. Invest*, *126*, 639–652.

[139] Shao, Y., Ye, M., Jiang, X., Sun, W., Ding, X., Liu, Z., et al. (2014). Gastric juice long noncoding RNA used as a tumor marker for screening gastric cancer. *Cancer*, *120*, 3320–3328.

[140] Khaitan, D., Dinger, M. E., Mazar, J., Crawford, J., Smith, M. A., Mattick, J. S., et al. (2011). The melanoma-upregulated long noncoding RNA SPRY4-IT1 modulates apoptosis and invasion. *Cancer Res*, *71*, 3852–3862.

[141] Ling, H., Vincent, K., Pichler, M., Fodde, R., Berindan-Neagoe, I., Slack, F. J., et al. (2015). Junk DNA and the long non-coding RNA twist in cancer genetics. *Oncogene*, *34*, 5003–5011.

[142] Bhan, A. & Mandal, S. S. (2015). LncRNA HOTAIR: a master regulator of chromatin dynamics and cancer. *Biochim. Biophys. Acta*, *1856*, 151–164.

In: Research on New Generation …
Editor: Yildiz Dincer

ISBN: 978-1-53614-367-6

Chapter 9

MOLECULAR MARKERS FOR THYROID CANCER: CLINICAL APPLICATIONS

Sushela S. Chaidarun*[*]*, MD, PhD
Department of Medicine, Endocrine Section,
Geisel Medical School at Dartmouth,
Dartmouth-Hitchcock Medical Center, Lebanon, NH, US

ABSTRACT

Thyroid cancer is the most common type of cancer arising from an endocrine gland. Currently, there is an estimated incidence of 19.6/100,000 per year in the USA. The incidence of thyroid cancer, both in the US and worldwide, has increased over the last 4 decades. Diagnosis of thyroid cancer is usually obtained through fine needle aspiration biopsy of nodules exhibiting suspicious features on ultrasound. We will review the current recommendations to implement molecular genetic testing for thyroid nodules. The most common gene mutations harbored in the various differentiated thyroid cancers, including BRAF, RAS, RET/PTC, and the more recently discovered gene mutations for thyroid cancer, TERT mutations, will be reviewed. We will discuss the practice guidelines to implement molecular gene testing for thyroid nodules with indeterminate cytology as well as the new category of non-invasive follicular thyroid neoplasm with papillary like nuclear features (NIFTP) as an intermediate pathology which will not require aggressive therapy after diagnostic lobectomy. Lastly, we will discuss personalized targeted therapy for thyroid cancer.

Keywords: thyroid cancer, thyroid nodule, molecular marker, genetic test, indeterminate cytology, fine needle aspiration biopsy, tumor genotyping, targeted therapy, multi-targeted kinase inhibitor

---

[*] Corresponding Author Email: Sushela.s.chaidarun@hitchcock.org.

Introduction

Thyroid cancer is the most common type of cancer among endocrine tumors, the fouth common cancer of women in the US and the most common cancer of women in Korea after routine thyroid ultrasound surveillance was introduced. Currently there is an estimated incidence of 19.6 cases per 100,000 persons per year in the US with 64,300 new cases and 1,980 deaths annually, and the overall incidence of thyroid cancer worldwide has markedly increased over the last 4 decades [1].

Diagnosis of thyroid cancer is usually obtained through fine needle aspiration (FNA) biopsy of nodules exhibiting suspicious features on ultrasound. We will review current recommendations regarding molecular genetic testing for thyroid nodules, especially in the indeterminate categories with the goal of avoiding overdiagnosis of thyroid cancer and reduction of unnecessary surgery.

Thyroid nodules are very common but only 5-15% are malignant [2-4]. Though most malignant thyroid cancers can be identified pre-operatively by cytological analysis, approximately 20-30% are classified indeterminate [5, 6]. Subsequently, repeat FNA or thyroid surgery may be required for definitive diagnosis. In the US, the incidence of thyroid cancer has increased 3-4 folds since 1975, from 4.9 to 14.3-19.6 cases per 100,000 individuals, primarily due to increased diagnosis of papillary thyroid cancer. However mortality has remained stable (approximately 0.5 deaths per 100,000) [7]. This suggests overdiagnosis of thyroid cancer. The dramatic increases in diagnosis have been mainly due to new modes of screening and diagnosis of many indolent tumors that would not, if left alone, result in symptoms or death [1].

Thyroid malignancies are divided into papillary thyroid carcinomas (80%), follicular thyroid carcinomas (10-15%), medullary thyroid carcinomas (4%), anaplastic thyroid carcinomas (1-2%), primary thyroid lymphomas or sarcoma (rare), and metastatic cancer to the thyroid (rare). Papillary and follicular/Hurtle cell carcinomas are derived from thyroid epithelial follicular cells and are grouped together as differentiated thyroid carcinomas, while medullary thyroid carcinomas are derived from interstitial c cells of neuroendocrine origin. Anaplastic carcinomas are undifferentiated thyroid cancer, one of the most aggressive human cancers that is fortunately very rare.

FNA of the nodules with suspicious features on ultrasound is the most cost-effective diagnostic method. FNA is indicated for nodules larger than 1.0 cm with suspicious features (solid, hypoechoicity, microcalcifications, taller than wide shape, irregular margins, interrupted rim calcifications, extrathyroidal extension, pathologic lymph node) that are associated with cancer risk of 50-90% [2-4]. Low suspicioun nodules (isoechoic/hyperechoic solid, partially cystic, or spongiform appearance) may be biopsied when they are larger than 1.5-2.0 cm since there is a low risk of cancer (< 5-10%).

FNA is not recommended for a pure thyroid cyst, or hyperfunctiong nodule with a suppressed serum thyroid stimulating hormone (TSH) concentration suggestive of hyperthyroidism (i.e., no need to FNA a cyst or hot/hyperfunctiong thyroid nodule since it is almost always benign with less than 1-2% risk of cancer).

Selective FNA and cervical lymph node assessment should be based on thyroid ultrasound characteristics and high risk history (prior head and neck irradiation, exposure to radioactive fallout, rapid nodule growth, compressive symptoms such as difficulty swallowing or hoarseness, and family history of thyroid cancer). Familial thyroid cancer syndromes include multiple endocrine neoplasia syndrome type 2, Cowden syndrome, Carney complex, or familial adenomatous colonic polyposis are important recognize prior to FNA for early diagnosis of thyroid cancer [2-4].

The Bethesda system for reporting thyroid cytopathyology provides standardized diagnostic categories to classify thyroid nodules [5-6]. The 2017 updated version reaffirms 6 diagnostic categories that are unchanged since 2009: 1) Nondiagnostic or Unsatisfactory; 2) Benign; 3) Atypia of Undetermined Significance (AUS) or Follicular Lesion of Undetermined Significance (FLUS); 4) Follicular Neoplasm (FN) or Suspicious for a Follicular Neoplasm (SFN); 5) Suspicious for Malignancy (SM); and 6) Malignant.

Many patients with cytologically indeterminate nodules (Bethesda categories 3, 4 and 5), which is about 20-30% of all biopsies, undergo unnecessary diagnostic surgeries for a cancer risk of 15-30%, placing them at risk for potential surgical complications and lifelong levothyroxine replacement [8, 9]. The advent of molecular gene testing has drawn significant attention towards improved stratification of indeterminate lesions to benign or malignant [10-13].

Studies investigating the use of molecular testing in cytologically indeterminate nodules showed the detection of any pertinent molecular mutations in the indeterminate categories (AUS/FLUS, FN/SFN, SM) conferred the risk of histologic malignancy of 88, 87, and 95%, respectively [14] while the negative test for mutation would be very reassuring with a very high negative predictive value over 95% [8, 15].

The purpose of this chapter is to review the current recommendations to implement molecular genetic testing for thyroid nodules in clinical practice. We review the most common gene mutations harbored in the various thyroid cancers, including BRAF, RAS, RET/PTC, and the more recently discovered TERT gene mutations.

We will discuss the practice guidelines to implement molecular gene testing for cytologically indeterminate thyroid nodules as well as the new category of non-invasive follicular thyroid neoplasm with papillary like nuclear features (NIFTP) [16], as an intermediate pathology which will not require aggressive therapy after diagnostic lobectomy. Lastly, we will discuss personalized targeted therapy for thyroid cancer.

Molecular Testing for Thyroid Nodules

Current Recommendations for Thyroid Nodules and Fine Needle Biopsy

The American Association of Clinical Endocrinologists (AACE) and American Thyroid Association (ATA) have created guidelines for choosing which nodules require biopsy [2, 3]. Presently, cytological examination of cells collected by FNA is the most reliable method for evaluating thyroid nodules [3, 5]. In 2009, the Bethesda reporting system for classifying thyroid cytology was proposed by the National Cancer Institute in Bethesda, Maryland, USA and provides diagnostic categories with accompanying risk stratification [5]. According to data from the National Cancer Institute that were revised in 2017, FNA in the benign category have a 0-3% risk of malignancy, and thyroid FNAs in the malignant category have a 94-96% risk of malignancy [6]. However, 20-30% of thyroid FNA specimens fall into the category of indeterminate.

The indeterminate category include 3 subcategories: AUS/FLUS, (category III), FN/SFN (category IV), and SM (category V). The predictive probability of malignancy in each of these categories is 6-18%, 10-40% and 45-60%, respectively. Of the indeterminate thyroid nodules that are surgically resected, 15-30% are confirmed to be malignant, though the incidence varies at different institutions.

Subsequently, the majority of diagnostic surgeries are performed on benign nodules. The cost and benefit of surgery *versus* molecular testing therefore needs to be assessed. However, molecular testing is advantageous compared to unnecessary surgery in terms of morbidity because any surgical intervention is associated with risks, even in the hands of the most experienced thyroid surgeons.

When assessing the clinical utility of molecular testing, it is important to recognize that even the gold standard of thyroid cytologic analysis is limited by variations in interpretation and hence there is a role for molecular markers to increase the diagnostic sensitivity for malignancy [17]. The physician and patient ought to be aware that understanding of the role of molecular testing is growing, and is still in its infancy. Further research is required to validate these tests. Nonetheless, several large, prospective, multicenter studies have independently validated the utility of testing for markers of thyroid cancer with reproducible results [8, 18, 19].

The current molecular markers that have received the most attention in thyroid cancer include BRAF, NRAS, KRAS, HRAS point mutations and PAX8/PPARG and RET/PTC rearrangements, with the aim to identify (rule in) thyroid cancer [20]. These studies have found that a positive result carries a 77% positive predictive value and 97% negative predictive value for predicting thyroid cancer, though clinical validation has not been performed for nodules that are less than 1 cm in diameter [8, 21, 22]. More recent studies have focused on telomerase reverse transcriptase promoter mutations on chromosome 5 (TERT mutations) which are more prevalent in aggressive thyroid cancers [23].

Presently, the role of molecular testing for evaluation of thyroid cancer is mainly reserved for indeterminate nodules to help stratify these nodules and prevent unnecessary surgery or to prevent unnecessary repeat FNA if surgery is warranted.

Review of Molecular Mutations in Thyroid Cancer

The BRAF V600E point mutation is the most common genetic mutation detected in patients with papillary thyroid cancer (PTC) and occurs in approximately 40-60% of patients [24, 25]. In fact, the most aggressive cancers have a much higher frequency of this mutation, closer to 70-80% [24, 25]. The BRAF V600E mutation results in an amino acid substitution at position 600 from valine to glutamic acid with a resultant increase in kinase activity important for cellular proliferation [26]. BRAF is located on chromosome 7 and is the most potent activator of the mitogen-activated protein kinases (MAPK) pathway among the 3 forms of the RAF kinases [26]. BRAF mutations seem to be involved in tumor initiation as they are found in microscopic PTC [20]. Prevalence of the BRAF V600E mutation is much higher in classical PTC (45%) than the follicular variant of PTC (20%) and is rarely found in follicular thyroid cancer (1.4%) [27]. For the aggressive tall cell variant of PTC, it has been reported as 50% [27]. The BRAF V600E mutation is associated with advanced patient age, aggressive tumor phenotype with higher risk of recurrent and persistent disease. [24, 25]. Coexisting *BRAF* V600E and telomerase reverse transcriptase *(TERT)* promoter mutations occurred is 5-6% of PTC and showed a marked synergistic effect on tumor aggressiveness, including a sharply increased tumor recurrence and patient mortality [28-31].

Gene expression profiling of papillary thyroid cancer, including BRAF V600E mutation (40-60%), RET/PTC (15% of PTC, especially in young adults with early lymph node metastasis or radiation-related cancer), RAS mutation (15% of follicular variant PTC) helps to accurately classify the risk status of the cancers and further support that each of these oncogenes induces specific cancer phenotypes [26, 32]. Tumors associated with RET/PTC1 rearrangements have the conventional PTC histology while RET/PTC3 rearrangements are seen with solid variant sporadic PTC [33].

Although most PTC with BRAF mutations have a classic histology, approximately 50-80% of the tall cell variant PTCs have the BRAF T1799A point mutation resulting in BRAF V600E mutant protein; this histologic subtype is believed to be present more often in advanced disease [24, 34, 35]. PTCs with the BRAF mutation also have a higher recurrence rate, and the metastatic recurrences have diminished radio-iodine avidity [20, 25]. The mutant protein interferes with the unique ability of thyroid tissue to take up iodine (via the de-expression of the sodium iodine symporter), including radioactive iodine, which is the mainstay for treating patients who cannot be cured by surgery alone, causing a higher rate of cancer recurrence. Thus, alternate therapies have to be considered for aggressive metastatic PTC.

In contrast, follicular thyroid cancer (FTC) and follicular variant PTC mostly harbor RAS mutations (50%) or PAX8/PPARG rearrangements (30-35%) and less commonly, distinct BRAF mutations e.g., BRAF K601E [20, 26, 34]. FTC with RAS mutations have comparatively low frequency of lymph node metastases but greater predisposition to hematogenous pulmonary metastasis. The RAS mutation is found in a wide spectrum of follicular neoplasms, including 20-40% of benign follicular adenomas, 15% of follicular variant PTC, 30% of non-invasive NIFTP, and 40-50% of FTC. PAX8/PPARG rearrangements are also seen in 2-10% of benign follicular adenomas and 38% of follicular variant PTC, while PI3K/PTEN is found in <10% of FTC but indicates more aggressive FTC. Coexisting *RAS* and *TERT* promoter mutations also showed a marked synergistic effect on the aggressiveness of FTC [29].

Mutations associated with more aggressive types of thyroid cancer include TP53, CTNNB, PIK3CA, TERT mutations, which are commonly seen in poorly differentiated and anaplastic thyroid carcinomas. The TERT mutation alone was associated with advanced patient age, lymph node and distant metastasis, and has been reported less than 10% of PTC but more than 70% of anaplastic thyroid cancer. TP53 was found in less than 1% of PTC and more than 70% of anaplastic cancer.

Anaplastic thyroid carcinoma (also called undifferentiated thyroid carcinoma, accounting for ~1-2% of all thyroid cancers) is thought to sometimes develop from an existing papillary or follicular cancers which habors these additional mutations. This cancer often spreads quickly into the neck and to other parts of the body, is very difficult to treat, and has the worst prognosis.

The RET proto-oncogene mutation is a very specific and useful genetic marker for medullary thyroid cancer (MTC), which is rare <5% and is entirely different from other types of thyroid cancers as it is derived from c-cells (parafollicular cells) which produces calcitonin. Most cases of MTC are sporadic but up to 25% are familial, associated with other endocrine tumors (e.g., parathyroid adenoma and adrenal pheochromocytoma), so-called multiple endocrine neoplasia (MEN) types 2A and 2B. Thus, genetic analysis of RET is very helpful for both ruling in and ruling out this rare but important thyroid cancer type, not only for the definitive diagnosis of MTC for the patient but also for screening other members in the family for that specific mutation of the RET oncogene. Since the RET mutation is in the germline for MEN type 2, the genetic testing can be done on peripheral blood leucocyte DNA, whereas about 50% of the sporadic (no germline) MTC have a somatic mutation.

Overview of Available Genetic Tests for Thyroid Cancer

Currently, there are four commercially available molecular tests for thyroid nodules (Table 1), using mutational and/or gene expression analyses of the FNA samples. They all have sensitivity close to 90% and high negative predictive value (NPV) of 92-96%, but variable specificity i.e., 85-92% for ThyroSeq-v2 and ThyGenX/ThyraMIR, 74% for Rosetta, and 52% for Afirma, resulting in a higher positive predictive values (PPV) 70-78% for ThyroSeq and ThyGenX/ThyraMIR and a low PPV 37-50% for Afirma and Rosetta tests [10-13].

Table 1. Comparison of commercially available molecular tests for cytologically intermediate thyroid nodule

Test	Afirma GEC and the new GSC Classifier	ThyroSeq v.2, v.3 Next Generation Sequencing	ThyGenX/ ThyraMIR	Rosetta GX Reveal
Company	Veracyte	University of Pittsburgh through CBLPath	Interspace Diagnostics	Rosetta Genomics
Methods	Genomic Expression classifier (GEC) is based on mRNA of 167 genes and reported as benign or suspicious. Afirma Genomic Sequencing Classifier (GSC) also includes RNA sequencing for BRAF V600E and RET/PTC-1,-3 mutations.	Next-generation sequencing v.2 detects 14 gene mutations (AKT1, BRAF, HRAS, KRAS, NRAS, PIK3CA, PTEN, RET, TP53, TERT, CTNNB1, SHR, GNAS, and EIF1AX and 48 gene rearrangements (RET/ PTC, PAX-8/PPARG etc). The newest ThyroSeq v.3 detects 112 gene mutations plus GEC.	ThyGenX: Uses PCR for detection of 8 common gene mutations (BRAF, HRAS, NRAS, KRAS, PIK3CA) and rearrangements (RET/PTC1, RET/PTC3 and PAX8-PPARG) first. If initial mutation panel is negative then further testing for 10-gene microRNA classifier (ThyraMIR) will be done.	microRNA(24miRNA)-based diagnostic assay. First thyroid test that works on stained FNA smears containing at least 60 cells for analysis
Strengths	High NPV 94-96% "Rule Out" Test	High NPV 96%, PPV 68-78% "Rule in & rule Out"	High NPV 94%, PPV 74% "Rule in & rule Out"	High NPV 92% "Rule out" test
Limitations	Low PPV: 37-50% High false positive results	Risk stratification based on mutation and prognostic values Newer assay	Newer assay with limited experiences	Low PPV 43% Newer assay with limited performance.
Cost	~ $4,875-6,850	~ $3,200-6,000	~ $1,675-2,025 (ThyGenX) ~$3,300-6,024 (for both ThyGenX/ThyraMIR)	Not available in the US

Afirma GEC (Genomic Expression Classifier), and New Afirma GSC (Genomic Sequening Classifier) Tests

The Afirma genomic expression classifier is the first molecular test for thyroid nodules available for clinical use in the US, and was approved in 2010 [8]. It is based on the expression of 167 genes [8, 10]. Based on validation studies demonstrating high negative predictive value (NPV) among nodules with cytology of AUS/FLUS or FN/SFN, it has been designated as a "rule out" test to identify nodules that are benign. For nodules in the intermediate categories, Afirma demonstrated a high NPV of 95% and 94%, respectively [8, 10]. However, it has a low positive predictive value (PPV) of only 38% for AUS/FLUS and 37% for FN/SFN which renders high false positive results [10]. Its usefulness thus depends on local prevalence of the thyroid cancer. It also carries a high false positive rate for Hurthle cell lesions. The Afirma test can help identify medullary thyroid cancer by using a MTC gene expression classifier [10].

The Afirma specimens required two FNA passes in addition to the routine two FNA passes for cytopathologic assessment. The samples need to be stored in the refridgerator and transported with two icepacks by an overnight shipment for analysis. This temperature-sensitivity limits its use in some geographic areas.

Recently, the Afirma Genomic Sequencing Classifier (GSC) has expanded the genomic tests using RNA next generation sequencing for RET/PTC1, RET/PTC3 in addition to BRAF V600E and MTC, and also a new classifier model of RNA expression profiles to improve specificity from 50% to 68%, and sensitivity from 89 to 91%. The NPV of 96% and PPV of 50% were described at the sites with 24% prevalence of cancer, by using 191 of the original 210 indeterminate thyroid nodules [36]. The Afirma GSC provided more accurate results with less false positivity for Hurthle cell lesions than the Afirma GEC [37]. In the study using specimens from 324 consecutive patients, the Afirma GSC showed a significant increase in benign diagnoses (72% vs 47%) compared to Afirma GEC, and a reduction in false positive rate with suspicious results reduced to 28% from 53% with Afirma GEC [36]. These results will need to be validated further in multiple clinical centers.

ThyroSeq verion-2, 3 Next Generation Sequencing (ThyroSeq-v.2, v.3 NGS)

ThyroSeq Next Generation Sequencing is the first mutational test developed by the University of Pittsburgh Medical Center, and it is commercially available through CBLPath. ThyroSeq v.2 NGS assesses 14 point mutations and over 48 gene rearrangements or fusions [38]. It can detect BRAF and RAS point mutations as well as common rearrangements of RET/PTC and PAX8/PPARG. It is estimated that one of these mutations is present in approximately 70% of well differentiated thyroid cancers, although the histologists were not blinded to the molecular test results in the validation study. The ThyroSeq v.2 NGS oncogene panel assay is considered a "rule in" assay since thyroid nodules harboring these mutations have a high likelihood of cancer given the test's high PPV of 68-78% [10]. It is also useful as a "rule out" assay since the negative mutation result carries a very high NPV of 96% [10]. ThyroSeq v.2 NGS test also includes a 6 gene expression panel for medullary thyroid cancer and parathyroid tissues.

In another single-institution study of 182 patients with 190 intermediate nodule cytologies at Moffitt Cancer Center in Tampa, FL, the NPV was 91% but the PPV was much lower at 42% [39]. This is likely due to the many patients (21) found to have RAS mutations, since these mutations can be seen in non-malignant nodules, including follicular adenomas and non-invasive follicular thyroid neoplasm with papillary like nuclear features (NIFTP), causing higher false positive results.

The newest third generation of ThyroSeq v.3 has additional analyses to capture 112 genetic mutations, including some rarer mutations for thyroid cancer, and also gene expression classifier to better separate malignant lesions from benign lesions [40]. ThyroSeq v.3 is a DNA- and RNA-based next-generation sequencing assay that analyzes 5 classes of genetic alterations, including point mutations, insertions/deletions, gene fusions, copy number alterations, and abnormal gene expression. It was validated in 238 pathological tissue samples and 175 FNA samples with known surgical follow-up, and provides high accuracy (91-92%) for detecting all common types of thyroid cancer and parathyroid lesions. The analytical sensitivity (94-98%), specificity (82-89%), and robustness of the test have been successfully validated and indicate its suitability for clinical use [40]. ThyroSeq v.3 NGS became available commercially in 2018.

ThyGenX/ThyraMIR Tests (8-Gene Mutation Analysis Panel and 10-Gene MicroRNA Panel Classifier)

This test can help reduce the cost by half by testing for the common 8 gene point mutations or fusions (ThyGenX) first and if the initial mutation panel is negative, further testing for the 10-gene microRNA panel classifier (ThyraMIR) will be performed [10, 41, 42]. If ThyGenX gene panel mutation is positive for BRAF V600E mutation, there is over 95% likelihood of cancer and therefore ThyraMIR microRNA classifier is not required, which helps save the cost. If RAS mutation is identified on ThyGenX test, the ThyraMIR will be performed to evaluate if it shows a benign microRNA pattern suggestive of a benign follicular adenoma instead of a malignant pattern for follicular cancer or follicular variant PTC. If both ThyGenX and ThyraMIR tests are negative (i.e., double negative), it carries a high 94% likelihood that the negative results are truly benign. It is useful for the combined AUS/FLUS and FN/SFN lesions (intermediate thyroid cytology categories) with high NPV of 94% and PPV of 74%. It also includes MTC gene expression for medullary thyroid cancer.

ThyGenX/ThyraMIR test requires only an extra FNA pass and the sample can be stored at room temperature up to 30 days. This gives the benefit for the local cytopathologists to assess the FNA specimens before mailing the samples out for molecular testing only if the nodules show some atypia or in the intermediate cytology categories (AUS/FLUS or FN/SFN).

ThyGenX/ThyraMIR molecular test will soon be able to analze samples from direct extraction of DNA and RNA on the smear slides (without the need for additional FNA biopsy). The test performance will need to be validated.

Rosetta GX Reveal miRNA Test

This new molecular test has the benefit of avoiding an extra pass or repeat FNA biopsy as the extraction can be performed by using the existing FNA slides containing at least 60 cells for the analysis. It has high NPV of 91% but low PPV of 43-59%, and therefore can be used as a rule-out test [43]. It is available in Europe and is not yet commercially available in the US.

Clinical Applications

Practice Guidelines for Clinical Use of Thyroid Molecular Testing

For a nodule with indeterminate cytology (AUS/FLUS, FN/SFN or SMC), ATA recommends molecular testing (strong recommendation, low quality evidence) after counseling the patient regarding benefits and limitations of testing given that there are no long term outcomes data on the clinical and therapeutic implications of results [2].

The recently updated 2016 AACE guidelines recommend molecular testing to complement, not replace cytologic evaluation or clinical judgment (Grade A) [44]. Molecular testing may be considered for cytologically indeterminate nodules. They recommend testing for BRAF, RET/PTC, PAX8/PPRG and RAS mutations (Grade B). They opine that due to insufficient evidence and limited follow up studies, they do not recommend in favor of or against the use of gene expression classifiers for cytologically indeterminate nodules (Grade B).

The main benefits of molecular testing include reducing the diagnostic uncertaintly of cytologically indeterminate thyroid nodules and unnecessary surgeries but there are many factors that contribute to the overall probability of malignancy. Accordingly, the decision to use molecular tests would depend on clinical history, ultrasound features, cytological findings, and patient's preference, and not simply a reflex test that could lead to inappropriate use of molecular tests.

The cost of the molecular tests remains high, although it will mostly be covered by health insurances in the US with the maximal out-of-pocket cost of $300 for those without health insurance. The cost-effectiveness analysis and personalized approach would be critical to reduce the overall cost. There is no or little added benefit of using molecular tests if the patient has large nodules with compressive symptoms requiring surgery, or a high suspicion for malignancy because of a history of rapid growth with lymph node involvement, very suspicious ultrasound features, or if the nodule does not meet criteria for FNA and appears benign. We should use a molecular test if it will impact the management of the patients. The tests may be used more widely in the future if the cost is lowered.

Non-Invasive Follicular Thyroid Neoplasm with Papillary Like Nuclear Features (NIFTP): A New Thyroid Neoplasm Category

The two common subtypes of PTC, namely classic PTC and follicular variant PTC (FvPTC) diagnosed histologically by papillary architecture and follicular growth pattern respectively, and by distinctive nuclear features are well known. Among FvPTC, two major subtypes are described- infiltrative and encapsulated. Infiltrative FvPTC resembles classic PTC, while encapsulated FvPTC (EFvPTC) when invasive behaves like a follicular carcinoma [44].

There is a non-invasive variant of EFvPTC which lacked capsular/vascular invasion but due to nuclear features was previously diagnosed as a cancer. A multi-disciplinary team reviewed clinical and cytopathologic data of 268 tumors categorized as EFvPTC. This was a retrospective study of patients with thyroid nodules diagnosed as EFvPTC, including 109 patients with noninvasive EFvPTC observed for 10-26 years and 101 patients with invasive EFvPTC observed for 1-18 years [44]. At median follow-up of 13 years, clinical outcomes showed no evidence of disease in the group with non- invasive EFvPTC and molecular analysis showed they did not harbor the BRAF mutations seen in classic PTC [45, 46]. These data led to a reclassification of this subset of tumors as non-invasive follicular thyroid neoplasm with papillary-like nuclear features (NIFTP). This reclassification was intended to reduce overtreatment and to ameliorate psychological impact of a cancer diagnosis [45, 46].

NIFTP is not benign but it is also not an overt cancer since it carries an extremely low malignant potential, analogous to an "*in situ*" cancer. Diagnostic lobectomy would be adequate for diagnosis and treatment of a NIFTP lesion [47].

This intermediate classification will substantially reduce thyroid cancer prevalence and reduce overtreatment of indolent tumors. With NIFTP not classified as a cancer, the rate of malignancy in the Bethesda diagnostic categories would decrease substantially. The expected rate of decrease differs amongst authors. Strickland et al. [48] predicted in AUS/FLUS, FN/SN and SM, 17%, 8% and 41% reductions in risk of malignancy while Faquin et al. [49] predicted 13%, 15% and 23% reductions in these categories. Follow up studies will be required to re-assess pre- and post-test probabilities of malignancy by molecular testing with incorporation of this new classification.

It's notable that most NIFTP will be detected by FNA as atypia or suspicious for malignancy due to the classical papillary nuclear features, and molecular testing as "suspicious" on Afirma and Rosetta tests, or as "positive mutation" on ThyroSeq and ThyGenX/ThyraMIR tests. The NIFTP molecular profile results include the mutations of RAS, BRAF K601E, PPARG fusion and THADA fusion. The point mutation of BRAF V600E is essentially absent in NIFTP. Most NIFTP are therefore triaged for surgery due to the suspicious cytology or mutations on the molecular tests, and lobectomy is an appropriate treatment for NIFTP without the need of total or completion thyroidectomy, radioactive iodine ablation, or prolonged monitoring [47].

In April 2017, the World Health Organization codified NIFTP as a benign diagnosis, a move that will spare more than 45,000 patients worldwide each year from total thyroidectomies and a lifetime dependence on medication and regular checkups, along with the psychological burden of a cancer diagnosis for this tumor which does not behave invasively. The word "cancer" misrepresents the biological nature of this tumor and often drives doctors to overtreatment, exposing patients to unnecessary risk and expense.

Personalized Targeted Therapeutic Approach for Thyroid Cancer

The fact that most thyroid carcinomas are associated with alterations in the MAPK pathway (RET -> BRAF -> RAS) or the PI3K/AKT/mTOR pathway (RET -> RAS -> PTEN -> PIK3CA -> AKT1) offers adjuvant medical therapy that can be targeted to these multi-kinase pathways [20]. Targeted drugs are designed for a specific characteristic in cancer cells, such as a gene mutation or a protein, and specifically cells driven by the relevant signaling cascades. These drugs may kill the cells or may synergize with other therapies, such as chemotherapy or radioactive iodine (RAI) treatment. Kinase inhibitors target specific protein kinases (e.g., MAPK or PI3K/AKT/mTOR pathway) which regulate cell growth, while angiogenesis inhibitors which block vascular endothelial growth factor receptor (VEGFR) or platelet-derived growth factor receptors (PDGFR) can prevent tumors from establishing new vasculature.

Fortunately, thyroid cancer is usually very treatable and is often cured with surgery and, if indicated, RAI therapy. Only 15% of patients with differentiated thyroid cancer develop metastatic disease and shorter survival. The 10-year, median survival rate after the discovery of metastatic differentiated thyroid cancer is 42%, but the prognosis varies widely according to many factors, including age, histology, location and size of distant metastases, and whether the disease takes up radioactive iodine. For example, patients with RAI-refractory, progressive disease who develop macroscopic metastases in the lungs or bones have a poor prognosis, with a 10-year overall survival rate of approximately 10%. For patients with more advanced or refractory disease, effective multi-targeted kinase inhibitors (MKIs) are available as adjuvant treatment for differentiated thyroid cancer, but relatively little progress has been made against the rare and aggressive anaplastic thyroid cancer [50-57].

External radiation and chemotherapy (doxorubicin, bleomycin, and platinum compounds) may be used for advanced thyroid cancer but they confer only a minor benefit on tumor burden and overall survival with significant toxicity. The availability of multitargeted kinase inhibitors that can stabilize progressive metastatic disease is changing the standard approach to treating these patients, further limiting the role of traditional cytotoxic chemotherapy.

Research on the molecular mechanisms of thyroid cancer has underlined the role of oncogenic kinases, particularly in metastatic thyroid cancer. For example, *BRAF* mutations have been observed in up to 70% of dedifferentiated metastatic PTC-refractory to ^{131}I RAI treatment [36]. Alterations in the PI3K/AKT pathway have been found to be important in the tumorigenesis and dedifferentiation of follicular variant neoplasms, involving 31% of benign follicular adenomas, 24% of PTC, 55% of FTC and 58% of anaplastic thyroid cancers, suggestive of its important role in the progression of benign follicular lesions toward anaplastic thyroid cancer [58]. Because of the high incidence of thyroid cancer, this field has become a focus for the potential use of multi-kinase targeted therapies, especially the class of agents that inhibit kinases that signal for cellular growth, and tumor angiogenesis.

In the US, targeted therapy drugs approved for differentiated papillary or follicular thyroid cancer include Sorafenib (Nexavar®, a tyrosine kinase inhibitor targeting VEGFR 2-3, PDGFR, RET and BRAF) and Lenvatinib (Lenvima®, an oral tyrosine kinase inhibitor targeting VEGFR 1-3, FGFR 1-4, PDGFR, RET and kit). Targeted therapy drugs approved for medullary thyroid cancer include Vandetanib (Caprelsa®) and Cabozantinib (Cometriq®). Recently Vemurafenib and Dabrafenib, FDA-approved BRAF inhibitors used for treatment of melanoma or non-small cell lung cancer, were shown in many reports to be effective in patients with progressive, RAI-refractory, *BRAF*-mutant metastatic PTC with response rates of 30–40%. These drugs were also effective in some patients with anaplastic thyroid cancer [56-58]. In addition, pretreatment with Selumetinib, a MEK 1-2 inhibitor, for 4 weeks has been approved to enhance RAI uptake by RAI-refractory differentiated thyroid cancers [59].

These new therapies seem to be more potent in patients with PTC than with FTC or poorly differentiated carcinoma, those with a mutated *BRAF* gene, and in those with lung rather than bone metastases [60]. The DECISION study using Sorafenib vs placebo in 417 patients with RAI-refractory progressive differentiated thyroid cancer showed significant improvement in the primary outcome of progression-free survival of 10.8 vs 5.8 months. However, no complete response or changes in overall survival were observed (since crossover was allowed at progression and this occurred in 70% of the placebo group), and 18.8% of patients discontinued the medication due to side effects [51]. The SELECT trial using Lenvatinib showed very promising results in patients with RAI-refractory differentiated thyroid cancer with significant progression-free survival of 18.3 vs. 3.6 months compared to placebo (hazard ratio for progression or death, 0.21; 99% confidence interval, 0.14 to 0.31; P<0.001) [57]. The response rate was 64.8% in the Lenvatinib group (4 complete responses and 165 partial responses) and 1.5% in the placebo group (P<0.001). The median overall survival was not different, however, between the treatment groups. There was also a 14.2% medication discontinuation rate due to side effects which reduced quality of life.

The toxicities or adverse effects of these targeted therapy drugs primarily consist of hypertension (up to 67% of patients), diarrhea (60%), appetite loss and weight loss (50%), fatigue (up to 60%), nausea (40%) and skin disease (67%) including rash, hand and foot syndrome, and alopecia [50, 60, 61]. In the Lenvatinib group for the SELECT trial, 6 of 20 deaths that occurred during the treatment period were considered to be drug-related, among a total of 261 patients with RAI-refractory progressive differentiated thyroid cancer [62].

For unresectable locally advanced or metastatic meduallary thyroid cancer, Vandetanib is available in the US through a Risk Evaluation and Mitigation Strategy (REMS) program. It is an oral inhibitor that targets VEGFR, RET/PTC and the epidermal growth factor receptor (EGFR). In a randomized trial of 145 patients with metastatic PTC or FTC unresponsive to RAI therapy, fewer patients had disease progression after treatment, and the median progression-free survial was lengthened (11.1 vs 5.9 months) compared to placebo. However, there was no difference in partial response or overall survival. Therefore, Vandetanib is mainly used for metastatic MTC and rarely prescribed for patients with metastatic, RAI-refractory PTC or FTC who could not tolerate lenvatinib or sorafenib.

Researchers have explored the use of MEK 1-2 inhibitor-Selumetinib as a pretreatment to help restore RAI uptake in advanced widely metastatic RAI-refractory differentiated thyroid cancer [58]. However, its use in earlier-stage cancers has not been sufficiently studied, despite the fact that these tumors constitute the vast majority of thyroid cancers. Furthermore, cancer cells are smart and have escape pathways to vade the drug blockade, so additional blockade molecules such as ERBB3 in combination with MEK and BRAF inhibitors may be needed. Ongoing studies are under way *in vitro* and in phase II clinical trials.

More recently, advances in immunotherapy such as Pembrolizumab (a PD1 inhibitor) has been shown to elicit a transient response for advanced differentiated and anaplastic thyroid cancer [63]. Currently there are ongoing investigations to use this immune checkpoint therapy alone or in combination with a multi-kinase inhibitor, radiation therapy, or chemotherapy for the treatement of anaplastic thyroid cancer [64, 65].

CONCLUSION

The increasing incidence of thyroid cancer worldwide is largely associated with a higher number of screening examinations and imaging tests performed, leading to the discovery of more thyroid nodules. Clinical risk factors, ultrasound features, and selective fine needle aspiration biopsy of suspicious nodules greater than 1.0-1.5 cm remain the key to avoiding overdiagnosis of thyroid cancer, especially in small indolent low-risk tumors that would not, if left alone, result in symptoms or death. Despite the dramatic increase in the incidence of thyroid cancer, the mortality remains stable, and the burden of thyroid cancer diagnosis resulting in overtreatment that increases overall healthcare costs.

Molecular testing can be a useful tool in conjunction with clinical and ultrasound risk assessment of thyroid nodules, especially for patients with indeterminate thyroid cytology (Bethesda categories III, IV or V) to reduce the need for diagnostic surgery and guide further management of the nodules. However, there is no need for costly molecular testing if the ultrasound and/or FNA showed an entirely benign pattern or highly suspicious results. If patients prefer to have surgery, or develop hyperthyroidism due to hyperfunctiong nodule(s) (~5%), or experience compressive symptoms (~5%), then there is no need for a molecular test. The main purpose of molecular test for thyroid nodules with intermediate cytology is to avoid surgery, lifelong hypothyroidism, and possible injury to the parathyroid glands or recurrent laryngeal nerve. The risk of malignancy in these atypical categories is about 15-30% (i.e., 70-85% of the nodules with atypical cytology are benign). Molecular testing is advantageous compared to diagnostic surgery in terms of morbidity because any surgical intervention is associated with risks, even in the hands of the most experienced thyroid surgeons. Currently it is mainly a matter of cost that restrains widespread use of molecular testing, and it is expected to be less expensive in the near future.

Future classification of thyroid cancer will likely incorporate the genetic mutation information to help clarify the subtypes and stages of the tumor i.e., benign adenoma, intermediate category such as NIFTP (which is not a benign lesion but similar to an *in situ* cancer), differentiated thyroid cancer, more aggressive cancer type with certain mutations, or the worst anaplastic thyroid cancer. Molecular tests can help inform us about the cancer risk and also the prognosis of thyroid nodules. Personalized targeted therapeutic approaches are now available to use in advanced or metastatic RAI-refractory differentiated thyroid cancer.

Among differentiated thyroid cancer, identification of mutations that drive aggressive phenotypes, such as BRAF V600E, or the coexistence of BRAF or RAS with TERT mutations would warrant more aggressive treatment and closer surveillance due to a much poorer prognosis with a high risk of evolution to poorly differentiated or undifferentiated (anaplastic) thyroid cancers. Papillary thyroid cancer generally has an excellent prognosis but about 5-10% of patients have a particularly aggressive disease course, accounting for virtually all PTC-related mortality.

Although conventional chemotherapy with cytotoxic agents are occasionally used for the treatment of patients with progressive symptomatic thyroid cancer that is unresponsive or not amenable to surgery, radioiodine therapy, or external radiotherapy, complete remission is rare and long-term responses are uncommon.

The knowledge of the molecular pathways involved in the pathogenesis of thyroid cancer has made possible the development of new therapeutic targeted drugs which block the oncogenic kinases (BRAF V600E, RET/PTC) or signaling kinases (VEGFR, PDGFR) involved in cellular growth and proliferation. Some clinical trials have shown clearly the ability of multi-targeted kinase therapies (Sorafenib, Lenvatinib, Vandetanib, Cabozantinib, etc) in stabilizing the course of the disease. Unfortunately, eventual progression despite antiangiogenic blockade suggests emergence of alternate pathways to promote tumor growth and metastasis. Therefore, combination therapy with new immune checkpoint therapies may yield a longer-lasting response and several clinical trials are ongoing.

The availability of kinase inhibitors that can stabilize progressive metastatic disease has changed the approach to treating patients with disease that no longer responds to radioiodine and TSH-suppressive thyroid hormone therapy. However, these drugs are disease-modifying agents, usually tumoristatic rather than tumoricidal, and so far no significant improvement in overall survival has been observed. In addition, these new targeted therapies have significant toxicities and, therefore, it is important to limit the use of systemic agents to patients at risk for morbidity or mortality due to progressive metastatic disease. Molecular tests for BRAF and other genetic mutations in MAPK and PI3K/AKT/mTOR pathways can help predict the response to multi-kinase targeted therapies. Fortunately most thyroid cancers are curable using conventional therapy with no observed increase in overall mortality despite the marked increase of thyroid cancer incidence worldwide.

When the American Joint Committee on Cancer's Eighth Edition Cancer Staging Manual became effective for U.S. practice in early 2018, substantially more patients with thyroid cancer will meet the definition for stage I disease (80% instead of 60%) [66]. The age cutpoint for stage 1 thyroid cancer will change from 45 to 55 years old and the tumor size will be increased from 2cm to 4cm, but their survival prognosis will remain as good as it was under the previous definition (98-100% for disease-specific survival for stage 1 thyroid cancer). Fewer patients will be classified with advanced diseases which will help reduce overly aggressive treatment for indolent tumors. The future goals will be to (1) reduce unnecessary surgery or limit the extent of the surgery; (2) minimize the use of radioactive iodine and enhance its effect by pretreatment with the targeted molecules to increase radioiodine uptake in some aggressive tumor cell types even before they metastasize; and (3) use personalized multi-kinase targeted therapies for RAI-refractory metastatic thyroid cancers that harbor genetic mutations in the oncogenic and/or signaling kinase pathways.

REFERENCES

[1] Vaccarella, S., Franceschi, S., Bray, F., Wild, C. P., Plummer, M. and Dal Maso, M. (2016). Worldwide Thyroid-Cancer Epidemic? The Increasing Impact of Overdiagnosis. *N Engl J Med;* 375:614-617.

[2] Haugen, B. R., Alexander, E. K., Bible, K. C., Doherty, G. M., Mandel, S. J., Nikiforov, Y. E., et al. (2016). 2015 American Thyroid Association Management Guidelines for Adult Patients with Thyroid Nodules and Differentiated Thyroid Cancer: The American Thyroid Association Guidelines Task Force on Thyroid Nodules and Differentiated Thyroid Cancer. *Thyroid;* 26(1):1-133.

[3] Gharib, H., Papini, E., Garber, J. R., Duick, D. S., Harrell, R. M., Hegedus, L., et al. (2016). American Association of Clinical Endocrinologists, American College of Endocrinology, and Associazione Medici Endocrinologi Medical Guidelines for Clinical Practice for the Diagnosis and Management of Thyroid Nodules-2016 Update. *Endocr Pract*; 22(5):622-639.

[4] Durante, C., Giorgio, G., Lamartina, L., Filetti, S., Mandel, S. J. and Cooper, D. S. (2018). The Diagnosis and Management of Thyroid Nodules: A Review. *JAMA;* 319(9):914-924.

[5] Cibas, E. S. and Ali, S. Z. (2009). The Bethesda System for Reporting Thyroid Cytopathology. *Thyroid:* 19(11):1159-1165.

[6] Cibas, E. S. and Ali, S. Z. (2017). The 2017 Bethesda System for Reporting Thyroid Cytopathology. *Thyroid:* 27(11): 1341-1346.

[7] Davies, L. and Welch, H. G. (2014). Current thyroid cancer trends in the United States. *JAMA Otolaryngol Head Neck Surg;* 140(4):317-322.

[8] Alexander, E. K., Kennedy, G. C., Baloch, Z. W., Cibas, E. S., Chudova, D., Diggans, J., et al. (2012). Preoperative diagnosis of benign thyroid nodules with indeterminate cytology. *N Engl J Med*; 367(8):705-715.

[9] Lewis, C. A., Philip, D. and Chaidarun, S. S. (2017). Practical Perspectives for Management of Thyroid Nodules with Atypical Needle Biopsy. *Global J Otolaryngol.* https://juniperpublishers.com/gjo/pdf/GJO.MS.ID.555638.pdf .

[10] Nishino, M. (2016). Molecular cytopathology for thyroid nodules: A review of methodology and test performance. *Cancer Cytopathol;* 124(1):14-27.

[11] Lathief, S., Pothuloori, A., Liu, X. and Chaidarun, S. S. (2016). Advances and Practical Use of the Molecular Markers for Thyroid Cancer. *Adv Cell Mol Otolaryngol*, 4: 33948 - http://dx.doi.org/10.3402/acmo.v4.33948 .

[12] Vargas-Salas, S., Martínez, J. R., Urra, S., Domínguez, J. M., Mena, N., Uslar, T., et al. (2018). Genetic testing for indeterminate thyroid cytology: review and meta-analysis. *Endocr Relat Cancer*; 25(3):R163-R177.

[13] Zhang, M. and Lin, O. (2016). Molecular testing of thyroid nodules. *Arch Pathol Lab Med;* 140:1138-1344.

[14] Nikiforov, Y. E., Ohori, N. P., Hodak, S. P., Carty, S. E., LeBeau, S. O., Ferris, R. L., et al. (2011). Impact of mutational testing on the diagnosis and management of patients with cytologically indeterminate thyroid nodules: a prospective analysis of 1056 FNA samples. *J Clin Endocrinol Metab;* 96(11):3390-3397.

[15] Nikiforov, Y. E., Carty, S. E., Chiosea, S. I., Coyne, C., Duvvuri, U., Ferris, R. L., et al. (2015). Impact of the Multi-Gene ThyroSeq NextGeneration Sequencing Assay on Cancer Diagnosis in Thyroid Nodules with Atypia of Undetermined Significance/Follicular Lesion of Undetermined Significance Cytology. *Thyroid;* 25(11):1217-1223.

[16] Baloch, Z. W., Seethala, R. R., Faquin, W. C., Papotti, M. G., Basolo, F., Fadda, G., et al. (2016). Noninvasive follicular thyroid neoplasm with papillary-like nuclear features (NIFTP): A changing paradigm in thyroid surgical pathology and implications for thyroid cytopathology. *Cancer Cytopathol;* 124(9):616-620.

[17] Ferris, R. L., Baloch, Z., Bernet, V., Chen, A., Fahey, T. J.,3rd, Ganly, I., et al. (2015). American Thyroid Association Statement on Surgical Application of Molecular Profiling for Thyroid Nodules: Current Impact on Perioperative Decision Making. *Thyroid;* 25(7):760-768.

[18] Nikiforov, Y. E., Carty, S. E., Chiosea, S. I., Coyne, C., Duvvuri, U., Ferris, R. L., et al. (2014). Highly accurate diagnosis of cancer in thyroid nodules with follicular neoplasm/suspicious for a follicular neoplasm cytology by ThyroSeq v2 next-generation sequencing assay. *Cancer;* 120(23):3627-3634.

[19] Santhanam, P., Khthir, R., Gress, T., Elkadry, A., Olajide, O., Yaqub, A., et al. (2016). Gene expression classifier for the diagnosis of indeterminate thyroid nodules: a meta-analysis. *Med Oncol;* 33(2):14-015-0727-3.

[20] Fagin, J. A. and Wells, S. A., Jr. (2016). Biologic and Clinical Perspectives on Thyroid Cancer. *N Engl J Med;* 375(11):1054-1067.

[21] Yarchoan, M., LiVolsi, V. A. and Brose, M. S. (2015). BRAF mutation and thyroid cancer recurrence. *J Clin Oncol*; 33(1):7-8.

[22] Nikiforov, Y. E., Carty, S. E., Chiosea, S. I., Coyne, C., Duvvuri, U., Ferris, R. L., et al. (2015). Impact of the Multi-Gene ThyroSeq Next-Generation Sequencing Assay on Cancer Diagnosis in Thyroid Nodules with Atypia of Undetermined Significance/Follicular Lesion of Undetermined Significance Cytology. *Thyroid;* 25(11):1217-1223.

[23] Liu, R. and Xing, M. (2016). TERT promoter mutations in thyroid cancer. *Endocr Relat Cancer;* 23(3):R143-55.

[24] Xing, M., Alzahrani, A. S., Carson, K. A., Viola, D., Elisei, R., Bendlova, B., et al. (2013). Association between BRAF V600E mutation and mortality in patients with papillary thyroid cancer. *JAMA;* 309(14):1493-1501.

[25] Xing, M., Alzahrani, A. S., Carson, K. A., Shong, Y. K., Kim, T. Y., Viola, D., et al. (2015). Association between BRAF V600E mutation and recurrence of papillary thyroid cancer. *J Clin Oncol;* 33(1):42-50.

[26] Xing, M., Haugen, B. R. and Schlumberger, M. (2013). Progress in molecular-based management of differentiated thyroid cancer. *Lancet;* 381(9871):1058-1069.

[27] Tufano, R. P., Teixeira, G. V., Bishop, J., Carson, K. A. and Xing, M. (2012). BRAF mutation in papillary thyroid cancer and its value in tailoring initial treatment: a systematic review and meta-analysis. *Medicine* (Baltimore);91(5):274-286.

[28] Jin, L., Chen, E., Dong, S., Cai, Y., Zhang, X., Zhou, Y., et al. (2016). BRAF and TERT promoter mutations in the aggressiveness of papillary thyroid carcinoma: a study of 653 patients. *Oncotarget;* 7(14):18346-18355.

[29] Liu, R. and Xing, M. (2016). TERT promoter mutations in thyroid cancer. *Endocr Relat Cancer;* 23(3):R143-55.

[30] Liu, R., Bishop, J., Zhu, G., Zhang, T., Ladenson, P. W. and Xing, M. (2017). Mortality Risk Stratification by Combining BRAF V600E and TERT Promoter Mutations in Papillary Thyroid Cancer: Genetic Duet of BRAF and TERT Promoter Mutations in Thyroid Cancer Mortality. *JAMA Oncol;* 3(2):202-208.

[31] Moon, S., Song, Y. S., Kim, Y. A., Lim, J. A., Cho, S. W., Haln, S., et al. (2017). Effects of Coexistent $BRAF^{V600E}$ and TERT promoter mutations on poor clinical outcomes in Papillary thyroid cancer: A metaanalysis. *Thyroid;* 27(5):651-660.

[32] Duick, D. S. (2012). Overview of molecular biomarkers for enhancing the management of cytologically indeterminate thyroid nodules and thyroid cancer. *Endocr Pract;* 18(4):611-615.

[33] Tallini, G. and Asa, S. L. (2001). RET oncogene activation in papillary thyroid carcinoma. *Adv Anat Pathol;* 8(6):345-354.

[34] Xing, M. (2015). BRAF Mutation and Thyroid Cancer Recurrence. *J Clin Oncol;* 33(22):2482-2483.

[35] Xing, M. (2005). BRAF mutation in thyroid cancer. *Endocr Relat Cancer;* 12(2):245-262.

[36] Afirma, G. S. C. (2017). Afirma Genomic Sequensing Classifier-The Next Generationn Afirma Test. *Veracyte performance validation on 191 of the original 210 indeterminate thyroid nodules from the NEJM study*. www.veracyte.com.

[37] Duh, Q. Y., Angell, T. E., Babiarz, J., Barth, N., Blevin, T., et al. (2017). Development and Validation of Classifiers to Enhance the Afirma Genomic Sequencing Classifier Performance Among Hurthle Cell Specimens. Abtract (oral

presentation) at *American Thyroid Association 87th Annual Meeting* in Victoria, BC, Canada, Oct 18-22.

[38] Nikiforova, M. N., Wald, A. I., Roy, S., Durso, M. B., Nikiforov, Y. E. (2013). Targeted next-generation sequencing panel (ThyroSeq) for detection of mutations in thyroid cancer. *J Clin Endocrinol Metab;* 98(11):E1852-60.

[39] Valderrabano, P., Khazai, L., Leon, M. E., Thompson, Z. J., Ma, Z., Chung, C. H., et al. (2017). Evaluation of ThyroSeq v2 performance in thyroid nodules with indeterminate cytology. *Endocr Relat Cancer;* 24(3):127-136.

[40] Nikiforova, M. N., Mercurio, S., Wald, A. I., Moura, B. D., Callenberg, K., Santana-Santos, L, et al. (2018). Analytical performance of the ThyroSeq V3 Genomic Classifier for cancer diagnosis in Thyroid nodules. *Cancer;* Jan 18. doi: 10.1002/cncr.31245. [Epub ahead of print].

[41] Labourier, E., Shifrin, A., Busseniers, A. E., Lupo, M. A., Manganelli, M. L., Andruss, B., et al. (2015) Molecular testing for miRNA, mRNA, and DNA on fine-needle aspiration improves the preoperative diagnosis of thyroid nodules with indeterminate cytology. *J Clin Endocrinol Metab;* 100: 2743–2750.

[42] Wylie, D. Beaudenon-Huibregtse, S., Haynes, B. C., Giordano, T. J., Labourier, E. (2017). *J Pathol Clin Res;* 2(2):93-103.

[43] Lithwick-Yanai, G., Dromi, N., Shtabsky, A., Morgenstern, S., Strenov, Y., Feinmesser, M., Kravtsov, V., et al. (2017). Multicentre validation of a microRNA-based assay for diagnosing indeterminate thyroid nodules utilising fine needle aspirate smears. *J Clin Pathol;* 70:500–507.

[44] Gharib, H., Papini, E., Garber, J. R., Duick, D. S., Harrell, R. M., Hegedus, L., et al. (2016). American Association of Clinical Endocrinologists, American College of Endocrinology, and Associazione Medici Endocrinologi Medical Guidelines for Clinical Practice for the Diagnosis and Management of Thyroid Nodules-2016 Update. *Endocr Pract;* 22(5):622-639.

[45] Nikiforov, Y. E., Seethala, R. R., Tallini, G., Baloch, Z. W., Basolo, F., Thompson, L. D., et al. (2016). Nomenclature Revision for Encapsulated Follicular Variant of Papillary Thyroid Carcinoma: A Paradigm Shift to Reduce Overtreatment of Indolent Tumors. *JAMA Oncol;* 2(8):1023-1029.

[46] Baloch, Z. W., Seethala, R. R., Faquin, W. C., Papotti, M. G., Basolo, F., Fadda, G., et al. (2016). Noninvasive follicular thyroid neoplasm with papillary-like nuclear features (NIFTP): A changing paradigm in thyroid surgical pathology and implications for thyroid cytopathology. *Cancer Cytopathol;* 124(9):616-620.

[47] Agrawal, N., Abbott, C. E., Liu, C., et al. (2017). Non-invasive Follicular Tumor with Papillary-like Nuclear Features (NIFTP): Not a Tempest in a Teapot. *Endocr Pract;* 23:451-457.

[48] Strickland, K. C., Howitt, B. E., Marqusee, E., Alexander, E. K., Cibas, E. S., Krane, J. F., et al. (2015). The Impact of Noninvasive Follicular Variant of

Papillary Thyroid Carcinoma on Rates of Malignancy for Fine-Needle Aspiration Diagnostic Categories. *Thyroid;* 25(9):987-992.

[49] Faquin, W. C., Wong, L. Q., Afrogheh, A. H., Ali, S. Z., Bishop, J. A., Bongiovanni, M., et al. (2016). Impact of reclassifying noninvasive follicular variant of papillary thyroid carcinoma on the risk of malignancy in The Bethesda System for Reporting Thyroid Cytopathology. *Cancer Cytopathol;* 124(3):181-187.

[50] Antonelli, A., Fallahi, P., Ferrari, S. M., Ruffilli, I., Santini, F., Minuto, M., Galleri, D., Miccoli, P. (2011). New Targeted Therapies for Thyroid Cancer. *Current Genomics;* 12: 626-631.

[51] Brose, M. S., Nutting, C. M., Jarzab, B., et al. (2014). Sorafenib in radioactive iodine-refractory, locally advanced or metastatic differentiated thyroid cancer: a randomised, double-blind, phase 3 trial. *Lancet;* 384:319-328.

[52] Carr, L. L., Mankoff, D. A., Goulart, B. H., et al. (2010). Phase II study of daily sunitinib in FDG-PET-positive, iodine-refractory differentiated thyroid cancer and metastatic medullary carcinoma of the thyroid with functional imaging correlation. *Clin Cancer Res;* 16:5260-5268.

[53] Bible, K. C., Suman, V. J., Molina, J. R., et al. (2010). Efficacy of pazopanib in progressive, radioiodine-refractory, metastatic differentiated thyroid cancers: results of a phase 2 consortium study. *Lancet Oncol;* 11:962-972.

[54] Locati, L. D., Licitra, L., Agate, L., et al. (2014). Treatment of advanced thyroid cancer with axitinib: Phase 2 study with pharmacokinetic/pharmaco dynamic and quality-of-life assessments. *Cancer;* 120:2694-2703.

[55] Hoftijzer, H., Heemstra, K. A., Morreau, H., et al. (2009). Beneficial effects of sorafenib on tumor progression, but not on radioiodine uptake, in patients with differentiated thyroid carcinoma. *Eur J Endocrinol;* 161:923-931.

[56] Marten, K. A. and Gudena, V. K. (2015). Use of Vemurafenib in Anaplastic Thyroid Carcinoma: A Case Report. *Cancer Biol Ther;* 16(10): 1430–1433.

[57] Subbiah, V., Kreitman, R. J., Wainberg, Z. A., Cho, J. Y., Schellens, J. H. M., Soria, J. C., et al. (2018). Dabrafenib and Trametinib Treatment in Patients with Locally Advanced or Metastatic BRAF V600-Mutant Anaplastic Thyroid Cancer. *J Clin Oncol* (Rapid Communication); 36 (1):7-13.

[58] Saji, M. and Ringel, M. D. (2010). The PI3K-Akt-mTOR pathway in initiation and progression of thyroid tumors. *Mol. Cell Endocrinol;* 321:20–28.

[59] Ho, A. L., Grewal, R. K., Leboeuf, R., Sherman, E. J., Pfister, D. G., Deandreis, D., et al. (2013). Selumetinib-Enhanced Radioiodine Uptake in Advanced Thyroid Cancer. *N Engl J Med;* 368(7): 623–632.

[60] Castro, M. R., Bergert, E. R., Goellner, J. R., Hay, I. D. and Morris, J. C. (2001). Immunohistochemical analysis of sodium iodide symporter expression in metastatic differentiated thyroid cancer: correlation with radioiodine uptake. *J. Clin. Endocrinol. Metab;* 86:5627–5632.

[61] Schlumberger, M., Tahara, M., Wirth, L. J., Robinson, B., Brose, M., Elisei, R., Habra, M. A., et al. (2015). Lenvatinib versus Placebo in Radioiodine-refractory Thyroid Cancer. *N Engl J Med;* 372:621-630.

[62] Cabanillas, M. E., Patel, A., Danysh, B. P., Dadu, R., Kopetz, S. and Falchook, G. (2015). BRAF inhibitors: Experience in Thyroid Cancer and General Reviw of Toxicity. *Horm Cancer;* 6(1): 21–36.

[63] Bastman, J. J., Serracino, H. S., Zhu, Y., Koenig, M. R., Mateescu, V., Sams, S. S., Davies, K. D., et al. (2016). Tumor-Infiltrating T Cells and the PD-1 Checkpoint Pathway in Advanced Differentiated and Anaplastic Thyroid Cancer. *J Clin Endocrinol Metab;* 101(7): 2863-2873.

[64] Khan, S. A. (2018). Phase II Trial of Pembrolizumab in Metastatic or Locally Advanced Anaplastic/Undifferentiated Thyroid Cancer. UT Southwestern Medical Center, Dallas, TX. *ClinicalTrials.gov* ID: NCT02688608. Oct 2016- 2019.

[65] Chintakuntlawar, A. (2018). Pembrolizumab, Chemotherapy, and Radiation Therapy With or Without Surgery in Treating Patients with Anaplastic Thyroid Cancer. Mayo clinic, Rochester, MN. *ClinicalTrials.gov* ID: NCT03211117. Aug 2017-Dec-2019.

[66] Tuttle, R. M., Haugen, B. and Perrier, N. D. (2017). Updated American Joint Committee on Cancer/Tumor-Node-Metastasis Staging System for Differentiated and Anaplastic Thyroid Cancer (Eighth Edition): What Changed and Why? *Thyroid;* 27(6):751-756.

About the Editor

Yildiz Dincer (PhD) is Professor in Istanbul University Cerrahpasa Medical Faculty, Department of Medical Biochemistry, Istanbul, Turkey. She gained her MSc and PhD degree in Biochemistry in the same department. She has been in Imperial Research Fund-Clare Hall Laboratories, UK as a fellow in 1999, and gained experience in the field of DNA repair by working in the team of Dr. Tomas Lindahl who shared the 2015 Nobel Prize in Chemistry for work on DNA repair. Dr. Dincer maintains an active research program studying on oxidative DNA damage, DNA repair/apoptosis, antioxidants and epigenetics. Dr. Dincer has published many articles and book chapters. She has served as the editor of various edited collections on topics such as iron deficiency, chemical carcinogens, cancer biochemistry, chemotherapeutics, epigenetics and Alzheimer's Disease.

INDEX

A

B

C

D

E

F

G

H

N

O

P

Q

R

S

T

U

V

W

X

Z

Brain Cancer, Tumor Targeting and Cervical Cancer

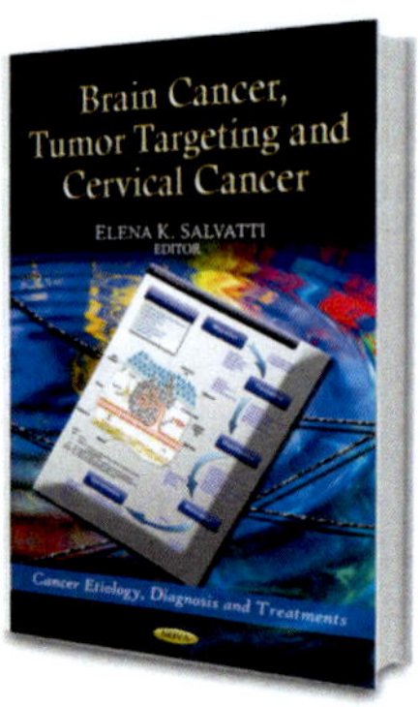

Editor: Elena K. Salvatti

Series: Cancer Etiology, Diagnosis and Treatments

Book Description: This book gathers current research from across the globe in the study of brain cancer, cervical cancer and tumor targeting. Some topics discussed in this compilation include emerging targets in cancer therapy and novel nanotechnology-based therapeutic strategies that have been changing the paradigm in cancer treatment; laparascopy and robotics in cervical cancer treatment; the history and current status of glioma therapy and glioma gene therapy; congenital tumors of the central nervous system; a review of tumors of the paranasal sinuses and real-time tumor targeting in external beam radiotherapy.

Hardcover ISBN: 978-1-61122-738-3
Retail Price: $130

Tumor Suppressors

Editor: Susan D. Nguyen

Series: Cell Biology Research Progress

Book Description: This book presents topical research data in the study of tumor suppressors, including Rab GTPases as potential tumor suppressors; the regulation of neutrophil function by tumor suppressor PTEN; the role of BRIT1/MCPH1 in genome maintenance and tumor suppression; epigenetic silencing of tumor suppressors; and the relationship between MicroRNA and tumor suppressors.

Hardcover ISBN: 978-1-61761-986-1
Retail Price: $170